Springer Proceedings in Mathematics & Statistics

Volume 299

Springer Proceedings in Mathematics & Statistics

This book series features volumes composed of selected contributions from workshops and conferences in all areas of current research in mathematics and statistics, including operation research and optimization. In addition to an overall evaluation of the interest, scientific quality, and timeliness of each proposal at the hands of the publisher, individual contributions are all refereed to the high quality standards of leading journals in the field. Thus, this series provides the research community with well-edited, authoritative reports on developments in the most exciting areas of mathematical and statistical research today.

More information about this series at http://www.springer.com/series/10533

Vladas Sidoravicius
Editor

Sojourns in Probability Theory and Statistical Physics - II

Brownian Web and Percolation, A Festschrift for Charles M. Newman

NYU SHANGHAI
上海纽约大学

Springer

Editor
Vladas Sidoravicius
NYU Shanghai
Shanghai, China

ISSN 2194-1009 ISSN 2194-1017 (electronic)
Springer Proceedings in Mathematics & Statistics
ISBN 978-981-15-0297-2 ISBN 978-981-15-0298-9 (eBook)
https://doi.org/10.1007/978-981-15-0298-9

Mathematics Subject Classification (2010): 60-XX, 82-XX, 60J65, 82B43

This Springer imprint is published by the registered company Springer Nature Singapore Pte Ltd.
The registered company address is: 152 Beach Road, #21-01/04 Gateway East, Singapore 189721, Singapore

Nautsense, a painting by Serena Newman, Buzios, Brazil, August 2010

Preface

This three-volume set, entitled Sojourns in Probability Theory and Statistical Physics, constitutes a Festschrift for Chuck Newman on the occasion of his 70th birthday. In these coordinated volumes, Chuck's closest colleagues and collaborators pay tribute to the immense impact he has had on these two deeply intertwined fields of research. The papers published here include original research articles and survey articles, on topics gathered by theme as follows:

Volume 1: Spin Glasses and Statistical Mechanics
Volume 2: Brownian Web and Percolation
Volume 3: Interacting Particle Systems and Random Walks

Our colleague Vladas Sidoravicius conceived the idea for this Festschrift during the conference on Probability Theory and Statistical Physics that was hosted on 25–27 March 2016 by the NYU-ECNU Institute of Mathematical Sciences at NYU Shanghai. This conference brought together more than 150 experts to discuss frontier research at the interface between these two fields, and it coincided with Chuck's 70th birthday. After the conference, Vladas approached various of Chuck's colleagues with invitations to contribute. Papers flowed in during the Fall of 2016 and the Spring of 2017. The Festschrift suffered delays in 2018, and then on 23 May 2019, Vladas passed away unexpectedly. Following discussions in June 2019 with NYU Shanghai and Springer Nature, we offered to assume editorial responsibility for bringing the volumes to completion.

We gratefully acknowledge Vladas's investment in these volumes, and we recognise that his presence in our community worldwide will be sorely missed. We offer our thanks to Julius Damarackas (NYU Shanghai) for his detailed preparation of the articles in these volumes.

Chuck has been one of the leaders in our profession for nearly 50 years. He has worked on a vast range of topics and has collaborated with and inspired at least three generations of mathematicians, sharing with them his deep insights into mathematics and statistical physics and his views on key developments, always leavened with his acute and captivating sense of humour. We wish him and his family many fruitful years to come.

July 2019

Federico Camia
Geoffrey Grimmett
Frank den Hollander
Daniel Stein

Contents

Differential Geometry for Model Independent Analysis of Images and Other Non-Euclidean Data: Recent Developments

Rabi Bhattacharya[1(✉)] and Lizhen Lin[2]

[1] Department of Mathematics, The University of Arizona, Tucson, AZ, USA
rabi@math.arizona.edu

[2] Department of Applied and Computational Mathematics and Statistics, The University of Notre Dame, Notre Dame, IN, USA
lizhen.lin@nd.edu

In celebration of Chuck's 70th Birthday

Abstract. This article provides an exposition of recent methodologies for nonparametric analysis of digital observations on images and other non-Euclidean objects. Fréchet means of distributions on metric spaces, such as manifolds and stratified spaces, have played an important role in this endeavor. Apart from theoretical issues of uniqueness of the Fréchet minimizer and the asymptotic distribution of the sample Fréchet mean under uniqueness, applications to image analysis are highlighted. In addition, nonparametric Bayes theory is brought to bear on the problems of density estimation and classification on manifolds.

Keywords: Fréchet means · Image analysis · Nonparametric inference on manifolds · Nonparametric Bayes · Statistics on manifolds · Stratified spaces

1 Introduction

Historically, directional statistics, that is, statistics on spheres, especially S^2, have been around for a long time, and there is a great deal of literature on it (See the books by Watson 1983; Mardia and Jupp 2000; Fisher et al. 1987). Much of that was inspired by a seminal paper by Fisher (1953) proving beyond any reasonable doubt that the earth's magnetic poles had shifted over geological times. Indeed, the two sets of data that he analyzed, one from the Quaternary period and the other from recent times (1947–48), showed an almost reversal of the directions of the magnetic poles. In addition to this first scientific demonstration of a phenomenon conjectured by some paleontologists, such studies of magnetic poles in fossilized remanent magnetism had an enormous impact on tectonics,

V. Sidoravicius (Ed.): *Sojourns in Probability Theory and Statistical Physics - II*, PROMS 299, pp. 1–43, 2019.
https://doi.org/10.1007/978-981-15-0298-9_1

essentially validating the theory of continental drift (Irving 1964; Fisher et al. 1987). There are other important applications of directional statistics, such as designing of windmills based on wind directions, etc. Fisher's example is presented in Sect. 8, in comparison with the nonparametric method highlighted in this article.

The advancement of imaging technology and increase in computing prowess have opened up a whole new vista of applications. Medical imaging, for example, is now an essential component of medical practice. Not only have MRIs (magnetic resonance imaging) become routine for diagnosing a plethora of diseases, there are more advanced techniques such as the DTI (diffusion tensor imaging) which measures diffusion coefficients of water molecules in tiny voxels along nerve fibers in the cortex of the brain in order to understand or monitor diseases such as Parkinson's and Alzheimer's (Goodlett et al. 2006; Kindlmann et al. 2007; Morra et al. 2000). Beyond medicine, there are numerous applications to morphometrics (Bookstein 1991), graphics, robotics, and machine vision (Aggarwal et al. 2004; Ma et al. 2005; Veeraraghavan et al. 2005).

Images are geometric objects and their precise mathematical descriptions and identifications in different fields of applications are facilitated by the use of differential geometry. Kendall (1984) and Bookstein (1991) were two pioneers in the geometric description and statistical analysis of images represented by landmarks on two or three dimensional objects. The spaces of such images, or shapes, are differential manifolds, or *stratified spaces* obtained by gluing together manifolds of different dimensions. In the following sections these spaces are described in detail. Much of the earlier statistical analysis on differential manifolds were parametric in nature, where a distribution Q on a manifold M is assumed to belong to a finite dimensional parametric family; that is, Q is assumed to have a density (with respect a standard distribution, e.g., the volume measure on M) which is specified except for the value of a finite dimensional parameter θ lying in an open subset Θ of an Euclidean space. The statistician's task is then to estimate the parameter (or test for its belonging to a particular subset of Θ), using observed data. There are standard methodologies for estimation (say, the maximum likelihood estimator, MLE), or testing (such as the likelihood ratio test) that one may try to use. Of course, it still requires a great deal of effort to analytically compute these statistical indices and their (approximate) distributions on specific manifolds. A reasonably comprehensive account of these for the shape spaces of Kendall, or similar manifolds, may be found in Dryden and Mardia (1998).

The focus of the present article is a model independent, or nonparametric, methodology for inference on general manifolds. As a motivation consider the problem of discriminating between two distributions on an Euclidean space based on independent samples from them. In parametric inference one would use a density (with respect to a sigma-finite measure) which is specified except for a finite dimensional parameter as described above. One may use one of a number of standard asymptotically efficient procedures to test if the two distributions have different parameter values (See, e.g., Hotelling 1931; Goodall 1991).

If the statistician is not confident about this parametric model, or any other, one popular method is to test for the differences between the means of the two distributions by using the two sample means. When the sample sizes are reasonably large then the difference between the sample means is asymptotically normal with mean given by the difference between the population means. If the observations are from a normal distribution with the mean as the unknown parameter then this test is optimal in an appropriate sense (Bhattacharya et al. 2016, pp. 296–300; Lehmann 1959, p. 93, 94). But used in other parametric models the test is not, in general, optimal and may even be inconsistent; that is, there may be many pairs of distributions $Q_1 \neq Q_2$ whose means are the same. However, when the components or coordinates of the distributions are such that the differences between Q_1 and Q_2 are reasonably expected to manifest in shifts of the mean vector, this widely used nonparametric test is quite effective, especially since with large sample sizes the asymptotic distribution is normal. Turning now to distributions Q on non-Euclidean metric spaces S, one has an analogue of the mean given by the minimizer, if unique, of the average (with respect to Q) of the squared distance from a point. This is the so called *Fréchet mean* introduced by Fréchet (1948), although physicists probably had used the notion earlier in specific physical contexts for the distribution Q of the mass of a body, calling it the center of mass. Of course it is in general a non-trivial matter to find out broad conditions for the *uniqueness of the Fréchet minimizer* and, in the case of uniqueness, to derive the *(asymptotic) distribution of the sample Fréchet mean*. These allow one to obtain proper confidence regions for the Fréchet mean of Q and critical regions for tests for detecting differences in means of distributions on M (Bhattacharya and Patrangenaru 2002, 2003, 2005). The theory of Fréchet means is presented in Sect. 2 (uniqueness and consistency), and in Sect. 4 (asymptotic distributions). The main results in Sects. 2 and 4 are presented with complete proofs. Section 4 plays a central role for inference in the present context, and it contains some improvements of earlier results.

It has been shown in data examples that the nonparametric procedures based on Fréchet means often greatly outperform their parametric counterparts (See Bhattacharya and Bhattacharya 2012). Misspecification of the model is a serious issue with parametric inference, especially for distributions on rather complex non-Euclidean spaces.

In this article two types of images and their analysis are distinguished. The greater emphasis is on landmarks based shapes introduced by Kendall (1984) and Bookstein (1991). This looks at a k-*ad* or a set of k properly chosen points, not all the same, on an m-dimensional image (usually $m = 2$ or 3), $k > m$, such as an MRI scan of a section of the brain for purposes of diagnosing a disease, or a scan of some organ of a species for purposes of morphometrics. In order to properly compare images taken from different distances and angles using perhaps different machines, the *shape of a* k-*ad* is defined modulo translation, scaling and rotation. The resulting shapes comprise *Kendall's shape spaces.* In addition, one may consider *affine shapes* which are invariant under all affine transformations appropriate in scene recognition; similarly, *projective shapes* invariant under projective transformations are often used for robotic vision. The precise mathematical (geometric) descriptions of these kind of images are presented in Sect. 3. Section 5 provides the asymptotic theory of tests and confidence regions on manifolds, based on the asymptotic distribution theory developed in Sect. 4.

Section 7 considers briefly the second type of images, namely, the actual geometric shape of a compact two-dimensional surface or a three dimensional body. Here the shape space is infinite dimensional and may be viewed as a *Hilbert manifold* (Ellingson et al. 2013). For purposes of diagnostics such as described above, this is probably not to be preferred in comparison with the finite dimensional landmarks based shapes considered by Kendall, because of the curse of dimensionality. The Hilbert manifolds then are better suited for purposes of machine vision. However, for that task a more effective methodology seems to be one which builds on the exciting inquiry of Kac (1966): Can one hear the shape of a drum? It turns out that for two-dimensional compact Riemannian manifolds such as compact surfaces, the spectrum of the Laplace–Beltrami operator identifies the manifold in most cases, although there are exceptions. In three and higher dimensions, on the other hand, iso-spectral manifolds are not so rare (Milnor 1964; Gordon et al. 1992; Zelditch 2000). Still, computer scientists and other researchers in machine vision have successfully implemented algorithms to identify two and three-dimensional images by the spectrum of their Laplacians, sometimes augmented by their eigen-functions (Demmel et al. 1999; Gotsman et al. 2003; Jain and Zhang 2007; Shamir 2006; Reuter et al. 2009). A mathematical breakthrough was achieved by Jones et al. (2008), who proved that indeed compact manifolds are determined by this augmentation.

Section 6 is devoted to another very important statistical problem: *nonparametric classification via density estimation, and nonparametric regression* on manifolds. In particular, we emphasize Ferguson's nonparametric Bayes theory of using *Dirichlet process priors* for this endeavor (Ferguson 1973, 1974).

Section 8 provides a number of applications of the theory of Fréchet means, including Fisher's example mentioned above, but focusing on two-sample problems on landmarks based shape spaces such as those introduced by Kendall (1984, 1989).

The appendix, provides a ready access to some notions in Riemannian geometry used in the text.

Remark 1. This paper is a survey of recent developments in the field. Some of the material presented in this article has appeared previously, and is collected here for the reader's convenience. Sections 2 and 3 have some overlaps with Bhattacharya (2013) and Bhattacharya and Oliver (2019); Sect. 4 is taken in part from Bhattacharya and Lin (2017); Sects. 5 and 6 contain material previously announced in Bhattacharya (2013); the material in Sect. 7 may be found in Bhattacharya and Oliver (2019); Sect. 8 brings together examples appearing in Bhattacharya (2013), Bhattacharya et al. (2016), Bhattacharya and Lin (2017), and Bhattacharya and Oliver (2019). The reader is referred to the original sources for further details and proofs.

2 Existence of the Fréchet Mean on Non-Euclidean Spaces.

Let (S, ρ) be a metric space and Q a probability measure on it. The *Fréchet function* of Q is defined as

$$F(p) = \int \rho^2(p, q)Q(\mathrm{d}q), p \in S. \tag{1}$$

If F is finite at some p then it is finite on S. The set $C(Q)$ of minimizers of F is called the Fréchet mean set. If the minimizer is unique, i.e., $C(Q)$ is a singleton, then it is called the *Fréchet mean* of Q, and one says that the *Fréchet mean* of Q exists. We will often use the topological condition

$$\textit{All closed bounded subsets of } S \textit{ are compact.} \tag{2}$$

When S is a Riemannian manifold and $\rho = \rho_g$ is the geodesic distance on it, then (2) is equivalent to the completeness of S, by the Hopf–Rinow Theorem (Do Carmo 1992, pp. 146–149).

Let $X_1, \ldots, X_n$ be a random sample from Q, i.e., X_j are i.i.d. with common distribution Q, defined on a probability space $(\Omega, \mathcal{F}, P)$. Denote by F_n the Fréchet function of the empirical $Q_n = (1/n)\sum_{1\le j\le n} \delta_{X_j}$, where δ_x is the point mass at x. Also let $B^\epsilon = \{p \in S : \rho(p, B) < \epsilon\}$ for $B \subset S$.

Theorem 1 (Bhattacharya and Patrangenaru 2003**).** *Assume* (2) *and that the Fréchet function F of Q is finite. Then (a) $C(Q)$ is nonempty and compact, and (b) for each $\epsilon > 0$, there exists a random positive integer $N = N(\omega; \epsilon)$ and a P-null set Γ such that $\forall n \ge N(\omega; \epsilon)$,*

$$C(Q_n) \subset (C(Q))^\epsilon \textit{ for every } \omega \notin \Gamma. \tag{3}$$

(c) In particular, if the Fréchet mean of Q, say μ, exists, then every measurable selection μ_n from $C(Q_n)$, converges almost surely to μ. In this case μ_n is called the sample Fréchet mean.

Proof. First assume S is compact. Then (a) is obvious. To prove (b), it is enough to show that $\delta_n = \max\{\mid F_n(p) - F(p) \mid : p \in S\} \to 0$ almost surely as $n \to \infty$. To see this let $\lambda = \min\{F(p) : p \in S\} = F(q)\ \forall q \in C(Q)$. If $(C(Q))^\epsilon = S$, then (3) holds with $N = 1$ (for every ω). Assume $(C(Q))^\epsilon$ is not S, and write $M_1 = S\backslash(C(Q))^\epsilon$. There exists $\theta(\epsilon) > 0$, such that $\min\{F(p) : p \in M_1\} = \lambda + \theta(\epsilon)$. Also, there exists $\epsilon_1 > 0$, $\epsilon_1 \le \epsilon$, such that $F(p) \le \lambda + \theta(\epsilon)/4\ \forall\ p \in (C(Q))^{\epsilon_1}$. Since $\delta_n \to 0$ a.s., there exists $N = N(\omega)$ such that $\forall n \ge N$, $F_n(p) < \lambda + \theta(\epsilon)/3$ $\forall p \in (C(Q))^{\epsilon_1}$ and $F_n(p) > \lambda + \theta(\epsilon)/2\ \forall p \in M_1$, so that $C(Q_n) \subset (C(Q))^\epsilon$, proving (3). In order to show that $\delta_n \to 0$ a.s. first note that, irrespective of Q, $|F(p) - F(p')| \le c\rho(p, p')$ where $c = 2\max\{\rho(q, q') : q, q' \in S\}$. Given any $\delta > 0$, $|F(p) - F(p')| < \delta/4$ if $\rho(p, p') < \eta = \delta/4c$. Let $q_1, \ldots, q_k$ be such that the balls $B(q_i : \eta)$ with radius η and center q_i cover S. Then $|F(p) - F(q_i)| < \delta/4\ \forall p \in B(q_i : \eta)(i = 1, \ldots, k)$. The same is true with Q replaced by Q_n.

By the strong law of large numbers (SLLN), there exists $N_1 = N_1(\omega;\delta)$ such that $|F_n(q_i) - F(q_i)| < \delta/2\ \forall n \geq N_1$ $(i = 1, \ldots k)$, outside a P-null set. It follows that, outside a P-null set, $|F_n(p) - F(p|)| < |F_n(p) - F_n(q_i)| + |F_n(q_i) - F(q_i)| + |F(q_i) - F(p)| < \delta\ \forall p \in B(q_i : \eta)$ $(i = 1, \ldots, k)$, provided $n \geq N_1$.

Consider now the non-compact case, but assuming (2). Let $\lambda = \inf\{F(p) : p \in S\}$. This infimum is attained in S. To see this, let p_k $(k = 1, 2, \ldots)$ be such that $F(p_k) \to \lambda$ as $k \to \infty$. Since $\rho(p, q) \leq \rho(p, x) + \rho(q, x)\ \forall p, q, x$, one has

$$\rho(p, q) \leq \int \rho(p, x) Q(\mathrm{d}x) + \int \rho(q, x) Q(\mathrm{d}x) \leq F^{1/2}(p) + F^{1/2}(q), \forall p, q \in S. \quad (4)$$

Letting $p = p_1$ and $q = p_k$, one obtains $\limsup_k \rho(p_k, p_1) < \infty$. Hence the sequence $\{p_k\}$ is bounded, and its closure is compact. Therefore, there exists p^* such that $F(p^*) = \lambda$. Thus $C(Q)$ is nonempty and closed. If q is any point in $C(Q)$ then taking $p = p^*$ and $q \in C(Q)$ in (4), one has $\rho(p^*, q) \leq 2\lambda^{1/2}$. That is $C(Q) \subset B(p^*, \lambda^{1/2})$. Thus part (a) is proved. To prove part (b), one has, using Q_n for Q and a fixed point p^* for q in $C(Q)$ in (4), the inequality $F_n^{1/2}(p) \geq \rho(p, p^*) - F_n^{1/2}(p^*)$, $\forall p$. Fix a $\delta > 0$. Consider the compact set $M_1 = \{q : \rho(q, p^*) \leq 2(\lambda + \delta)^{1/2} + \lambda^{1/2}\}$. Then for $p \in S \backslash M_1$, one has $F_n(p) \geq [2(\lambda + \delta)^{1/2} + \lambda^{1/2} - F_n^{1/2}(p^*)]^2 > \lambda + \delta$, $F_n(p^*) < \lambda + \delta$ for all sufficiently large $n \geq N_1 = N_1(\omega)$ except for ω lying in a P-null set, in view of the SLLN. Hence $C(Q_n) \subset M_1$ for $n \geq N_1$. Applying the result in the compact case (with $S = M_1$), one arrives at (b). Part (c) is an immediate consequence of part (b). □

For compact metric spaces S, part (c) of Theorem 1 follows from Ziezold (1977).

Remark 2. Theorem 1 extends to more general Fréchet functions, including $F(p) = \int \rho^\alpha(p, q) Q(\mathrm{d}q)$, $\alpha \geq 1$.

Remark 3. Relation (3) does not imply that the sets $C(Q)$ and $C(Q_n)$ are asymptotically close in the Hausdorff distance. Indeed, in many examples $C(Q_n)$ may be a singleton, while $C(Q)$ is not. See, e.g., Bhattacharya and Patrangenaru (2003), Remark 2.6, where it is shown that whatever be the absolutely continuous distribution Q on S^1, $C(Q_n)$ is almost surely a singleton; in particular, this is the case when Q is the uniform distribution for which $C(Q) = S^1$. In view of this, and for asymptotic distribution theory considered later, it is important to find broad conditions on Q for the *existence of the Fréchet mean* (as the unique minimizer of the Fréchet function).

Let $S = M$ be a *differentiable manifold* of dimension d – a topological space which is metrizable as a separable metric space such that (i) every $p \in M$ has an open neighborhood U_p with a homeomorphism $\psi_p : U_p \to B_p$, where B_p is an open subset of $\mathbb{R}^d$, and (ii) (compatibility condition) if $U_p \cap U_q$ is nonempty, then the map $\psi_p \circ \psi_q^{-1} : \psi_q(U_p \cap U_q) \to \psi_p(U_p \cap U_q)$ is a C^∞ – a common example is the sphere $S^d = \{x \in \mathbb{R}^{d+1} : |x| = 1\}$; one may take p as the north pole $(0, 0, \ldots, 0, 1)$ and q as the south pole $(0, 0, \ldots, 0, -1)$, $U_p = S^d \backslash \{q\}$, $U_q = S^d \backslash \{p\}$, and ψ_p and ψ_q are the stereographic projections on $S^d \backslash \{q\}$ and $S^d \backslash \{p\}$, respectively, onto $\mathbb{R}^d$. Or, one may take $2d$ open hemispheres U_p of S^d with poles whose coordinates are all zeros, except for $+1$ or -1 at the i-th coordinate ($i = 1, \ldots, d$), each mapped diffeomorphically onto the open unit disc in $\mathbb{R}^d$. There are infinitely many distances which metrize the topology of M. The two most common are (1) the Euclidean distance under an embedding, and (2) the geodesic distance when M is endowed with a metric tensor. For the first, recall that a smooth (C^∞) map $J : M \to E^N$ is an embedding into an Euclidean space E^N, if (a) J is one-to-one and $M \to J(M)$ is a homeomorphism with $J(M)$ given the relative topology of E^N, and (b) the differential $\mathrm{d}_p J$ on the tangent space $T_p(M)$ into the tangent space of E^N at $J(p)$ is one-to-one. The Euclidean distance on $J(M)$ (transferred to M via J^{-1}) is called the *extrinsic distance* ρ_J on M. The embedding is said to be closed if $J(M)$ is closed. For S^d one may, for example, take J to be the inclusion map of S^d into $\mathbb{R}^{d+1}$, and the extrinsic distance is the chord distance.

Theorem 2 (Extrinsic Fréchet Mean on a Manifold, (Bhattacharya and Patrangenaru 2005)**).** *Let M be a differentiable manifold and Q a probability measure on it. If J is a closed embedding of M into an Euclidean space E^N, and the Fréchet function of Q is finite with respect to the induced Euclidean distance on $J(M)$, then the (extrinsic) Fréchet mean exists as the unique minimizer of the Fréchet function if and only if there is a unique point $\mu_{J,E}$ in $J(M)$ closest to the Euclidean mean m of the (push forward) distribution $Q_J = Q \circ J^{-1}$ on E^N, and then the extrinsic mean is $J^{-1}\mu_{J,E}$.*

Proof. For a point $c \in J(M)$, writing $|y|^2 = \sum_{i=1}^N (y^{(i)})^2$ for the usual squared Euclidean norm on E^N,

$$\begin{aligned}\int_{J(M)} |c-y|^2 Q_J(\mathrm{d}y) &= \int_{E^N} |c-y|^2 Q_J(\mathrm{d}y) \\ &= \int_{E^N} |m-y|^2 Q_J(\mathrm{d}y) + |c-m|^2.\end{aligned}$$

This is minimized with respect to c, by setting c to be the point in $J(M)$ closest to m, if there is only one such point, and the minimizer is not unique otherwise. □

Example 1 (Extrinsic Mean on the Sphere S^d). Let the inclusion map on S^d into $\mathbb{R}^{d+1}$ be the embedding J. Then the mean m of Q_J on $\mathbb{R}^{d+1}$ lies inside the unit ball $B(0; 1)$ in $\mathbb{R}^{d+1}$ unless Q is degenerate at a point $m \in S^d$. If Q is nondegenerate, the closest point to m in S^d is $m/|m|$ unless $m = 0$ (i.e., m lies at the center of the unit ball). Thus (the image of) the extrinsic mean is $\mu_{J,E} = m/|m|$. If $m = 0$, then $C(Q) = S^d$. If Q is degenerate at m, then m is the extrinsic mean. Taking Q to be the empirical Q_n, the sample Fréchet mean is $\bar{X}/|\bar{X}|$, if $\bar{X}$ is not the origin in $\mathbb{R}^{d+1}$. If $\bar{X} = 0$, then $C(Q_n) = S^d$.

Theorem 2 allows one in many important cases of interest in image analysis to find analytic characterizations for the existence of the extrinsic mean (i.e., as the unique minimizer of the Fréchet function) and computable formulas for its computation. This will be discussed in Sect. 3.

Unfortunately, on a Riemannian manifolds M with metric tensor g there is no good analog of Theorem 2 for the *intrinsic mean* of Q, – the minimizer of the Fréchet function under the geodesic distance ρ_g. The pioneering work by Karcher (1977) followed by generalizations and strengthening, most notably, by Kendall (1990), Le (2001) and Afsari (2011) hold under support restrictions on Q, which are untenable for general statistical inference. The recent results of Afsari (2011) are the sharpest among these, which we state below (for the Fréchet function (1)) without proof. For the terminology used in the statement we refer to the Appendix on Riemannian geometry. Recall that the support of a probability measure Q on a metric space is the smallest closed set D such that $Q(D) = 1$.

Theorem 3 (Intrinsic Mean on a Riemannian Manifold, (Afsari 2011)). *On a complete Riemannian manifold (M, g), there exists an intrinsic Frećhet mean of Q, as the unique minimizer of the Frećhet function* (1) *with the geodesic distance $\rho = \rho_g$, if the support of Q is contained in a geodesic ball of radius less than $r^* = (1/2)\min\{\mathrm{inj}(M), \pi/\sqrt{\bar{C}}\}$. Here $\mathrm{inj}(M)$ is the injectivity radius of M; and $\bar{C}$ is the supremum of sectional curvatures of M, if positive, or zero otherwise.*

Remark 4. If the Riemannian manifold M is complete, simply connected and has non-positive curvature and the Fréchet function of Q is finite, then the intrinsic mean of Q exists (as the unique minimizer of F). An important generalization of this is to the so called *metric spaces of non-positive curvature*, or the NPC spaces, which include many interesting metric spaces which are not manifolds. Such spaces were introduced by Alexandrov (1957) and further developed by Reshetnyak (1968) and Gromov (1981). See Sturm (2003) for a fine exposition.

Example 2. Let $M = S^2$. Then it has constant sectional curvature 1, and its injectivity radius is π. Thus if Q has support contained in an open hemisphere, then the Fréchet mean of Q under the geodesic distance exists. To see that one cannot relax this support condition in general, consider the uniform distribution on the equator. Then the minimum expected squared distance is attained at both the North and South poles (say, (0, 0, 1), and (0, 0, −1)), so that $C(Q)$ has two points.

Remark 5. For purposes of statistical inference the support condition in Theorem 3 is restrictive, but as Example 2 shows one cannot dispense with the support condition without some further conditions on the nature of Q. In statistical practice a reasonable assumption is that the distribution is absolutely continuous. In S^1 under the assumption that Q has a continuous density (with respect to the arc length measure on intervals, i.e., the Lebesgue measure on $[0, 2\pi)$) a necessary and sufficient condition, which applies broadly, was obtained in Bhattacharya (2007) and may be found in Bhattacharya and Bhattacharya (2012), pp. 31–33, 73–75.

3 Geometry of Shape Spaces

In this section, we describe the geometry of some well-known shape spaces including Kendall type similarity shape spaces, affine shape spaces and projective shape spaces. Some of the description in this section is a partial repetition of Sect. 6 in one of the co-author's earlier paper (Bhattacharya 2013).

3.1 Kendall's Similarity Shape Space Σ_m^k

The similarity shape of a k-ad $x = (x_1, \cdots, x_k)$ in $\mathbb{R}^m$, not all points the same, is its orbit under the group generated by translations, scaling and rotations. Writing $\bar{x} = (x_1 + \cdots + x_k)/k$, $<\bar{x}> = (\bar{x}, \cdots, \bar{x})$, the effect of translation is removed by looking at $(x_1 - \bar{x}, \cdots, x_k - \bar{x}) = x - <\bar{x}>$, which lies in the $mk - m$ dimensional hyperplane L of $\mathbb{R}^{mk}$ made up of $m \times k$ matrices with the m row sums all equal to zero. To get rid of scale, one looks at $u = (x - <\bar{x}>)/|x - <\bar{x}>|$, where $|.|$ is the usual norm in $\mathbb{R}^{mk}$. This translated and scaled k-ad is called the *preshape* of the k-ad. It lies on the unit sphere in L, and is isomorphic to $S^{m(k-1)-1}$. An alternative representation of the preshape, which we use, is obtained as $p = xH/|xH|$, where H is the $k \times (k-1)$ *Helmert matrix* comprising $k-1$ column vectors forming an orthonormal basis of $1^{\perp}$, namely, the subspace of $\mathbb{R}^k$ orthogonal to $(1, \cdots, 1)'$. A standard H has the j-th column given by $(a(j), \cdots, a(j), -ja(j), 0, \cdots, 0)'$, where the first j elements equal $a(j) = [j(j+1)]^{-1/2}$ $(j = 1, \cdots, k-1)$. Then p is an $m \times (k-1)$ matrix of norm one. The *shape* $\sigma(x) = \sigma(p)$ of x is then identified with the orbit of p under all rotations:

$$\sigma(x) = \sigma(p) = \{Ap : A \in SO(m)\}, \tag{5}$$

where $SO(m) = \{\mathrm{A} : AA' = I_m, \det(A) = 1\}$ is called the *special orthogonal group* acting on $\mathbb{R}^m$. The set of all shapes $\sigma(p)$ is Kendall's *similarity shape* space Σ_m^k.

If $m = 2$, $k > 2$, the action of $SO(2)$ on the preshape sphere S^{2k-3} is *free*, i.e., no $A \in SO(2)$ other than the identity has a fixed point and each orbit of a point in S^{2k-3} has an orbit of dimension one, namely the dimension of $SO(2)$. Since each $A \in SO(2)$ is an isometry of S^{2k-3} endowed with the geodesic distance, it follows that $\Sigma_2^k = S^{2k-3}/SO(2)$ is a Riemannian manifold. For $m > 2$, $k > m$, however, the action of $SO(m)$ on $S^{m(k-1)-1}$ is not free. For example, for $m = 3$, each collinear k-ad in $S^{3(k-1)-1}$ is invariant under all rotations in $\mathbb{R}^3$ around the line of the k-ad. Σ_3^k is then a disjoint union of two Riemannian manifolds, not complete, one comprising of the orbits of collinear k-ads under rotation by elements of $SO(2)$ other than those that keep it fixed (except for the identity). The other comprises of orbits under $SO(3)$ of all non-collinear k-ads in $S^{3(k-1)-1}$. Σ_3^k is then a stratified space with two strata. More generally, Σ_m^k, $m > 2$ $(k > m)$, is a *stratified space* with $m - 1$ strata. See Kendall et al. (1999), Chapter 6, for a complete description of the intrinsic geometry of Σ_m^k. Also see Huckemann et al. (2010) for intrinsic analysis of more general stratified spaces of the form $M = N/\mathcal{G}$, where N is a Riemannian manifold and $\mathcal{G}$ is a Lie group of isometries acting on N.

Intrinsic Geometry of Σ_2^k. For the case $m = 2$, it is convenient to regard a k-ad $x = ((x_1, y_1), \cdots, (x_k, y_k))$ as a k-tuple $z = (z_1, \cdots, z_k)$ of numbers $z_1 = x_1 + iy_1, \cdots, z_k = x_k + iy_k$ in the complex plane $\mathbb{C}$, and let $p = (z - <\bar{z}>)/|z - <\bar{z}>|$. Then the shape of p, or z, is identified with the orbit

$$\sigma(z) = \sigma(p) = \{e^{i\theta} p : \theta \in (-\pi, \pi]\}. \tag{6}$$

One may equivalently, consider the shape as the orbit $\{\lambda(z - <\bar{z}>) : \lambda \in \mathbb{C}\}$. That is, after Helmertization, the shape of x, or z, is identified with a complex line passing through the origin in $\mathbb{C}^{k-1}$. The shape space is then identified with the *complex projective space* $\mathbb{C}P^{k-2}$, of (real) dimension $2k - 4$. We will, however, use the representation $\Sigma_2^k = \mathbb{C}S^{k-1}/G$, where $G = \{e^{i\theta} : \theta \in (-\pi, \pi]\}$ is a 1-dimensional compact group ($G \simeq S^1$) of isometries of the preshape sphere $\mathbb{C}S^{k-1} = \{p = (p_1, \cdots, p_{k-1}) \in \mathbb{C}^{k-1} : |p| = 1\}$, which is isomorphic to S^{2k-3}. Recall that the metric tensor on $S^{2k-3} \simeq \mathbb{C}S^{k-1}$ is that inherited from the inclusion map into $\mathbb{R}^{2(k-1)} = \{(x_1, y_1, x_2, y_2, \cdots, x_{k-1}, y_{k-1}) : (x_j, y_j) \in \mathbb{R}^2\ \forall j\} \simeq \mathbb{C}^{k-1} = \{(z_1, z_2, \cdots, z_{k-1}) : z_j = x_j + iy_j \in \mathbb{C}\ \forall j\}$. That is, the inner product at the tangent space $T_p\mathbb{C}S^k$ is $\langle \tilde{v}, \tilde{w} \rangle = \mathrm{Re}(vw^*)$, when $\tilde{v}$, $\tilde{w}$ are expressed as complex $1 \times (k-1)$ matrices (row vectors) in $\mathbb{C}S^{k-1}$, satisfying $\mathrm{Re}(p\tilde{v}^*) = 0 = \mathrm{Re}(p\tilde{w}^*)$. The projection map is then $\pi : p \to \sigma(p)$. The *vertical subspace* V_p is obtained by differentiating the curve $\theta \to e^{i\theta} p$, say at $\theta = 0$, yielding ip. That is, $V_p = \{cip : c \in \mathbb{R}\}$. Thus the *horizontal subspace* is $H_p = \{\tilde{v} : \mathrm{Re}(p\tilde{v}^*) = 0, \mathrm{Re}((ip)\tilde{v}^*) = 0\} = \{\tilde{v} : p\tilde{v}^* = 0\}$. The *geodesics* $\gamma(t; \sigma(p), v)$ for $v = (d_p\pi)\tilde{v}$ (for $\tilde{v}$ in H_p), and the exponential map $\mathrm{Exp}_{\sigma(p)}$ on Σ_2^k are specified by this isometry between $T_{\sigma(p)}(\Sigma_2^k)$ and H_p for all shapes $\sigma(p)$ (See the Appendix A). Thus, identifying vectors v in H_p with vectors v in $T_{\sigma(p)}(\Sigma_2^k)$, one obtains

$$\begin{aligned} T_{\sigma(p)}(\Sigma_2^k) &= \{v = (d_p\pi)\tilde{v} : \forall v \ \text{ such that } p\tilde{v}^* = 0\}, \\ \mathrm{Exp}_{\sigma(p)}\, 0 &= \sigma(p), \\ \mathrm{Exp}_{\sigma(p)}\, v &= \sigma(\cos(|\tilde{v}|)p + \sin(|\tilde{v}|)\tilde{v}/|\tilde{v}|)\ (v \neq 0,\ p\tilde{v}^* = 0), \\ \gamma(t; \sigma(p), v) &= \sigma((\cos t|\tilde{v}|)p + (\sin t|\tilde{v}|)\tilde{v}/|\tilde{v}|),\ (t \in \mathbb{R}, p\tilde{v}^* = 0),\ v \neq 0. \end{aligned} \tag{7}$$

Denoting by ρ_{gs} and ρ_g the geodesic distances on $\mathbb{C}S^{k-1}$ and Σ_2^k, respectively, and recalling that (See Example 10.1 and Kendall et al. 1999, p. 114) $\rho_{gs}(p, q) = \arccos(\mathrm{Re}\, pq^*)$, one has

$$\begin{aligned} \rho_g(\sigma(p), \sigma(q)) &= \inf\{\rho_{gs}(p, q) : p \in O_u, q \in O_w\} \\ &= \inf\{\arccos(\mathrm{Re}\, e^{i\theta}\, pq^*) : \theta \in [0, 2\pi)\} \\ &= \arccos(|pq^*|) \in [0, \pi/2]. \end{aligned} \tag{8}$$

It follows that the geodesics are periodic with period π, and the cut locus of $\sigma(p)$ is $\{\sigma(q) : \text{all } q \ \text{ such that} \arccos(|pq^*|) = \pi/2\}$, and that the injectivity radius of Σ_2^k is $\pi/2$. The inverse exponential map is given by $\mathrm{Exp}_{\sigma(p)}^{-1}(\sigma(q)) = v$, where $v = (\mathrm{d}_p\pi)\tilde{v}$ ($\tilde{v} \in H_p$), and $\tilde{v}$ satisfies (Use (A.3) with the representation of S^{2k-3} as $\mathbb{C}S^{k-1}$)

$$\begin{aligned}\tilde{v} &= \mathrm{Exp}_p^{-1}(q\,\mathrm{e}^{i\theta}) \qquad (9)\\ &= [\arccos(\mathrm{Re}(pq^*\,\mathrm{e}^{-i\theta})](1-[\mathrm{Re}(pq^*\,\mathrm{e}^{-i\theta})]^2)^{-1/2}(q\,\mathrm{e}^{i\theta} - (pq^*\,\mathrm{e}^{-i\theta})p),\end{aligned}$$

where θ is so chosen as to minimize $\rho_{gs}(p, q\,\mathrm{e}^{i\theta}) = \arccos(\mathrm{Re}(pq^*\,\mathrm{e}^{-i\theta}))$. That is, $(pq^*\,\mathrm{e}^{-i\theta}) = |pq^*|$, or $\mathrm{e}^{i\theta} = pq^*/|pq^*|$ (for $pq^* \neq 0$, i.e., for $\sigma(q)$ not in $C(\sigma(p))$). Hence, writing $\rho = \rho_g(\sigma(p), \sigma(q))$, $\rho \neq 0$, one has

$$\begin{aligned}\tilde{v} &= [\arccos(|pq^*|)](1-|pq^*|^2)^{-1/2}\{(pq^*/|pq^*|)q - |pq^*|p\} \qquad (10)\\ &= [\rho/\sin\rho]\{q\,\mathrm{e}^{i\theta} - (\cos\rho)p\} \qquad (\mathrm{e}^{i\theta} = pq^*/\cos\rho).\end{aligned}$$

This horizontal vector $\tilde{v}$ ($\in H_p$) represents $\mathrm{Exp}_{\sigma(p)}^{-1}(\sigma(q)) = v$.

The sectional curvature of Σ_2^k at a section generated by two orthonormal vector fields $\tilde{W}_1$ and $\tilde{W}_2$ is $1 + 3\cos^2\phi$ where $\cos\phi = \langle U_1, iU_2\rangle$, U_1 and U_2 being the horizontal lifts of $\tilde{W}_1$ and $\tilde{W}_2$ (See Do Carmo 1992).

Extrinsic Geometry of Σ_2^k Induced by an Equivariant Embedding. As mentioned in Sect. 2, no broad sufficient condition is known for the existence of the intrinsic mean (i.e., of the uniqueness of the minimizer of the corresponding Fréchet function). The extrinsic mean, on the other hand, is unique for most Q, and is generally computable analytically. However, for an extrinsic analysis to be very effective one should choose a good embedding which retains as many geometrical features of the shape manifold as possible. Let Γ be a Lie group acting on a differentiable manifold M, and denote by $GL(N, \mathbb{R})$ the *general linear group* of nonsingular transformations on a Euclidean space E^N of dimension N onto itself. An embedding J on M into E^N is said to be Γ-*equivariant* if there exists a group homomorphism $\Phi : \gamma \to \phi_\gamma$ of Γ into $GL(N, \mathbb{R})$ such that $J(\gamma p) = \phi_\gamma(Jp)$ $\forall p \in M$, $\gamma \in \Gamma$. Often, when there is a natural Riemannian structure on M, Γ is a group of isometries of M. Consider the so-called *Veronese–Whitney embedding* J of Σ_2^k into the (real) vector space $S(k-1, \mathbb{C})$ of all $(k-1)\times(k-1)$ Hermitian matrices $B = B^*$, defined by

$$J\sigma(p) = p^*p \qquad [\sigma(p) = \{\mathrm{e}^{i\theta}p, \theta \in [0, 2\pi), p \in \mathbb{C}S^{k-1}\}]. \qquad (11)$$

The Euclidean inner product on $S(k-1, \mathbb{C})$, considered as a real vector space, is given by $\langle B, C\rangle = \mathrm{Re}(\mathrm{Trace}(BC^*))$. Let $SU(k-1)$ denote the *special unitary group* of all $(k-1)\times(k-1)$ unitary matrices A (i.e., $A^*A = I$, $\det(A) = 1$) acting on $S(k-1, \mathbb{C})$ by $B \to A^*BA$. Then the embedding (11) is Γ-equivariant, with $\Gamma = \{\gamma_A : A \in SU(k-1)\}$ and the group action on Σ_2^k given by: $\gamma_A\sigma(p) = \sigma(pA)$. For $J\sigma(pA) = A^*p^*pA = \phi(\gamma_A)(J\sigma(p))$, say, where the group homomorphism on Γ onto $SU(k-1)$ is given by $\gamma_A \to \phi(\gamma_A) : \phi(\gamma_A)B = A^*BA$. Note that $SU(k-1)$ is a group of isometries of $S(k-1, \mathbb{C})$. In the notation defining equivariance, one lets $S(k-1, \mathbb{C}) = E^N$ ($N = (k-1)^2$), $SU(k-1)$ is a subgroup of $GL(N, \mathbb{R})$.

To compute the extrinsic mean of Q on Σ_2^k, let $Q_J = Q \circ J^{-1}$ be the probability induced on $S(k-1, \mathbb{C})$ by the map J in (11), and let μ_J denote its Euclidean mean. By Theorem 1, the (image of the) extrinsic mean of Q is given by the orthogonal projection P on $J(\Sigma_2^k)$.

Proposition 1 (Bhattacharya and Patrangenaru 2003). *The image under J of the extrinsic mean of Q comprises all elements of the form w^*w where w^* is a normalized (column) eigenvector with the largest eigenvalues of μ_J. In particular, the extrinsic mean of Q exists if and only if the largest eigenvalue of μ_J is simple.*

Proof. Let T be a $(k-1)\times(k-1)$ unitary matrix such that $T\mu_J T^* = D \equiv \text{diag}(\lambda_1, \lambda_2, \ldots, \lambda_{k-1})$ where $\lambda_1 \le \lambda_2 \le \cdots \le \lambda_{k-1}$ are the ordered eigenvalues of μ_J. Then the columns of T^* form a complete orthonormal set of eigenvectors of μ_J. By relabelling the landmarks, if necessary, we may assume that the ith column of T^* is an eigenvector with eigenvalue λ_i. Write $\|A\|^2 = \text{Trace}(AA^*)$ as the square of the Euclidean norm of A. Then for elements w^*w of $J(\Sigma_2^k) = \{w^*w, w \in \mathbb{C}S^{k-1}\}$, denoting $v = wT^*$, one has

$$\begin{aligned}\|w^*w - \mu_J\|^2 &= \|Tw^*wT^* - T\mu_J T^*\| = \|v^*v - D\|^2 \\ &= \sum_{i,j} |\bar{v}_i v_j - \lambda_i \delta_{ij}|^2 = 1 + \sum_i \lambda_i^2 - 2\sum_i \lambda_i |v_i|^2\end{aligned}$$

which is minimized over $J(\Sigma_2^k)$ by taking $v_{k-1} = 1$ and $v_i = 0\ \forall i < k-1$, i.e., by taking $w^* = T^*v^*$ be any normalized eigenvector of μ_J with the largest eigenvalue. □

A *size-and-shape similarity shape* $s\sigma(z)$ is defined for Helmertized k-ads $z = (z_1, \cdots, z_{k-1})$ as its orbit under $SO(m)$. An equivariant embedding for it is $s\sigma(z) \to z^*z/|z|$, on the *size-and-shape-similarity shape* space $S\Sigma_2^k$ into $S(k-1, \mathbb{C})$.

3.2 Reflection Similarity Shape Space $R\Sigma_m^k$, $m > 2$, $k > m$

For $m > 2$, let $\tilde{N}S^{m(k-1)-1}$ be the subset of the centered preshape sphere $S^{m(k-1)-1}$ whose points p span $\mathbb{R}^m$, i.e., which, as $m \times k$ matrices, are of full rank. We define the *reflection similarity shape* of the k-ad as

$$r\sigma(p) = \{Ap : A \in O(m)\}\ (p \in \tilde{N}S^{m(k-1)-1}), \tag{12}$$

where $O(m)$ is the set of all $m \times m$ *orthogonal matrices* $A : AA' = I_m$, $\det(A) = \pm 1$. The set $\{r\sigma(p) : p \in \tilde{N}S^{m(k-1)-1}\}$ is the *reflection similarity shape space* $R\Sigma_m^k = \tilde{N}S^{m(k-1)-1}/O(m)$. Since $\tilde{N}S^{m(k-1)-1}$ is an open subset of the sphere $S^{m(k-1)-1}$, it is a Riemannian manifold. Also $O(m)$ is a compact Lie group of isometries acting on $S^{m(k-1)-1}$. Hence there is a unique Riemannian structure on $R\Sigma_m^k$ such that the projection map $p \to r\sigma(p)$ is a Riemannian submersion.

We next consider a useful embedding of $R\Sigma_m^k$ into the vector space $S(k, \mathbb{R})$ of all $k \times k$ real symmetric matrices (See Bandulasiri and Patrangenaru 2005; Bandulasiri et al. 2009; Dryden et al. 2008; and Bhattacharya 2008). Define

$$J(r\sigma(p)) = p'p\ (p \in \tilde{N}S^{m(k-1)-1}), \tag{13}$$

with p an $m\times(k-1)$ matrix with norm one. Note that the right side is a function of $r\sigma(p)$. Here the elements p of the preshape sphere are Helmertized. To see that this is an embedding, we first show that J is one- to-one on $R\Sigma_m^k$ into $S(k-1,\mathbb{R})$. For this note that if $J(r\sigma(p))$ and $J(r\sigma(q))$ are the same, then the Euclidean distance matrices $((|p_i-p_j|))_{1\leq i\leq j\leq k-1}$ and $((|q_i-q_j|))_{1\leq i\leq j\leq k-1}$ are equal. Since p and q are centered, by geometry this implies that $q_i = Ap_i (i = 1,\cdots,k-1)$ for some $A \in O(m)$, i.e., $r\sigma(p) = r\sigma(q)$. We omit the proof that the differential d_pJ is also one-to-one. It follows that the embedding is equivariant with respect to a group action isomorphic to $O(k-1)$.

Proposition 2 (Bhattacharya 2008**).** *(a) The projection of $\tilde{\mu}$ into $J(R\Sigma_m^k)$ is given by*

$$P_{J(R\Sigma_m^k)}(\tilde{\mu}) = \left\{A\colon A = \sum_{j=1}^{m}\left(\lambda_j - \bar{\lambda} + \frac{1}{m}\right)U_jU_j{}'\right\} \tag{14}$$

where $\lambda_1 \geq \ldots \geq \lambda_k$ are the ordered eigenvalues of $\tilde{\mu}$, $U_1,\ldots,U_k$ are corresponding orthonormal (column) eigenvectors and $\bar{\lambda} = \sum_{j=1}^{m}\lambda_j/m$. (b) The projection set is a singleton and Q has a unique extrinsic mean μ_E iff $\lambda_m > \lambda_{m+1}$. Then $\mu_E = \sigma(F)$ where $F = (F_1,\ldots,F_m)'$, $F_j = \sqrt{\lambda_j - \bar{\lambda} + \frac{1}{m}}\,U_j$.

For a detailed proof see Bhattacharya (2008), or Bhattacharya and Bhattacharya (2012), p. 114, 115.

For $m > 2$, a *size-and-reflection shape* $sr\sigma(z)$ of a Helmertized k-ad z in $\mathbb{R}^m$ of full rank m is given by its orbit under the group $O(m)$. The space of all such shapes is the size-and-reflection shape space $SR\Sigma_m^k$. An $O(k-1)$-equivariant embedding of $SR\Sigma_m^k$ into $S(k-1,\mathbb{R})$ is: $J(sr\sigma(z)) = z'z/|z|$.

3.3 Affine Shape Space $A\Sigma_m^k$

Let $k > m+1$. Consider the set of all k-ads in $\mathbb{R}^m$, with full rank m as $m\times k$ matrices. The affine shape of a k-ad x may be identified with its orbit under all affine transformations:

$$\sigma(x) = \{Ax + c : A \in GL(m,\mathbb{R}), c \text{ an } m\times k \text{ matrix}\}. \tag{15}$$

If the k-ad is centered as $u = x - \lt\bar{x}\gt$, then the affine shape of x, or of u, is given by

$$\sigma(x) = \sigma(u) = \{Au : A \in GL(m,\mathbb{R})\},\ (u \text{ centered } k\text{-ad of rank } m). \tag{16}$$

The space of all such affine shapes is the *affine shape space* $A\Sigma_m^k$. Note that two Helmertized k-ads u and v (as $m\times(k-1)$ matrices of full rank) have the same shape if and only if the rows of u and v span the same m-dimensional subspace of $\mathbb{R}^{k-1}$. Hence we can identify $A\Sigma_m^k$ with the Grasmannian $G_m(k-1)$, namely, the set of all m-dimensional subspaces of $\mathbb{R}^{k-1}$ (Sparr 1992). For the Grassmann manifold, refer to Boothby (1986), p. 63, 168, 362, 363. For extrinsic analysis on $A\Sigma_m^k \simeq G_m(k-1)$, consider the embedding of $A\Sigma_m^k$ into $S(k-1, \mathbb{R})$ given by

$$J(\sigma(u)) = FF', \tag{17}$$

where $F = (f_1 \cdots f_m)$ is a $(k-1) \times m$ matrix and $\{f_1, \cdots, f_m\}$ is an orthonormal basis of the m-dimensional subspace L, say, of $\mathbb{R}^{k-1}$ spanned by the rows of u. Note that the $(k-1) \times (k-1)$ matrix FF' is idempotent and is the matrix of orthogonal projection of $\mathbb{R}^{k-1}$ onto L. It is independent of the orthonormal basis chosen. The embedding is $O(k-1)$-equivariant under the group action $\sigma(u) \to \sigma(uO)$ $(O \in O(k-1))$ on $A\Sigma_m^k$, with $O(k-1)$ acting on $S(k, \mathbb{R})$ by $A \to OAO'$.

Proposition 3 (Sugathadasa 2006**).** *The projection of $\tilde{\mu}$ into $J(A\Sigma_m^k)$ is given by*

$$P(\tilde{\mu}) = \left\{ \sum_{j=1}^{m} U_j U_j' \right\} \tag{18}$$

where $U = (U_1, \ldots, U_k) \in SO(k)$ is such that $\tilde{\mu} = U\Lambda U'$, $\Lambda = \mathrm{diag}(\lambda_1, \ldots, \lambda_k)$, $\lambda_1 \geq \ldots \geq \lambda_k = 0$. The extrinsic mean μ_E exists if and only if $\lambda_m > \lambda_{m+1}$, and then $\mu_E = \sigma(F')$ where $F = (U_1, \ldots, U_m)$.

For a proof see Bhattacharya and Bhattacharya (2012), p. 140, 141.

3.4 Projective Shape Space $P\Sigma_m^k$

First, recall that the real projective space $\mathbb{R}P^m$ is the space of all lines through the origin in $\mathbb{R}^{m+1}$. Its elements are $[p] = \{\lambda p : \lambda \in \mathbb{R}\backslash\{0\}\}$ for all $p \in \mathbb{R}^{m+1}\backslash\{0_{R^{m+1}}\}$. It is also conveniently represented as the quotient S^m/G where G is the two-point group $\{e, -e\}$, e being the identity map and $-ep = -p$ $(p \in S^m)$. That is, a line through p is identified with $\{p/|p|, -p/|p|\}$ $(p \in \mathbb{R}^{m+1}\backslash\{0_{R^{m+1}}\})$. As a consequence, there is a unique Riemannian metric tensor on $\mathbb{R}P^m = S^m/G$ such that $p \to \{p, -p\}$ is a Riemannian submersion, with $\langle u, v\rangle_{\mathbb{R}P^m} = u'v$ for all vectors u, v in $T_{[p]}\mathbb{R}P^m$. The geodesic distance is given by $\rho_g([p],[q]) = \arccos(|p'q|) \in [0, \pi/2]$, and the cut locus of $[p]$ is $\mathrm{Cut}([p]) = \{[q] : \cos(|p'q|) = \pi/2\}$, so that the injectivity radius of $\mathbb{R}P^m$ is $\pi/2$. Its sectional curvature is constant $+1$ (as it is of S^m). The exponential map of $T_{[p]}\mathbb{R}P^m$ (and its inverse on $\mathbb{R}P^m\backslash(\mathrm{Cut}([p]))$ can be easily expressed in terms of those for the sphere S^m. We will use [ ] for both representations.

The so-called *Veronese–Whitney embedding* of $\mathbb{R}P^m$ into $S(m+1, \mathbb{R})$ is given by

$$J([p]) = pp^t, \ (p = (p_1, \cdots, p_{m+1})' \in S^m). \tag{19}$$

It is clearly $O(m+1)$-equivariant, with the group action on $\mathbb{R}P^m$ as: $A[p] = [Ap]$ $(A \in O(m+1))$.

Turning to landmarks based projective shapes, assume $k > m+2$. *A frame of* $\mathbb{R}P^m$ is a set of $m+2$ ordered points $([p_1], \cdots, [p_{m+2}])$ such that every subset of $m+1$ of these points spans $\mathbb{R}P^m$, i.e., every subset of $m+1$ points of $\{p_1, \cdots, p_{m+2}\}$ spans $\mathbb{R}^{m+1}$. The *standard frame* of $\mathbb{R}P^m$ is

$$([e_1], [e_2], \cdots, [e_{m+1}], [e_1 + e_2 + \cdots + e_{m+1}]),$$

where e_i $(\in \mathbb{R}^{m+1})$ has 1 in the ith position and zeros elsewhere. A k-ad $y = (y_1, \cdots, y_k) = ([p_1], \cdots, [p_k]) \in (\mathbb{R}P^m)^k$ is in *general position* if there exist $i_1 < i_2 < \cdots < i_{m+2}$ such that $(y_{i_1}, \cdots, y_{i_{m+2}})$ is a frame of $\mathbb{R}P^m$. A *projective transformation* α on $\mathbb{R}P^m$ is defined by

$$\alpha[p] = [Ap], \ (p \in \mathbb{R}^{m+1} \backslash \{0\}) \tag{20}$$

where $A \in GL(m+1, \mathbb{R})$. The usual operation of matrix multiplication on $GL(m+1, \mathbb{R})$ then leads to a corresponding group of projective transformations on $\mathbb{R}P^m$. This is the *projective group* $PGL(m)$. Note that, for a given A in $GL(m+1, \mathbb{R})$, cA determines the same element of $PGL(m)$ for all $c \neq 0$. The *projective shape* of a k-ad $y = (y_1, \cdots, y_k) = ([p_1], \cdots, [p_k]) \in (\mathbb{R}P^m)^k$ in general position is its orbit under $PGL(m)$:

$$\sigma(y) = \{\alpha y \equiv (\alpha[p_1], \cdots, \alpha[p_k]) : \alpha \in PGL(m)\}, \tag{21}$$
$$(y = ([p_1], \cdots, [p_k] \ \text{ in general position}).$$

The *projective shape space* $PG\Sigma_m^k$ is the set of all projective shapes of k-ads in general position. Following Mardia and Patrangenaru (2005) and Patrangenaru et al. (2010), we will consider a particular dense open subset of $PG\Sigma_m^k$. Fix a set of $m+2$ indices $I = \{i_j : j = 1, \cdots, m+2\}$, $1 \leq i_1 < i_2 < \cdots < i_{m+2} \leq k$. Define $PG_I\Sigma_m^k$ as the set of shapes $\sigma(y)$ in $PG\Sigma_m^k$, $y = (y_1, \cdots, y_k) = ([p_1], \cdots, [p_k])$, such that every subset of $m+1$ points of $\{[p_{i_j}], j = 1, \cdots, m+2\}$ spans $\mathbb{R}P^m$.

The shape space $PG_I\Sigma_m^k$ (with $I = \{1, 2, \cdots, m+2\}$) may be identified with $(\mathbb{R}P^m)^{k-m-2}$ (See Mardia and Patrangenaru 2005).

4 Asymptotic Distribution Theory for Fréchet Means

This section is devoted to the asymptotic distribution theory of sample Fréchet means, which lies at the heart of statistical inference based on Fréchet means. We first present a result which is broadly applicable to distributions on manifolds as well as more general locally Euclidean spaces such as *stratified spaces*. The basic idea behind it is rather simple. Suppose a probability Q on a metric space (S, ρ) has a Fréchet mean μ. Assume also the sample Fréchet mean μ_n converges to it (a.s. or in probability), which is true in particular under the topological assumption (2). If, in local coordinates, μ and μ_n are expressed as ν and ν_n in an open subset of $\mathbb{R}^s$ for some s, then the Fréchet function F_n of Q_n, expressed in local coordinates as $\tilde{F}_n$, say, satisfies a first order condition: grad $\tilde{F}_n(\nu_n) = 0$. A Taylor expansion of the left side around ν, one expresses $\nu_n - \nu$ approximately as $-\Delta^{-1}(\nu)\text{grad}\tilde{F}_n(\nu)$, where Δ is the Hessian of $\tilde{F}$ at ν. Since grad $\tilde{F}_n(\nu)$ is the average of n s-dimensional i.i.d. random vectors, the classical CLT is applied to show that $\sqrt{n}[\nu_n - \nu]$ is asymptotically normal. Here is the precise statement. For a detailed proof see Bhattacharya and Lin (2017), Theorem 3.3. A slightly weaker version appears in Bhattacharya and Lin (2013).

Let (S, ρ) be a metric space and Q a probability measure on its Borel σ-field. As before, define the *Fréchet function* of Q as

$$F(p) = \int \rho^2(p, q) Q(\mathrm{d}q) \ (p \in S). \tag{22}$$

Assume that F is finite on S and has a unique minimizer $\mu = \operatorname{argmin}_p F(p)$. Then μ is called the *Fréchet mean* of Q (with respect to the distance ρ). Under broad conditions, the Fréchet sample mean μ_n of the empirical distribution $Q_n = \dfrac{1}{n}\sum_{j=1}^n \delta_{Y_j}$ based on independent S-valued random variables Y_j $(j = 1, \ldots, n)$ with common distribution Q is a consistent estimator of μ. That is, $\mu_n \to \mu$ almost surely, as $n \to \infty$. Here μ_n may be taken to be any measurable selection from the (random) set of minimizers of the Fréchet function of Q_n, namely, $F_n(p) = \dfrac{1}{n}\sum_{j=1}^n \rho^2(p, Y_j)$ (See Theorem 1).

The following assumptions are used in the proof of Theorem 4.

(A1) The Fréchet mean μ of Q is unique.
(A2) $\mu \in G$, where G is a measurable subset of S, and there is a homeomorphism $\phi : G \to U$, where U is an open subset of $\mathbb{R}^s$ for some $s \geq 1$ and G is given its relative topology on S. The function

$$x \mapsto h(x; q) := \rho^2(\phi^{-1}(x), q) \tag{23}$$

is twice continuously differentiable on U, for every q outside a Q-null set.
(A3) $P(\mu_n \in G) \to 1$ as $n \to \infty$.
(A4) Let $\mathrm{D}_r h(x; q) = \partial h(x; q)/\partial x_r, r = 1, \ldots, s$. Then

$$E|\mathrm{D}_r h(\phi(\mu); Y_1)|^2 < \infty, \ E|\mathrm{D}_{r,r'} h(\phi(\mu); Y_1)| < \infty \text{ for } r, r' = 1, \ldots, s. \tag{24}$$

(A5) Let $u_{r,r'}(\epsilon; q) = \sup\{|\mathrm{D}_{r,r'} h(\theta; q) - \mathrm{D}_{r,r'} h(\phi(\mu); q)| : |\theta - \phi(\mu)| < \epsilon\}$. Then

$$E|u_{r,r'}(\epsilon; Y_1)| \to 0 \text{ as } \epsilon \to 0 \text{ for all } 1 \leq r, r' \leq s. \tag{25}$$

(A6) The matrix $\Lambda = [E\mathrm{D}_{r,r'} h(\phi(\mu); Y_1)]_{r,r'=1,\ldots,s}$ is nonsingular.

Remark 6. Observe that

$$Eh(x, Y_1) = F(\phi^{-1}(x)) = E\mathrm{D}_r h(x, Y_1) = \mathrm{D}_r F(\phi^{-1}(x)), 1 \leq r \leq s, x \in U.$$

Also, $E\mathrm{D}_r h(\phi(\mu), Y_1) = \mathrm{D}_r F(\phi^{-1}(x)) \mid_{x=\phi(\mu)} = 0$, $1 \leq r \leq s$, since $F(\phi^{-1}(x))$ attains a minimum at $x = \phi(\mu)$.

Theorem 4 (Bhattacharya and Lin 2017**).** *Under assumptions* (A1)–(A6),

$$n^{1/2}[\phi(\mu_n) - \phi(\mu)] \xrightarrow{\mathcal{L}} N(0, \Lambda^{-1} C \Lambda^{-1}), \ \text{as } n \to \infty, \tag{26}$$

where C is the covariance matrix of $\{\mathrm{D}_r h(\phi(\mu); Y_1), r = 1, \ldots, s\}$.

Proof. The function $x \to F_n(\phi^{-1}x) = \frac{1}{n}\sum_{j=1}^n h(x, Y_j)$ on U attains a minimum at $\phi(\mu_n) \in U$ for all sufficiently large n (almost surely). For all such n one therefore has the first order condition

$$\nabla F_n(\phi^{-1}\nu_n) = \frac{1}{n}\sum_{j=1}^n \nabla h(\nu_n, Y_j) = 0, \tag{27}$$

where $\nu = \phi(\mu)$, $\nu_n = \phi(\mu_n)$ (column vectors in U). Here ∇ is the gradient $(\mathrm{D}_1, \ldots, \mathrm{D}_r)$. A Taylor expansion yields

$$0 = \frac{1}{n}\sum_{j=1}^n \nabla h(\nu_n, Y_j) = \frac{1}{n}\sum_{j=1}^n \nabla h(\nu, Y_j) + \Lambda_n(\nu_n - \nu) \tag{28}$$

where Λ_n is the $s \times s$ matrix given by

$$\Lambda_n = \frac{1}{n}\sum_{j=1}^n [\mathrm{D}_{r,r'} h(\theta_{n,r,r'}, Y_j)]_{r,r'=1,\ldots,s}, \tag{29}$$

and $\theta_{n,r,r'}$ lies on the line segment joining ν_n and ν. We will show that

$$\Lambda_n \to \Lambda \text{ in probability,} \quad \text{as } n \to \infty. \tag{30}$$

Fix $r, r' \in \{1, \ldots, s\}$. For $\delta > 0$, write $Eu_{r,r'}(\delta, Y_1) = \gamma(\delta)$. There exists $n = n(\delta)$ such that $P(|\nu_n - \nu| > \delta) < \delta$ for $n > n(\delta)$. Now

$$E\left|\left[\frac{1}{n}\sum_{j=1}^n \mathrm{D}_{r,r'} h(\nu_n, Y_j) - \frac{1}{n}\sum_{j=1}^n \mathrm{D}_{r,r'} h(\nu, Y_j)\right] \cdot 1_{[|\nu_n - \nu| \le \delta]}\right|$$
$$\le E\frac{1}{n}\sum_{j=1}^n u_{r,r'}(\delta, Y_j) = Eu_{r,r'}(\delta, Y_1) = \gamma(\delta) \to 0 \tag{31}$$

as $\delta \to 0$. Hence, by Chebyshev's inequality for first moments, for $n > n(\delta)$ one has for every $\epsilon > 0$,

$$P\left(\left|\frac{1}{n}\sum_{j=1}^n \mathrm{D}_{r,r'} h(\nu_n, Y_j) - \frac{1}{n}\sum_{j=1}^n \mathrm{D}_{r,r'} h(\nu, Y_j)\right| > \epsilon\right) \le \delta + \gamma(\delta)/\epsilon \to 0 \tag{32}$$

as $\delta \to 0$. This shows that

$$\left[\frac{1}{n}\sum_{j=1}^n \mathrm{D}_{r,r'} h(\nu_n, Y_j) - \frac{1}{n}\sum_{j=1}^n \mathrm{D}_{r,r'} h(\nu, Y_j)\right] \to 0; \text{ in probability as } n \to \infty. \tag{33}$$

Next, by the strong law of large numbers,

$$\frac{1}{n}\sum_{j=1}^n \mathrm{D}_{r,r'} h(\nu, Y_j) \to E\mathrm{D}_{r,r'} h(\nu, Y_1) \text{ almost surely,} \quad \text{as } n \to \infty. \tag{34}$$

Since (32)–(34) hold for all r, r', (30) follows. The set of symmetric $s \times s$ positive definite matrices is open in the set of all $s \times s$ symmetric matrices, so that (30) implies that Λ_n is nonsingular with probability going to 1 and $\Lambda_n^{-1} \to \Lambda^{-1}$ in probability, as $n \to \infty$. Note that $E\nabla h(\nu, Y_1) = 0$ (see Remark 6). Therefore, using (A4), by the classical CLT and Slutsky's Lemma, (28) leads to

$$\sqrt{n}(\nu_n - \nu) = \Lambda_n^{-1} \left[-(1/\sqrt{n}) \frac{1}{n} \sum_{j=1}^{n} \nabla\, h(\nu, Y_j) \right] \xrightarrow{\mathcal{L}} N(0, \Lambda^{-1} C \Lambda^{-1}), \qquad (35)$$

as $n \to \infty$. □

For the case of the *extrinsic mean*, let M be a d-dimensional differentiable manifold, and $J : M \to E^N$ an embedding of M into an N-dimensional Euclidean space. Assume that $J(M)$ is closed in E^N, which is always the case, in particular, if M is compact. The extrinsic distance $\rho_{E,J}$ on M is defined as $\rho_{E,J}(p, q) = |J(p) - J(q)|$ for $p, q \in M$, where $| \cdot |$ denotes the Euclidean norm of E^N. The image μ in $J(M)$ of the *extrinsic mean* $\mu_{E,J}$ is then given by $\mu = P(m)$, where m is the usual mean of $Q \circ J^{-1}$ thought of as a probability on the Euclidean space E^N, and P is the orthogonal projection defined on an N-dimensional neighborhood V of m into $J(M)$ minimizing the Euclidean distance between $p \in V$ and $J(M)$. If the projection P is unique on V then the projection $\mu_n = P(m_n)$ of the Euclidean mean $m_n = \sum_{j=1}^{n} J(Y_j)/n$ on $J(M)$ is, with probability tending to one as $n \to \infty$, unique and lies in an open neighborhood G of $\mu = P(m)$ in $J(M)$. Theorem 4 immediately implies the following result of Bhattacharya and Patrangenaru (2003) (Also see Bhattacharya and Bhattacharya (2012), Proposition 4.3). Assume that P is uniquely defined in a neighborhood of the N-dimensional Euclidean mean m of $Q \circ J^{-1}$. Let ϕ be a diffeomorphism on a neighborhood G of $\mu = P(m)$ in $J(M)$ onto an open set U in $\mathbb{R}^d$. Then, using the notation of (26),

$$\sqrt{n}\, [\phi(\mu_n) - \phi(\mu)] = \sqrt{n}\, [\phi(P(m_n)) - \phi(P(m))] \xrightarrow{\mathcal{L}} N(0, \Lambda^{-1} C \Lambda^{-1}),$$

as $n \to \infty$.

One may, in particular, choose (U, ϕ) to be a coordinate neighborhood of $\mu = P(m)$ in $J(M)$. In Bhattacharya and Patrangenaru (2003), however, ϕ is chosen to be the linear orthogonal projection on G into the tangent space $T_\mu J(M)$.

For a more computable expression of the limit, let X_j, $1 \le j \le n$, be i.i.d, M-valued observations with common distribution Q, and $Y_j = J(X_j)$, $1 \le j \le n$.

In a neighborhood V of m, the differential $\mathrm{d}_y P$ maps $T_j E^N \approx E^N$. One expresses $\mathrm{d}_{P(m)} e_j = \sum_{i=1}^{d} b_{ij} F_i$,

$$\mathrm{d}_m(\bar{Y} - m) = \sum_{j=1}^{N} \sum_{i=1}^{d} b_{ji} (\bar{Y} - m)(j) F_i$$
$$= \sum_{i=1}^{d} \left(\sum_{j=1}^{N} b_{ji} (\bar{Y} - m)^{(j)} \right) F_i \qquad (F_i = F_i(P(m))).$$

Thus one arrives at the following result.

Proposition 4. *Assume the projection P is uniquely defined and is continuously differentiable in a neighborhood V of $m = EY_j$, and $E|Y_j|^2 < \infty$. Then*

$$\sqrt{n}\mathrm{d}_m P(\bar{Y} - m) \xrightarrow{d} N(0, \Sigma) \tag{36}$$

where $\Sigma = B'CB$ with $b = ((b_{ji}))$ and C is the $N \times N$ covariance matrix of Y_j.

Corollary 1 (CLT for Intrinsic Means-I). *Let (M, g) be a d-dimensional complete Riemannian manifold with metric tensor g and geodesic distance ρ_g. Suppose Q is a probability measure on M with intrinsic mean μ_I, and that Q assigns zero mass to a neighborhood, however small, of the* cut locus *of μ_I. Let $\phi = \mathrm{Exp}_{\mu_I}^{-1}$ be the inverse exponential, or log-, function at μ_I defined on a neighborhood G of $\mu = \mu_I$ onto its image U in the tangent space $T_{\mu_I}(M)$. Assume that the assumptions* (A4)–(A6) *hold. Then, with $s = d$, the CLT* (26) *holds for the intrinsic sample mean $\mu_n = \mu_{n,I}$, say.*

Remark 7. In addition to providing a CLT for manifolds (of dimension d), Theorem 4 applies to many stratified spaces which are manifolds of different dimensions s glued together. See Bhattacharya and Lin (2017) for a simple derivation of a CLT for the so called Open Book model, originally due to Hotz et al. (2013). Another stratified space to which Theorem 4 applies is Σ_m^k, $m > 2$, $k > m$, described in Sect. 3 (see Remark 15).

Remark 8. For manifolds of dimension d (i.e., $s = d$), Theorem 4 and Corollary 1 improve upon Theorem 2.3 and 5.3 in Bhattacharya and Bhattacharya (2012) (and earlier results in Bhattacharya and Patrangenaru 2005).

We now turn to the derivation of the asymptotic distribution of sample intrinsic Fréchet means on Riemannian manifolds which does not require the support restriction of Corollary 1.

For the case of the circle S^1, necessary and sufficient conditions for the existence of the intrinsic mean was established in Bhattacharya (2007), under the assumption of a continuous density with respect to the uniform distribution. The result, along with a central limit theorem for the sample intrinsic mean, also appears in Bhattacharya and Bhattacharya (2012), pp. 31–34, 72–75. A proof of the CLT was also obtained independently in McKilliam et al. (2012). Some additional results, especially for distributions with discontinuous density may be found in the recent article Hotz and Huckemann (2015).

Proposition 5. *Suppose a complete orientable d-dimensional Riemannian manifold (M,g) has the property that the image $D \subset T_pM$ of $M \backslash \mathrm{Cut}(p)$ under the map $\log_p = \mathrm{Exp}_p^{-1}$ is the same for all $p \in M$, and the push forward of the volume measure on D under the $\log_p$ map is also the same for all p. Assume that the intrinsic mean μ_I of a probability Q on M exists, and that Q is absolutely continuous in a neighborhood W of $\mathrm{Cut}(\mu_I)$ with a density f on W which is twice continuously differentiable. Assume also that the first and second derivatives of $p \to f(\mathrm{Exp}_p v)$, in local coordinates, are bounded for p in a neighborhood of μ_I by functions $f_i(v)$, $f_{ij}(v)$ such that, for a sufficiently small $\epsilon > 0$,*

$$\int_{\{R-\epsilon<|v|<R\}} |v|^2 f_i(v)m(\mathrm{d}v) < \infty,$$

$$\int_{\{R-\epsilon<|v|<R\}} |v|^2 f_{ij}(v)m(\mathrm{d}v) < \infty, \quad (i,j = 1,\ldots,d),$$

where $m(\mathrm{d}v)$ is the push forward on T_pM of the volume measure by the map $\log_{\mu_I}$. Then there exists a neighborhood of μ_I in which the Fréchet function (1) with $\rho = \rho_g$, is twice continuously differentiable.

Proof. First note that there exist $r > 0$ and $\epsilon > 0$, both sufficiently small and a geodesic ball B_r with center μ_I and radius $r > 0$ such that $(\mathrm{Cut}(B_r))^\epsilon \subset W$, where $\mathrm{Cut}(B_r) = \cup\{\mathrm{Cut}(p) : p \in B_r\}$ and A^ϵ is the ϵ-neighborhood of a set $A \subset M$. For $p \in B_r$,

$$F(p) = \int_{\{q:|\log_p q|<R-\epsilon\}} \rho_g^2(p,q)Q(\mathrm{d}q) + \int_{\{q:R-\epsilon<|\log_p q|<R\}} \rho_g^2(p,q)Q(\mathrm{d}q). \quad (37)$$

The first integral in (37) is clearly twice continuously differentiable. The second integral may be expressed as

$$\int_{\{R-\epsilon<|v|<R\}} |v|^2 f(\mathrm{Exp}_p v)m(\mathrm{d}v), \quad (38)$$

where $R = \rho_g(p, \mathrm{Cut}(p))$. □

Remark 9. We conjecture that the conclusion of Proposition 5 holds much more generally and, in particular, for all compact orientable Riemannian manifolds, if Q has a twice continuously differentiable density.

The following result is an immediate consequence of Proposition 5.

Corollary 2. *Let $M = S^d$ with the usual Riemannian metric tensor, and Q a probability measure on it. (a) Then the Fréchet function is twice continuously differentiable if Q has a twice continuously differentiable density. (b) If Q is absolutely continuous in a neighborhood of the cut locus $\mathrm{Cut}(p)$ of a point p, with a twice continuously differentiable density there, then the Fréchet function is twice continuously differentiable in a neighborhood of p.*

For the statement of the next result we continue to use the notation of Corollary 1.

Theorem 5 (Bhattacharya and Lin 2017). *Suppose that the intrinsic mean μ of Q exists, and that Q is absolutely continuous in a neighborhood W of the cut locus of μ with a continuous density with respect to the volume measure. Assume also that (i) $Q(\mathrm{Cut}(B(\mu;\epsilon))) = O(\epsilon^{d-c})$, $\epsilon \to 0$, for some c, $0 \le c < d$, (ii) on some neighborhood V of $\nu = \phi(\mu) = 0$ the function $\theta \to F\left(\phi^{-1}(\theta)\right)$ is twice continuously differentiable with a nonsingular Hessian $\Lambda(\theta)$, and (iii)* (A4) *holds with $\phi(\mu)$ replaced by θ, $\forall \theta \in V$. Then, if $d > c+2$, one has the CLT* (26) *for the sample intrinsic mean μ_n.*

Proof. One may take the neighborhood V of $\nu = 0$ sufficiently small such that $\mathrm{Cut}(\phi^{-1}(V)) \subset W$. Then $Z_n(\theta) := n^{-1}\sum_{1\le j\le n} \mathrm{grad}\, h(\theta, Y_j)$ is well defined for $Y_j \notin \mathrm{Cut}(\phi^{-1}\theta)$, $j = 1, \dots, n$, that is, with probability one, provided $\theta \in V$, since $Q(\mathrm{Cut}(\phi^{-1}\theta)) = 0$. By the classical CLT, $Z_n(0) := n^{-1}\sum_{1\le j\le n} \mathrm{grad}\, h(0, Y_j)$ is of the order $O_p(n^{-1/2})$. Let B_n be the ball in $T_\mu M$ with center $\nu = \phi(\mu) = 0$ and radius $n^{-1/2}\log n$. By hypothesis, the probability that $Y_j \in \mathrm{Cut}(\phi^{-1}(B_n))$ is $O((n^{-1/2}\log n)^{d-c})$. For $\phi^{-1}(B_n)$ is the geodesic ball $B(\mu; n^{-1/2}\log n)$, hence the probability that the set $\{Y_j : j = 1, \dots, n\}$ intersects $\mathrm{Cut}(\phi^{-1}(B_n))$ is $O(n(n^{-1/2}\log n)^{d-c}) = o(1)$ if $d > c+2$. Therefore, with probability converging to 1, one may use a Taylor expansion of $Z_n(\theta)$ in B_n,

$$Z_n(\theta) = Z_n(\nu) + \Lambda_n(\theta)(\theta - \nu), \quad (\theta \in B_n), \ (\nu = 0), \tag{39}$$

where $\Lambda_n(\theta)$ is the $d \times d$ matrix whose (r, r') element is

$$\frac{1}{n}\sum_{1\le j\le n} \mathrm{D}_{r,r'} h(\theta(n; r, r', Y_j), Y_j)$$

with $\theta(n; r, r', Y_j)$ lying on the line segment joining θ and $\nu = 0$. By hypothesis (ii), with probability converging to one as $n \to \infty$, $\Lambda_n(\theta)$ is nonsingular for all large n ($\theta \in B_n$) since its difference (in norm) from the Hessian $\Lambda(\theta)$ goes to zero as $n \to \infty$, by the strong law of large numbers. Also, with probability going to 1, the function $\theta \to H_n(\theta) = 0 - \Lambda_n(\theta)^{-1} Z_n(\nu)$ maps $\bar{B}_n$ into itself, where $\bar{B}_n$ is the closure of B_n. For this argument recall that $Z_n(0) = O_p(n^{-1/2})$ by the classical CLT. By the *Brouwer fixed point theorem* (Milnor 1965), $H_n(\theta)$ has a fixed point. Let ν_n denote a measurable selection from the set of fixed points in $\bar{B}_n$. It follows that ν_n converges in probability to ν, and satisfies the first order condition (27). Note that ν_n is the sample intrinsic mean, since the Fréchet function is strictly convex in a neighborhood of ν. The CLT now follows as in the last sentence and relation (35) of the proof of Theorem 4. □

Corollary 3. *Suppose Q on $M = S^d$ ($d > 2$) has an intrinsic mean μ and is absolutely continuous on a neighborhood W of* $\mathrm{Cut}(\mu)$ *with a continuous density on W. Then the CLT for the sample intrinsic mean holds.*

Remark 10. It follows from the proof of Theorem 2.2, specialized to S^d, that the Hessian Λ is positive definite.

Remark 11. Although it is curious that the proof of Theorem 5 does not hold for $d = 2$, the authors expect that a proof of Corollary 3 for the case $d = 2$ may be given using polar coordinates. For the moment, the CLT for S^2 is derived only under the support restriction of Corollary 1.

Remark 12. Suppose $\mathcal{G}$ is a Lie group of isometries on S^d, $d > 2$. Then the projection $\pi : S^d \to S^d/\mathcal{G}$ is a *Riemannian submersion* on S^d onto its quotient space $M = S^d/\mathcal{G}$ (Gallot et al. 1990, pp. 63–65, 97–99). Let Q be a probability measure on S^d with a twice continuously differentiable density and a Karcher or intrinsic mean μ. Let $\tilde{\mu}$ be the projection of μ and $\tilde{\mu}$ that of the sample intrinsic (or Karcher) mean μ_n. Then, in local coordinates, the differential of the Fréchet function on M vanishes at $\tilde{\mu}$, because π is smooth and the differential of the Fréchet function on S^d vanishes at μ. The delta method provides a CLT for the corresponding sample Fréchet mean $\tilde{\mu}_n$ in local coordinates. However, $\tilde{\mu}$ ($\tilde{\mu}_n$) are unlikely to be the intrinsic (respectively, sample intrinsic) mean of $\tilde{Q}$ (respectively, $\tilde{Q}_n$) obtained from Q (respectively, Q_n) by the projection map.

Remark 13. One may also explore the opposite route for a probability $\tilde{Q}$ on M with a density and a unique intrinsic/Karcher mean $\tilde{\mu}$ and a probability Q, among a fairly large family of distributions with smooth densities on S^d whose projection on M is $\tilde{Q}$, such that Q satisfies the hypothesis of Corollary 3 with $\pi(\mu) = \tilde{\mu}$. One may then apply the CLT on S^d to derive one on $S^d/\mathcal{G}$. As an example consider the antipodal map $g(p) = -p$, and $\mathcal{G} = \{g, \text{identity}\}$. Let $\tilde{Q}$ be a probability on $M = S^d/\mathcal{G} = \mathbb{R}P^d$ (the real projective space) thought of as a probability on the upper hemisphere vanishing smoothly at the boundary, and with a unique intrinsic mean $\tilde{\mu} = \{\mu, -\mu\}$, where μ is the Karcher mean of Q (restricted to the hemisphere). This opens a way for CLT's on Kendall's shape spaces as well.

Remark 14. As indicated in Remark 12, one of the significances of a CLT on S^d is that it may provide a route to intrinsic CLTs on $S^d/\mathcal{G}$, the space of orbits under a *Lie group $\mathcal{G}$ of isometries* of S^d. Such spaces include the so-called *axial spaces* (or *real projective spaces* $\mathbb{R}P^d$), and Kendall type shape spaces which are important in *shape-based image analysis*. For the latter spaces, S^d is the so-called *preshape sphere*. Observe that the hypothesis (i) of Theorem 5 may not hold in all such quotient spaces. For example, on $\mathbb{R}P^d$ one only has the order $O(\epsilon)$ in hypothesis (i) in Theorem 5, since the cut locus of the a point in $\mathbb{R}P^d$ is isomorphic to $\mathbb{R}P^{d-1}$. For Kendall's planar shape space, identified as the *complex projective space* $\mathbb{C}P^{k-2}$, of dimension $d = 2k - 4$, the volume measure of Cut $(B(\mu;\epsilon))$ is $O(\epsilon^2)$, since the cut locus of a point of $\mathbb{C}P^{k-2}$ is isomorphic to $\mathbb{C}P^{k-3}$. For these facts one may refer to Gallot et al. (1990), Section 2.114, p. 102, 103.

Remark 15. For $m > 2$, $k > m$, Σ_m^k is a stratified space in the intrinsic topology. But the projection $S^{m(k-1)-1} \to \Sigma_m^k$ (see (5)) is continuous and Σ_m^k is a compact metric space. Hence Theorem 4 still applies to this stratified space.

5 Intrinsic and Extrinsic Analysis and Curvature

In this section we provide explicit expressions of asymptotic dispersions for the intrinsic CLT on general Riemannian manifolds, and relate this to curvature, with applications to the sphere and planar shape spaces. We also provide an example of detailed extrinsic analysis for the planar shapes.

For intrinsic analysis, consider the function $h(z, y) = \rho_g^2(\mathrm{Exp}_p\, z, \mathrm{Exp}_p\, y)$ for z, y in T_pM, with an appropriate choice of p. One first needs to express explicitly the quantities $\mathrm{D}_r h(z, y)$, $\mathrm{D}_r\mathrm{D}_s h(z, y)$ in normal coordinates at p, i.e., at $z = 0 \equiv \mathrm{Exp}_p^{-1} p$. For this let $\gamma(s)$ be a (constant speed) geodesic starting at p, and $m \in M$. Define the *parametric surface* $c(s, t) = \mathrm{Exp}_m(t\, \mathrm{Exp}_m^{-1} \gamma(s))$, $s \in [0, \epsilon)$, $\epsilon > 0$ small. Note that $c(s, 0) = m$ for all s, $c(s, 1) = \gamma(s)$, and that, for all fixed $s \in [0, \epsilon)$, $t \to c(s, t)$ is a geodesic starting at m and reaching $\gamma(s)$ at $t = 1$. Writing $T(s, t) = (\partial/\partial t)c(s, t)$, $S(s, t) = (\partial/\partial s)c(s, t)$, one then has $S(s, 0) = 0$, $S(s, 1) = \dot{\gamma}(s)$. Also, $\langle T(s, t), T(s, t)\rangle$ does not depend on t and, therefore,

$$\rho_g^2(\gamma(s), m) = \int_0^1 \langle T(s, t), T(s, t)\rangle \mathrm{d}t. \tag{40}$$

Differentiating this with respect to s and recalling the symmetry

$$(\mathrm{D}/\partial s)T(s, t) = (\mathrm{D}/\partial t)S(s, t)$$

on a parametric surface (See Do Carmo 1992, p. 68, Lemma 3.4), and

$$(\mathrm{D}/\partial t)T(s, t) = 0,$$

one has

$$\begin{aligned}(\mathrm{d}/\mathrm{d}s)\rho_g^2(\gamma(s), m) &= 2\int_0^1 \langle (\mathrm{D}/\partial s)T(s, t), T(s, t)\rangle \mathrm{d}t \\ &= 2\int_0^1 \langle (\mathrm{D}/\partial t)S(s, t), T(s, t)\rangle \mathrm{d}t \\ &= 2\int_0^1 (\mathrm{d}/\mathrm{d}t)\langle S(s, t), T(s, t)\rangle \mathrm{d}t \\ &= 2\langle S(s, 1), T(s, 1)\rangle = -2\langle \dot{\gamma}(s), \mathrm{Exp}_{\gamma(s)}^{-1} m\rangle.\end{aligned} \tag{41}$$

Setting $s = 0$ in (41) and letting $\dot{\gamma}(0) = v_r$, with $\{v_r : r = 1, \cdots, d\}$ an orthonormal basis of T_pM, one shows that the normal coordinates y_r of m (i.e., the coordinates of $y = \mathrm{Exp}_p^{-1} m$ with respect to $\{v_r : r = 1, \cdots, d\}$) satisfy

$$-2y^r \equiv -2\langle \mathrm{Exp}_p^{-1} m, v_r\rangle = [(\mathrm{d}/\mathrm{d}s)\rho_g^2(\gamma(s), m)]_{s=0}. \tag{42}$$

From this one gets

$$\mathrm{D}_r h(0, y) = -2y^r (r = 1, \cdots, d). \tag{43}$$

If $Q(\mathrm{Cut}(p)) = 0$, then writing $\tilde{Q}$ for the distribution induced from Q by the map Exp_p^{-1} on T_pM, the Fréchet function and its *gradient* in local coordinates may be expressed as

$$F(q) = \int \rho_g^2(q,m)Q(\mathrm{d}m) = \int h(z,y)\tilde{Q}(\mathrm{d}y) = \tilde{F}(z), \tag{44}$$

where $z = \mathrm{Exp}_p^{-1} q$ and $\mathrm{D}_r\tilde{F}(z) = -2\int y^r\tilde{Q}(\mathrm{d}y)$. Since a (local) minimum of this is attained at $q = \mu_I$, $\tilde{F}$ must satisfy a first order condition $\mathrm{D}_r\tilde{F}(z) = 0$ at $z = \nu$. In particular, letting $p = \mu_I$ and, consequently, $\nu = 0$, one has $\int \mathrm{D}_r h(0,y)\tilde{Q}(\mathrm{d}y) = 0$, so that (43) yields

$$\int y^r\tilde{Q}(\mathrm{d}y) = 0 \ (r = 1, \cdots, d), \ (\tilde{Q} = Q\circ\phi^{-1}, \phi = \mathrm{Exp}_{\mu_I}^{-1}). \tag{45}$$

By Theorem 5, the asymptotic distribution of the sample intrinsic mean μ_n is that of $\phi^{-1}(\nu_n)$, where $\phi = \mathrm{Exp}_p^{-1}$, and

$$\sqrt{n}(\nu_n - \nu) \simeq \Lambda^{-1}\left[(1/\sqrt{n})\sum_{1\le j\le n}\mathrm{D}h(\nu, Y_j)\right], \tag{46}$$

$$(\Lambda_{rs} = E\mathrm{D}_r\mathrm{D}_s h(\nu, Y_1), 1 \le r,s \le d),$$

with $Y_j = \phi(X_j)$, where X_j are i.i.d. with distribution Q. By (43), the right side of (46) simplifies to $\Lambda^{-1}[-2(1/\sqrt{n})\sum_{1\le j\le n} Y_j]$, if $p = \mu_I$ (and $\nu = 0$).

For an analytical study of the Hessian Λ of the Fréchet function, one derives from (55) the relation

$$\frac{\mathrm{d}^2}{\mathrm{d}s^2}\rho_g^2(\gamma(s),m) = 2\langle \mathrm{D}_sT(s,1), S(s,1)\rangle = 2\langle \mathrm{D}_tS(s,1), S(s,1)\rangle, \tag{47}$$

$$\left(\mathrm{D}_s = \frac{\mathrm{D}}{\partial s}, \mathrm{D}_t = \frac{\mathrm{D}}{\partial t} \text{ covariant derivatives}\right). \tag{48}$$

Using the theory of Jacobi fields (Do Carmo 1992, p. 111) the following relations may be derived. Let C denote the supremum of all sectional curvatures of M and let

$$f(t) = \begin{cases} 1 & \text{if } C = 0, \\ (\sqrt{C}t)\cos(\sqrt{C}t)/\sin(\sqrt{C}t) & \text{if } C > 0, \\ \sqrt{-C}t\cosh(\sqrt{-C}t)/\sinh(\sqrt{-C}t) & \text{if } C < 0. \end{cases} \tag{49}$$

Also let t_0 be the supremum of all t such that $f(t) > 0$. For $d\times d$ symmetric matrices A, B, the order relation $A \ge B$ means $A - B$ is nonnegative definite.

Theorem 6 (Bhattacharya and Bhattacharya 2008**).** *Assume* $|Y_1| = |\log_{\mu_I} X_1| \le t_0$ *a.s. In addition, if the hypotheses for the CLT in Corollary 1 or Theorem 5 hold, one has*

$$\Lambda = ((\Lambda_{ij})) \ge \left(\left(2E\left(\frac{1 - f|Y_1|}{|Y_1|^2}Y_1^iY_1^j + f(|Y_1|)\delta_{ij}\right)\right)\right), \tag{50}$$

with equality if the sectional curvature is consistent.

Remark 16. It is simple to check that on $M = S^d$, the Hessian Λ given by the right side of (50), with $C = 1$, is nonsingular.

Remark 17. Kendall and Le (2011) obtained the exact expression for the Hessian for the intrinsic Fréchet mean on the important case of the planar shape space Σ_2^k, which has a constant holomorphic curvature.

Remark 18. Note that the relations in (42) provide the gradient of the intrinsic Fréchet function $F(p)$ in normal coordinates around p.

Example 3 (Confidence region for the intrinsic mean of Q on the sphere S^d). Let μ_I be the intrinsic mean of Q on S^d. Given n i.i.d. observations $X_1, \cdots, X_n$ on S^d with common distribution Q, let μ_n be the intrinsic sample mean. Write $\phi = \mathrm{Exp}_{\mu_I}^{-1}$, and $\phi_p = \mathrm{Exp}_p^{-1}$, so that $\phi_{\mu_I} = \phi$. By Theorem 4,

$$\sqrt{n}[\phi(\mu_n) - \phi(\mu_I)) = \sqrt{n}\phi(\mu_n) \to N(0, \Lambda^{-1}\tilde{\Sigma}\Lambda^{-1}) \text{ in distribution as } n \to \infty, \tag{51}$$

where the $d \times d$ matrices Λ and $\tilde{\Sigma}$ are given by

$$\begin{aligned} \tilde{\Sigma} &= 4\,\mathrm{Cov}(\phi(X_1)), \\ \Lambda_{rs} &= 2E[(1 - (X_1^t\mu_I)^2)^{-1}\{1 - (1 - (X_1^t\mu_I)^2)^{-1/2} \\ &\quad \cdot (X_1^t\mu_I)\arccos(X_1^t\mu_I)\}(X_1^t\nu_r)(X_1^t\nu_s) \\ &\quad + (1 - (X_1^t\mu_I)^2)^{-1/2} \cdot (X_1^t\mu_I)(\arccos(X_1^t\mu_I)))\delta_{rs}], 1 \le r, s \le d. \end{aligned} \tag{52}$$

Here $\{\nu_r : 1 \le r \le d\}$ is an orthonormal basis of $T_{\mu_I}S^d$. A confidence region for μ_I, of asymptotic level $1 - \alpha$, is then given by

$$\{p \in S^d : n\phi_p(\mu_n)^t\hat{\Lambda}_p\hat{\tilde{\Sigma}}_p^{-1}\hat{\Lambda}_p\phi_p(\mu_n) \le \chi_d^2(1-\alpha)\}, \tag{53}$$

where Λ_p, $\tilde{\Sigma}_p$ are obtained by replacing μ_I by p in the expressions for Λ and $\tilde{\Sigma}$ in (52). The "hat" ($\hat{}$) indicates that the expectations are computed under the empirical Q_n, rather than Q. It would be computationally simpler to choose a particular $p = p_0$, say, and let $\phi = \mathrm{Exp}_{p_0}^{-1}$. Then one can construct a simpler confidence region:

$$\{p \in S^d : n[\phi(\mu_n) - \phi(p)]^t\hat{\Lambda}_{p_0}\tilde{\Sigma}_{p_0}^{-1}\hat{\Lambda}_{p_0}[\phi(\mu_n)) - \phi(\mu_p)] \le \chi_d^2(1-\alpha)\}. \tag{54}$$

Example 4 (Inference on the planar shape space Σ_2^k). To apply Theorem 5, we use (46) where $\phi = \mathrm{Exp}_{\sigma(p)}^{-1}$ and p is a suitable point in $\mathbb{C}S^{k-1}$. To derive a computable expression for Λ, write the geodesic γ in the parametric surface $c(s,t)$ as $\gamma = \pi \circ \tilde{\gamma}$, where $\tilde{\gamma}$ is a geodesic in $\mathbb{C}S^{k-1}$ starting at $\tilde{\mu} \in \pi^{-1}\{\mu_I\}$. Then, with $\tilde{T}(s,1) = (\mathrm{d}_{\gamma(s)}\pi^{-1})T(s,1)$,

$$\begin{aligned} (\mathrm{d}/\mathrm{d}s)\rho_g^2(\gamma(s), m) &= 2\langle T(s,1), \dot{\gamma}(s)\rangle = 2\langle \tilde{T}(s,1), \dot{\tilde{\gamma}}(s)\rangle, \\ (\mathrm{d}^2/\mathrm{d}s^2)\rho_g^2(\gamma(s), m) &= 2\langle \mathrm{D}_s\tilde{T}(s,1), \dot{\tilde{\gamma}}(s)\rangle. \end{aligned} \tag{55}$$

The final inner products are in $T\mathbb{C}S^{k-1}$, namely, $\langle\tilde{v},\tilde{w}\rangle = \mathrm{Re}(\tilde{v}\tilde{w}^*)$. Note that $\tilde{T}(s,1) = -\operatorname{Exp}^{-1}_{\tilde{\gamma}(s)} q$, $q \in \pi^{-1}m$, may be expressed by (9) and (10) as

$$\tilde{T}(s,1) = -(\rho(s)/\sin\rho(s))[\mathrm{e}^{i\theta(s)}\, q - (\cos\rho(s))\tilde{\gamma}(s)], \tag{56}$$

where $\rho(s) = \rho_g(\gamma(s), m)$ and $\mathrm{e}^{i\theta(s)} = (1/\cos\rho(s))\tilde{\gamma}(s)q^*$. The covariant derivative $\mathrm{D}_s\tilde{T}(s,1)$ is the projection of $(\mathrm{d}/\mathrm{d}s)\tilde{T}(s,1)$ onto $H_{\tilde{\gamma}(s)}$. Since $\langle\tilde{\mu},\dot{\tilde{\gamma}}(0)\rangle = 0$, (55) then yields

$$[(\mathrm{d}^2/\mathrm{d}s^2)\rho_g^2(\gamma(s),m)]_{s=0} = 2\langle[(\mathrm{d}/\mathrm{d}s)\tilde{T}(s,1)]_{s=0},\dot{\tilde{\gamma}}(0)\rangle. \tag{57}$$

Differentiating (56) one obtains

$$\begin{aligned}[(\mathrm{d}/\mathrm{d}s)\tilde{T}(s,1)]_{s=0} = {} & [(\mathrm{d}/\mathrm{d}s)(\rho(s)\cos\rho(s))/\sin\rho(s))]_{s=0}\tilde{\mu} \\ & + [(\rho(s)\cos\rho(s))/\sin\rho(s))]_{s=0}\dot{\tilde{\gamma}}(0) \\ & - [(\mathrm{d}/\mathrm{d}s)(\rho(s)/(\cos\rho(s))(\sin\rho(s))]_{s=0}(\tilde{\mu}q^*)q \\ & - [\rho(s)/(\cos\rho(s))(\sin\rho(s))]_{s=0}(\dot{\tilde{\gamma}}(0)q*)q.\end{aligned} \tag{58}$$

From (55), $2\rho(s)\dot{\rho}(s) = 2\langle\tilde{T}(s,1),\tilde{\gamma}'(s)\rangle$, which along with (56) leads to

$$[(\mathrm{d}/\mathrm{d}s)\rho(s)]_{s=0} = -(1/\sin r)\langle(\tilde{\mu}q^*/\cos r)q,\dot{\tilde{\gamma}}(0)\rangle,\ (r = \rho_g(m,\mu_I)). \tag{59}$$

One then gets (See Bhattacharya and Bhattacharya 2008, p. 93, 94)

$$\begin{aligned}\langle[(\mathrm{d}/\mathrm{d}s)\tilde{T}(s,1)]_{s=0},\dot{\tilde{\gamma}}(0)\rangle = {} & \{(r\cos r)/(\sin r)\}|\dot{\tilde{\gamma}}(0)|^2 \\ & - \{(1/\sin^2 r) - (r\cos r)/\sin^3 r\}(\mathrm{Re}(x))^2 \\ & + r/((\sin r)(\cos r))(\mathrm{Im}(x))^2, \\ & (x = \mathrm{e}^{i\theta}\, q\dot{\tilde{\gamma}}(0)^*, \mathrm{e}^{i\theta} = \tilde{\mu}q^*/\cos r).\end{aligned} \tag{60}$$

One can check that the right side of (60) depends only on $\pi(\tilde{\mu})$ and not any particular choice of $\tilde{\mu}$ in $\pi^{-1}\{\mu_I\}$.

Now let $\{\nu_1,\cdots,\nu_{k-2},i\nu_1,\cdots,i\nu_{k-2}\}$ be an orthonormal basis of $T_{\sigma(p)}\Sigma_2^k$ where we identify Σ_2^k with $\mathbb{C}P^{k-2}$, and choose the unit vectors

$$\nu_r = (\nu_r^1,\cdots,\nu_r^{k-1}), r = 1,\cdots,k-2,$$

to have zero imaginary parts and satisfy the conditions $p^*\nu_r = 0$, $\nu_r^t\nu_s = 0$ for $r \neq s$.

Suppose now that $\sigma(p) = \mu_I$, i.e., $\gamma(0) = \mu_I$. If $\dot{\gamma}(0) = v$, then $\gamma(s) = \operatorname{Exp}_{\mu_I}(sv)$, so that $\rho_g^2(\gamma(s),m) = h(sv,y)$ with $y = \operatorname{Exp}^{-1}_{\mu_I} m$. Then, expressing v in terms of the orthonormal basis,

$$[(\mathrm{d}^2/\mathrm{d}s^2)\rho_g^2(\gamma(s),m)]_{s=0} = [(\mathrm{d}^2/\mathrm{d}s^2)h(sv,y)]_{s=0} = \Sigma v_iv_j\mathrm{D}_i\mathrm{D}_jh(0,y). \tag{61}$$

Integrating with respect to Q now yields

$$\sum v_iv_j\Lambda_{ij} = E[(\mathrm{d}^2/\mathrm{d}s^2)\rho_g^2(\gamma(s),X)]_{s=0},\ (X \text{ with distribution } Q). \tag{62}$$

This identifies the matrix Λ from the calculations (57) and (60). To be specific, consider independent observations $X_1, \cdots, X_n$ from Q, and let

$$Y_j = \operatorname{Exp}_{\mu_I}^{-1} X_j (j = 1, \cdots, n).$$

In normal coordinates with respect to the above basis of $T_{\mu_I}\Sigma_2^k$, one has the following coordinates of Y_j:

$$\left(\operatorname{Re}(Y_j^1), \cdots, \operatorname{Re}(Y_j^{k-2}), \operatorname{Im}(Y_j^1), \cdots, \operatorname{Im}(Y_j^{k-2})\right) \in \mathbb{R}^{2k-4}. \tag{63}$$

Writing

$$\Lambda = \begin{pmatrix} \Lambda_{11} & \Lambda_{12} \\ \Lambda_{21} & \Lambda_{22} \end{pmatrix}$$

in blocks of $(k-2) \times (k-2)$ matrices, one arrives at the following expressions of the elements of these matrices, using (60)–(63). Denote $\rho_g^2(\mu_I, X_1) = h(0, Y_1)$ by ρ. Then

$$\begin{aligned}
(\Lambda_{11})_{rs} &= 2E[\rho(\cot\rho)\delta_{rs} - (1/\rho^2)(1-\rho\cot\rho)(\operatorname{Re} Y_1^r)(\operatorname{Re} Y_1^s) \\
&\quad + \rho^{-1}(\tan\rho)(\operatorname{Im} Y_1^r)(\operatorname{Im} Y_1^s)]; \\
(\Lambda_{22})_{rs} &= 2E[\rho(\cot\rho)\delta_{rs} - (1/\rho^2)(1-\rho\cot\rho)(\operatorname{Im} Y_1^r)(\operatorname{Im} Y_1^s) \\
&\quad + \rho^{-1}(\tan\rho)(\operatorname{Re} Y_1^r)(\operatorname{Re} Y_1^s)]; \\
(\Lambda_{12})_{rs} &= 2E[\rho(\cot\rho)\delta_{rs} - (1/\rho^2)(1-\rho\cot\rho)(\operatorname{Re} Y_1^r)(\operatorname{Im} Y_1^s) \\
&\quad + \rho^{-1}(\tan\rho)(\operatorname{Im} Y_1^r)(\operatorname{Re} Y_1^s)]; \\
(\Lambda_{21})_{rs} &= (\Lambda_{12})_{sr}.(r, s = 1, \cdots, k-2).
\end{aligned} \tag{64}$$

One now arrives at the CLT for the intrinsic sample mean μ_n by Theorem 5 and Corollary 1. An asymptotic two-sample test for $H0 : Q_1 = Q_2$ based on independent samples may be constructed from this in the usual manner (See Bhattacharya and Lin (2017))

We next turn to extrinsic analysis on Σ_2^k, using the embedding (11). Let μ_J be the mean of $Q \circ J^{-1}$ on $S(k-1, \mathbb{C})$, where J is the Veronese–Whitney embedding (11).

Assuming that the largest eigenvalue of μ_J is simple (see Proposition 1), one may now obtain the asymptotic distribution of the sample extrinsic mean $\mu_{n,E}$, namely, that of $J(\mu_{n,E}) = v_n^* v_n$, where v_n is a unit eigenvector of $\bar{\tilde{Y}} = \sum \tilde{Y}_j / n$ corresponding to its largest eigenvalue. Here $\tilde{Y}_j = J(Y_j)$, for i.i.d observations $Y_1, \cdots, Y_n$ on Σ_2^k. For this purpose, a convenient orthonormal basis (frame) of $T_p S(k-1, \mathbb{C}) \simeq S(k-1, \mathbb{C})$ is the following:

$$\begin{aligned}
\nu_{a,b} &= 2^{-1/2}(e_a e_b^t + e_b e_a^t) \text{ for } a < b, \nu_{a,a} = e_a e_a^t; \\
w_{a,b} &= i2^{-1/2}(e_a e_b^t - e_b e_a^t) \text{ for } b < a \ (a, b = 1, \cdots, k-1),
\end{aligned} \tag{65}$$

where e_a is the column vector with all entries zero other than the a-th, and the a-th entry is 1. Let $U_1, \cdots, U_{k-1}$ be orthonormal unit eigenvectors corresponding to the eigenvalues $\lambda_1 \le \cdots \le \lambda_{k-2} < \lambda_{k-1}$. Then choosing $T = (U_1, \cdots, U_{k-1}) \in SU(k-1)$ and $T\mu_J T^* = D = \mathrm{diag}(\lambda_1, \cdots, \lambda_{k-1})$, such that the columns of $T\nu_{a,b}T^*$ and $Tw_{a,b}T^*$ together constitute an orthonormal basis of $S(k-1, \mathbb{C})$. It is not difficult to check that the differential of the projection operator P satisfies

$$(\mathrm{d}_{\mu_J} P)Tv_{a,b}T^* = \begin{cases} 0 & \text{if } 1 \le a \le b < k-1, \\ & \text{or } a = b = k-1, \\ (\lambda_{k-1} - \lambda_a)^{-1} Tv_{a,k-1}T^* & \text{if } 1 \le a < k-1, b = k-1; \end{cases} \tag{66}$$

$$(\mathrm{d}_{\mu_J} P)Tw_{a,b}T^* = \begin{cases} 0 & \text{if } 1 \le a \le b < k-1, \\ (\lambda_{k-1} - \lambda_a)^{-1} Tw_{a,k-1}T^* & \text{if } 1 \le a < k-1. \end{cases}$$

To check these, take the projection of a linear curve $c(s)$ in $S(k-1, \mathbb{C})$ such that $\dot{c}(0)$ is one of the basis elements $v_{a,b}$, or $w_{a,b}$, and differentiate the projected curve with respect to s. It follows that $\{Tv_{a,k-1}T^*, Tw_{a,k-1}T^* : a = 1, \cdots, k-2\}$ form an orthonormal basis of $T_{P(\mu_J)} J(\Sigma_2^k)$. Expressing $\tilde{Y}_j - \mu_J$ in the orthonormal basis of $S(k-1, \mathbb{C})$, and $\mathrm{d}_{\mu_J} P(\tilde{Y}_j - \mu_J)$ with respect to the above basis of $T_{P(\mu_J)} J(\Sigma_2^k)$, one may now apply Proposition 4.

6 Nonparametric Bayes Estimation of Densities on a Manifold and the Problem of Classification

6.1 Density Estimation

Consider the problem of estimating the density q of a distribution Q on a Riemannian manifold (M, g) with respect to the volume measure λ on M. According to Ferguson (1973), given a finite non-zero base measure α on a measurable space $(\mathcal{X}, \Sigma)$, a random probability P on the class $\mathcal{P}$ of all probability measures on $\mathcal{X}$ has the *Dirichlet distribution* D_α if for every measurable partition $\{B_1, \ldots, B_k\}$ of $\mathcal{X}$, the D_α-distribution of $(P(B_1), \ldots, P(B_k)) = (\theta_1, \ldots, \theta_k)$, say, is Dirichlet with parameters $(\alpha(B_1), \ldots, \alpha(B_k))$. That is, $(P(B_1), \ldots, P(B_{k-1}))$ has the joint density

$$f(\theta_1, \ldots, \theta_{k-1}) = \mathrm{const}\left(\theta_1^{\alpha(B_1)-1} \ldots \theta_{k-1}^{\alpha(B_k-1)-1}\right)(1 - \theta_1 - \ldots - \theta_{k-1})^{\alpha(B_k)-1}$$

on $\{(\theta_1, \ldots, \theta_{k-1}) : \theta_i > 0\ \forall i,\ \theta_1 + \ldots \theta_{k-1} < 1\}$. If $\alpha(B) = 0$ for some B_j, then $P(B_j) = 0$ with probability 1. In the case $k = 2$, the D_α-distribution of $(P(B_1), P(B_2))$ is also called the *beta distribution*, denoted $\mathrm{beta}(\alpha(B_1), \alpha(B_2))$. Sethuraman (1994) gave a very convenient "stick breaking" representation of the random P. To define it, let $u_j (j = 1, \ldots)$ be an i.i.d. sequence of $\mathrm{beta}(1, \alpha(\mathcal{X}))$

random variables, independent of a sequence $Y_j (j = 1, \dots)$ having the distribution $G = \alpha/\alpha(\mathcal{X})$ on $\mathcal{X}$. Sethuraman's representation of the random probability with the Dirichlet prior distribution D_α is

$$P \equiv \sum w_j \delta_{Y_j}, \tag{67}$$

where $w_1 = u_1, w_j = u_j(1 - u_1) \dots (1 - u_{j-1})$ $(j = 2, \dots)$, and δ_{Y_j} denotes the Dirac measure at Y_j. As this construction shows, the Dirichlet distribution assigns probability one to the set of all discrete distributions on $\mathcal{X}$, and one cannot retrieve a density estimate from it directly. The Dirichlet priors constitute a conjugate family, i.e., the posterior distribution of a random P with distribution D_α, given observations $X_1, \dots, X_n$ from P is $D_{\alpha + \sum_{1 \le i \le n} \delta_{X_i}}$. A general method for Bayesian density estimation on a manifold (M, g) may be outlined as follows. Suppose that q is continuous and positive on M. First find a parametric family of densities $m \to K(m; \mu, \tau)$ on M where $\mu \in M$ and $\tau > 0$ are "location" and "scale" parameters, such that K is continuous in its arguments, $K(\cdot; \mu, \tau) d\lambda(\cdot)$ converges to δ_μ as $\tau \downarrow 0$, and the set of all "mixtures" of $K(\cdot; \mu, \tau)$ by distributions on $M \times (0, \infty)$ is dense in the set $C_\lambda(M)$ of all continuous densities on M in the supremum distance, or in $L^1(d\lambda)$. The density q may then be estimated by a suitable mixture. To estimate the mixture, use a prior D_β with full support on the set of all probabilities on the space $M \times (0, \infty)$ of "parameters" (μ, τ). A draw from the prior may be expressed in the form (67), where u_j are i.i.d. beta$(1, b)$ with $b = \beta(M \times (0, \infty))$, independent of $Y_j = (m_j, t_j)$, say, which are i.i.d. β/b on $M \times (0, \infty)$. The corresponding random density is then obtained by integrating the kernel K with respect to this random mixture distribution,

$$\sum w_j K(m; m_j, t_j). \tag{68}$$

Given M-valued (Q-distributed) observations $X_1, \dots X_n$, the posterior distribution of the mixture measure is Dirichlet D_{β_X}, where $\beta_X = \beta + \sum_{1 \le i \le n} \delta_{Z_i}$, with $Z_i = (X_i, 0)$. A draw from the posterior distribution leads to the random density in the form (68), where u_j are i.i.d. beta$(1, b+n)$, independent of (m_j, t_j) which are i.i.d. $\beta_X/(b+n)$. One may also consider using a somewhat different type of priors such as $D_\alpha \times \pi$ where D_α is a Dirichlet prior on M, and π is a prior on $(0, \infty)$, e.g., gamma or Weibull distribution.

Consistency (weak consistency) of the posterior is generally established by checking full Kullback–Liebler support of the prior D_β (See Ghosh and Ramamoorthi (2003), pp. 137–139). Strong consistency has been established for the planar shape spaces using the complex Watson family of densities (with respect to the volume measure or the uniform distribution on Σ_2^k) of the form $K([z]; \mu, \tau) = c(\tau) \exp |z * \mu|^2/\tau$ in Bhattacharya and Bhattacharya (2012) and Bhattacharya and Dunson (2010), where it has been shown, by simulation from known distributions, that, based on a prior $D_\beta \times \pi$ chosen so as to produce clusters close to the support of the observations, the Bayes estimates of quantiles and other indices far outperform the kernel density estimates (KDE) of Pelletier (2005), and also require much less computational time than the latter.

In moderate sample sizes, the nonparametric Bayes estimates perform much better than even the MLE (computed under the true model specification)!

6.2 Classification

Classification of a random observation to one of several groups is one of the most important problems in statistics. This is the objective in medical diagnostics, classification of subspecies and, more generally, this is the target of most image analysis. Suppose there are r groups or populations with a priori given relative sizes or proportions $\pi_i(i = 1, \ldots, r), \sum \pi_i = 1$, and densities $q_i(x)$ (with respect to some sigma-finite measure). Under $0-1$ loss function, the average risk of misclassification (i.e., the Bayes risk) is minimized by the rule: Given a random observation X, classify it to belong to group j if $\pi_j q_j(X) = \max\{\pi_i q_i(X) : i = 1, \ldots, r\}$. Generally, one uses sample estimates of π_i-s and q_i-s, based on random samples from the r groups (training data). Nonparametric Bayes estimates of q_i-s on shapes spaces perform very well in classification of shapes, and occasionally identify outliers and misclassified observations (See, Bhattacharya and Bhattacharya 2012 and Bhattacharya and Dunson 2010).

In a simulation study using samples of size 400 from a complex Watson distribution, Bhattacharya and Dunson (2010) found that the nonparametric Bayes estimate far outperformed the kernel density estimate KDE over a multitude of criteria. It also performed much better than the MLE of the correctly specified model! Here are the L^1 distances and the Kullback–Leibler divergences from f_0 (Table 1).

Table 1. L^1 distance and Kullback–Leibler divergence between the estimate and the true density

	NP Bayes	KDE	MLE
L^1	0.44	1.03	0.75
K-L	0.13	0.41	0.25

7 The Laplace–Beltrami Operator in Machine Vision

Mark Kac asked in a paper in 1966 in the American Mathematical Monthly: “Can one hear the shape of a drum?”. In other words, by listening to the frequencies of vibrations of a clamped drum, given by the eigenvalues of the Laplacian with Dirichlet (or zero) boundary condition, is it possible to reconstruct or identify the geometric shape of the drum? The origin of this question may be traced back to Hermann Weyl’s famous formula (Weyl 1911): For any bounded domain Ω in

$\mathbb{R}^d$ with a smooth boundary, the number $N(\lambda)$ of eigenvalues of the (negative) Laplacian $-\triangle$ which are less than λ has the asymptotic relation

$$N(\lambda) \sim \omega_d(2\pi)^{-d}\lambda^{d/2}\text{vol}(\Omega) \quad \text{as } \lambda \to \infty \quad (\omega_d = \text{vol of unit ball in } \mathbb{R}^d). \tag{69}$$

Here the relation $\sim$ indicates that the ratio of its two sides converges to 1 (as $\lambda \to \infty$). A similar formula holds for any d-dimensional compact Riemannian manifold (M, g) with or without boundary where $\triangle$ is the so-called *Laplace–Beltrami operator* (Chavel 1984; Rosenberg 1997), which may be expressed in a local chart given by u (on $B(0,r) \to U \subset M$) as

$$\triangle f = (\det\ g)^{-1/2} \sum_{1 \leq i,j \leq d} \partial_i g^{ij} (\det\ g)^{1/2} \partial_j f. \tag{70}$$

Weyl type spectral asymptotics for $-\triangle$ are given by

$$N(\lambda) \sim c(d)\lambda^{d/2}\text{vol}(\Omega) \quad \text{as} \quad \lambda \to \infty, \tag{71}$$

where $c(d)$ depends only on the dimension d.

There are many refinements of the estimates (69), (71) with an error term of the order $\lambda^{(d-1)/2}$. Although there are many spectral invariants of the manifold, it turns out unfortunately, that the answer to Kac's question is "no" (Milnor 1964). For two-dimensional surfaces, the answer is mostly "yes" outside a relatively small set of manifolds (Zelditch 2000). But in dimension 3 or higher the set of non-isometric manifolds with the same spectrum is not negligible. A natural question that arises is: if one uses eigenfunctions as well as eigenvalues of $-\triangle$ can one reconstruct or identify the manifold? One may find many interesting and important articles in computer science/machine vision journals where such reconstructions are displayed. But the mathematical question posed above is rigorously answered in the affirmative only by Jones et al. (2008), under only mild conditions on the manifold, such as uniform ellipticity of the Laplacian. The last mentioned authors actually construct coordinate patches covering M, and therefore the structure of the manifold, only using eigenvalues and eigenfunctions of $-\triangle$.

There are many numerical methods for the computation of eigenvalues and eigenfunctions carried out by computer scientists and applied for object identification and scene recognition (See, e.g. Reuter 2006). To economize the use of these features sometimes topological properties of M are also used. For example, see Dey et al. (2008), Dey and Li (2009) who use the first homology group to identify handles and holes in a closed bounded domain in 3D. For a more elaborate technique using algebraic topology, known as persistent homology, we refer to Carlsson (2009).

8 Examples and Applications

In this section we apply the theory to a number of data sets available in the literature.

Example 5 (Paleomagnetism). The first statistical confirmation of the shifting of the earth's magnetic poles over geological times, theorized by paleontologists based on observed fossilised magnetic rock samples, came in a seminal paper by R.A. Fisher (1953). Fisher analyzed two sets of data – one recent (1947–48) and another old (Quaternary period), using the so-called *von Mises–Fisher model*

$$f(x;\mu,\tau) = c(\tau)\exp\{\tau x^t \mu\}(x \in S^2),$$

Here $\mu(\in S^2)$, is the *mean direction*, extrinsic as well as intrinsic ($\mu = \mu_I = \mu_E$), and $\tau > 0$ is the concentration parameter. The maximum likelihood estimate of μ is $\hat{\mu} = \bar{X}/|\bar{X}|$, which is also our sample extrinsic mean. The value of the MLE for the first data set of $n = 9$ observations turned out to be $\hat{\mu} = \hat{\mu}_E = (0.2984, 0.1346, 0.9449)$, where (0,0,1) is the geographic north pole. Fisher's 95% confidence region for μ is $\{\mu \in S^2 : \rho_g(\hat{\mu},\mu) \leq 0.1536)\}$. The sample intrinsic mean is $\hat{\mu}_I = (0.2990, 0.1349, 0.9447)$, which is very close to $\hat{\mu}_E$. The nonparametric confidence region based on $\hat{\mu}_I$, as given by (53), and that based on the extrinsic procedure, are nearly the same, and both are about 10% smaller in area than Fisher's region. (See Bhattacharya and Bhattacharya (2012), Chapter 2).

The second data set based on $n = 29$ observations from the Quaternary period that Fisher analyzed, using the same parametric model as above, had the MLE $\hat{\mu} = \bar{X}/|\bar{X}| = (0.0172, -0.2978, -0.9545)$, almost antipodal of that for the first data set, and with a confidence region of geodesic radius 0.1475 around the MLE. Note that the two confidence regions are not only disjoint, they also lie far away from each other. This provided the first statistical confirmation of the hypothesis of shifts in the earth's magnetic poles, a result hailed by paleontologists (See Irving 1964). Because of difficulty in accessing the second data set, the nonparametric procedures could not be applied to it. But the analysis of another data set dating from the Jurassic period, with $n = 33$, once again yielded nonparametric intrinsic and extrinsic confidence regions very close to each other, and each about 10% smaller than the region obtained by Fisher's parametric method (See Bhattacharya and Bhattacharya 2012, Chapter 5, for details).

Example 6 (Brain scan of schizophrenic and normal patients). We consider an example from Bookstein (1991) in which 13 landmarks were recorded on a midsagittal two-dimensional slice from magnetic brain scans of each of 14 schizophrenic patients and 14 normal patients. The object is to detect the deformation, if any, in the shape of the k-ad due to the disease, and to use it for diagnostic purposes. The shape space is Σ_2^{13}. The intrinsic two-sample test has an observed value 95.4587 of the asymptotic chi-square statistic with 22 degrees of freedom, and a p-value 3.97×10^{-11}. The extrinsic test has an observed value 95.5476 of the chi-square statistic and a p-value 3.8×10^{-11}. It is remarkable, and reassuring, that completely different methodologies of intrinsic and extrinsic inference essentially led to the same values of the corresponding asymptotic chi-square statistics (a phenomenon observed in other examples as well). For details of these calculations and others we refer to Bhattacharya and Bhattacharya

(2012). This may also be contrasted with the results of parametric inference in the literature for the same data, as may be found in Dryden and Mardia (1998), pp. 146, 162–165. Using a isotropic Normal model for the original landmarks data, and after removal of "nuisance" parameters for translation, size and rotation, an F-test known as Goodall's F-test (See Bookstein 1991) gives a p-value 0.01. A Monte Carlo test based permutation test obtained by 999 random assignments of the data into two groups and computing Goodall's F-statistic, gave a p-value 0.04. A Hotelling's T^2 test in the tangent space of the pooled sample mean had a p-value 0.834. A likelihood ratio test based on the isotropic offset Normal distribution on the shape space has the value 43.124 of the chi-square statistic with 22 degrees of freedom, and a p-value 0.005.

Example 7 (Shapes of Gorilla Skulls). We consider another example in which two planar shape distributions via their extrinsic (and intrinsic) means are distinguished. A Bayesian nonparametric classifier is also built and applied.

In this data set, there are 29 male and 30 female gorillas and the eight landmarks are chosen on the midline plane of the 2D image of the skull. The data can be found in Dryden and Mardia (1998). It is of interest to study the shapes of the skulls and use that to detect differences in shapes between the sexes. This finds applications in morphometrics and other biological sciences.

To distinguish between the distribution of shapes of skulls of the two sexes, one may compare the sample extrinsic mean shapes or dispersions in shape as well as the intrinsic counterparts.

The value of the two sample test statistic which is also a chi-square statistic based on the asymptotic distribution for the sample Fréchet mean, for comparing the intrinsic mean shapes, and the asymptotic p-value for the chi-squared test are

$$T_{n1} = 391.63, \ p\text{-value} \ = P(\mathcal{X}_{12}^2 > 391.63) < 10^{-16}.$$

Hence we reject the null hypothesis that the two sexes have the same intrinsic mean shape. The test statistics (a chi-square statistic based on the central limit theorem) for comparing the extrinsic mean shapes, and the corresponding asymptotic p-values are

$$T_1 = 392.6, \ p\text{-value} \ = P(\mathcal{X}_{12}^2 > 392.6) < 10^{-16}.$$

Hence we reject the null hypothesis that the two sexes have the same extrinsic mean shape. We can also compare the mean shapes by pivotal bootstrap method using the test statistic T_2^* which is a bootstrap version of T_2. The p-value for the bootstrap test using 10^5 simulations turns out to be 0. In contrast, a parametric test carried out in Dryden and Mardia (1998), pp. 168–172, has a p-value 0.0001.

A Bayesian nonparametric classifier is next applied (see Bhattacharya and Dunson 2010) to predict gender. The shape densities for the two groups via non-parametric Bayesian methods are estimated which are used to derive the conditional distribution of gender given shape. 25 individuals of each gender are picked as a training sample, with the remaining 9 used as test data. Table 2 presents the estimated posterior probabilities of being female for each of the

Table 2. Posterior probability of being female for each gorilla in the test sample

gender	$\hat{p}([z])$	95% CI	$d_E([z_i], \hat{\mu}_1)$	$d_E([z_i], \hat{\mu}_2)$
F	1.000	(1.000, 1.000)	0.041	0.111
F	1.000	(0.999, 1.000)	0.036	0.093
F	0.023	(0.021, 0.678)	0.056	0.052
F	0.998	(0.987, 1.000)	0.050	0.095
F	1.000	(1.000, 1.000)	0.076	0.135
M	0.000	(0.000, 0.000)	0.167	0.103
M	0.001	(0.000, 0.004)	0.087	0.042
M	0.992	(0.934, 1.000)	0.091	0.121
M	0.000	(0.000, 0.000)	0.152	0.094

Here $\hat{p}([z])$ is estimated prob. of being female, given shape $[z]$; $d_E([z], \hat{\mu}_i)$ is extrinsic distance from the mean shape in group i, with $i = 1$ for females and $i = 2$ for males

gorillas in the test sample along with a 95% credible interval. For most of the gorillas, there is a high posterior probability of assigning the correct gender. There is misclassification only in the 3rd female and 3rd male. For the 3rd female, the credible interval includes 0.5, suggesting that there is insufficient information to be confident in the classification. However, for the 3rd male, the credible interval suggests a high degree of confidence that this individual is female. Perhaps this individual is an outlier and there is something unusual about the shape of his skull, with such characteristics not represented in the training data, or, alternatively, he was labeled incorrectly.

Example 8 (Corpus Callosum shapes of normal and ADHD children). We consider the third planar shape data set, which involve measurements of a group typically developing children and a group of children suffering the ADHD (Attention deficit hyperactivity disorder). ADHD is one of the most common psychiatric disorders for children that can continue through adolescence and adulthood. Symptoms include difficulty staying focused and paying attention, difficulty controlling behavior, and hyperactivity (over-activity). ADHD in general has three subtypes: (1) ADHD hyperactive-impulsive (2) ADHD-inattentive; (3) Combined hyperactive-impulsive and inattentive (ADHD-combined) Ramsay (2007). ADHD-200 Dataset (http://fcon_1000.projects.nitrc.org/indi/adhd200/) is a data set that record both anatomical and resting-state functional MRI data of 776 labeled subjects across 8 independent imaging sites, 491 of which were obtained from typically developing individuals and 285 in children and adolescents with ADHD (ages: 7–21 years old). The Corpus Callosum shape data are extracted using the CCSeg package, which contains 50 landmarks on the contour of the Corpus Callosum of each subject (see Huang et al. 2015). After quality control, 647 CC shape data out of 776 subjects were obtained, which included 404

(n_1) typically developing children, 150 (n_2) diagnosed with ADHD-Combined, 8 (n_3) diagnosed with ADHD-Hyperactive-Impulsive, and 85 (n_4) diagnosed with ADHD-Inattentive. Therefore, the data lie in the space Σ_2^{50}, which has a high dimension of $2 \times 50 - 4 = 96$.

We carry out *extrinsic two sample tests* between the group of typically developing children and the group of children diagnosed with ADHD-Combined, and also between the group of typically developing children and ADHD-Inattentive children. We construct test statistics that base on the asymptotic distribution of the extrinsic mean for the planar shapes.

The p-value for the two-sample test between the group of typically developing children and the group of children diagnosed with ADHD-Combined is 5.1988×10^{-11}, which is based on an asymptotic chi-square distribution, following Corollary 1 (See Bhattacharya and Lin 2017). The p-value for the test between the group of typically developing children and the group ADHD-Inattentive children is smaller than 10^{-50}.

Example 9 (Positive definite matrices with application to diffusion tensor imaging). Another important class of manifolds is $\text{sym}^+(p)$, the space of $p \times p$ positive definite matrices. In particular, when $p = 3$, $\text{sym}^+(3)$, the space of 3×3 positive definite matrices, has important applications in diffusion tensor imaging (DTI). DTI, is now an important tool for neuroimaging in clinical trials. It provides for the measurement of the diffusion matrix (3×3 positive definite matrices) of molecules of water in tiny voxels in the white matter of the brain. When there are no barriers, the diffusion matrix is isotropic, and in the presence of structural barriers in the brain white matter due to axon (nerve fiber) bundles and their myelin sheaths (electrically insulating layers) the diffusion is anisotropic, and DTI can be used to measure the anisotropic diffusion tensor. When a trauma occurs, due to an injury or a disease, this highly organized structure is disrupted and anisotropy decreases. Large scale DTI based studies have been used to investigate autism, schizophrenia, Parkinson's disease and Alzheimer's disease. The geometry of $\text{sym}^+(p)$ for general p is now described in the following.

Let $A \in \text{sym}^+(p)$ which follows a distribution Q. We first introduce the Euclidean metric of A, which is given by $\|A\|^2 = \text{Trace}(A)^2$. Since $\text{sym}^+(p)$ is an open convex subset of $\text{sym}(p)$, the space of all $p \times p$ symmetric matrices, the mean of Q with respect to the Euclidean distance is given by the Euclidean mean

$$\mu_E = \int AQ(\mathrm{d}A).$$

Another important metric for $\text{sym}^+(p)$ is the log-*Euclidean metric* (Arsigny et al. 2006). Let $J \equiv \log : \text{sym}^+(p) \to \text{sym}(p)$ be the inverse of the *exponential map* $B \to \mathrm{e}^B$, $\text{sym}(p) \to \text{sym}^+(p)$, which is the matrix exponential of B. Then J is a diffeomorphism. The log Euclidean distance is given by

$$\rho_{LE}(A_1, A_2) = \|\log(A_1) - \log(A_2)\|.$$

Note that J is an embedding on $\text{sym}^+(p)$ onto $\text{sym}(p)$ and, in fact, it is an equivariant embedding under the group action of $\text{GL}(p, \mathbb{R})$, the general linear group of $p \times p$ non-singular matrices. The extrinsic mean of Q under J is given by

$$\mu_{E,J} = \exp\left(\int (\log(A)) Q(\mathrm{d}A) \right).$$

We apply Theorem 4 to sample Fréchet means under both the Euclidean and log-Euclidean distances. In particular, we consider a diffusion tensor imaging (DTI) data set consisting of 46 subjects with 28 HIV+ subjects and 18 healthy controls. Diffusion tensors were extracted along the fiber tract of the splenium of the corpus callosum. The DTI data for all the subjects are registered in the same *atlas space* based on arc lengths, with 75 features obtained along the fiber tract of each subject. This data set has been studied in a regression setting in Yuan et al. (2012). On the other hand, we carry out two sample tests between the control group and the HIV+ group for each of the 75 sample points along the fiber tract. Therefore, 75 tests are performed in total. Two types of tests are carried out based on the Euclidean distance and the log-Euclidean distance.

The simple Bonferroni procedure for testing H_0 yields a p-value equal to 75 times the smallest p-value which is of order 10^{-7}. To identify sites with significant differences, the 75 p-values are ordered from the smallest to the largest with a *false discovery rate* of $\alpha = 0.05$, 58 sites are found to yield significant differences using the Euclidean distance, and 47 using the log-Euclidean distance (see Benjamini and Hochberg 1995).

Example 10 (Glaucoma detection – a match pair problem in 3D). Our final example is on the $3D$ reflection similarity shape space $R\Sigma_3^k$. To detect shape changes due to glaucoma, data were collected on twelve mature rhesus monkeys. One of the eyes of each monkey was treated with a chemical agent to temporarily increase the intraocular pressure (IOP). The increase in IOP is known to be a cause of glaucoma. The other eye was left untreated. Measurements were made of five landmarks in each eye, suggested by medical professionals. The data may be found in Bhattacharya and Patrangenaru (2005). A match pair test yielded an observed value 36.29 of the asymptotic chi-square statistic with 8 degrees of freedom. The corresponding p-value is 1.55×10^{-5} (See Bhattacharya and Bhattacharya 2012, Chapter 9). This provides a strong justification for using shape change of the inner eye as a diagnostic tool to detect the onset of glaucoma. An earlier computation using a different nonparametric procedure in Bhattacharya and Patrangenaru (2005) provided a p-value 0.058. Also see Bandulasiri et al. (2009) where a 95% confidence region is obtained for the difference between the extrinsic size-and-shape reelection shapes between the treated and untreated eyes.

Acknowledgement. The authors acknowledge support for this article from NSF grants DMS 1811317, CAREER 1654579 and IIS 1663870.

A Appendix on Riemannian Manifolds

Often the manifold M in applications has a natural Riemannian metric tensor g. That is, it is given an inner product $\langle , \rangle_p$ on the tangent space T_pM at p, which is smoothly defined. In local coordinates in U_p given by $\psi_p(\cdot) = x = (x_1, \ldots, x_d) \in B_p$, the functions $(g_{ij})(x) = \langle E_i, E_j \rangle_p$, with $E_i = \mathrm{d}\psi_p^{01}(\partial/\partial x_i)$ $(i, j = 1, \ldots, d)$, are smooth in B_p. This allows one to measure the length of a smooth arc γ joining any two points q, q' in U_p, namely, $\int_{[a,b]} |\mathrm{d}x(t)/\mathrm{d}t| \mathrm{d}t$, $\gamma(a) = q$, $\gamma(b) = q'$, $x(t) = \psi_p \circ \gamma(t)$. Here $|\mathrm{d}x(t)/\mathrm{d}t|^2 = \langle \mathrm{d}x(t)/\mathrm{d}t, \mathrm{d}x(t)/\mathrm{d}t \rangle_p$, with $\mathrm{d}x(t)/\mathrm{d}t$ expressed in the local frame $E_i (i = 1, \ldots, d)$. One may also write $\mathrm{d}x(t)/\mathrm{d}t$ as $\mathrm{d}\gamma(t)/\mathrm{d}t$. Using the compatibility condition (ii) above one now defines the length of a smooth arc joining any two points in M. The *geodesic distance* $\rho_g(p, q)$ between p and q is the minimum of lengths of all smooth arcs joining p and q. A standard parametrization of a curve is its arc length s: $s = \int_{[a,t]} |\mathrm{d}\gamma(u)/\mathrm{d}u| \mathrm{d}u$. In this parametrization of curves, one has $|\mathrm{d}\gamma(t)/\mathrm{d}t| = 1$. We will adopt this so called *unit speed* parametrization unless otherwise specified. The property of local minimization of arc lengths yields a first order condition on the velocity $\mathrm{d}\gamma(t)/\mathrm{d}t$ of the minimizing curve γ at t: the acceleration along γ is zero at every parameter join t. If M is a submanifold ((hyper) surface) of an Euclidean space $\mathbb{R}^N$, then the second derivative $\mathrm{d}^2\gamma(t)/\mathrm{d}t^2$ is well defined, but in general does not belong to the tangent space of M at $\gamma(t)$. By *'acceleration'* one means the orthogonal projection of the vector $\mathrm{d}^2\gamma(t)/\mathrm{d}t^2$ onto the tangent space of M at $\gamma(t)$. This projection is called the *covariant derivative* of the velocity and denoted $(\mathrm{D}/\mathrm{d}t)\mathrm{d}\gamma(t)/\mathrm{d}t$. The "zero acceleration" of a geodesic γ means $(\mathrm{D}/\mathrm{d}t)\mathrm{d}\gamma(t)/\mathrm{d}t = 0$. On a general differentiable manifold, which is not given explicitly as a submanifold, there is no "outside". The proper extension of the above notion of covariant derivative by Levi–Civita, using a notion known as *affine connection*, for all differentiable manifolds was a milestone in the development of differential geometry (See, e.g., Do Carmo 1992 Chapter 2).

In local coordinates the equation for a geodesic is a second order ordinary differential equation. By the standard existence theorem for ordinary differential equations, a geodesic γ is uniquely determined on a maximal interval (a, b) $(-\infty \leq a < b < \infty)$, given an initial point $\gamma(0) = p$ and an velocity $(\mathrm{d}\gamma(t)/\mathrm{d}t)_{t=0} = v$. According to a result of Hopf and Rinow (Do Carmo 1992, Chapter 7), the geodesics can be extended indefinitely, (i.e., $a = -\infty$ and $b = \infty$), i.e., it is *geodesically complete*, if and only if (M, ρ_g) is a complete metric space; this in turn is equivalent to the topological condition (2). In particular, all compact Riemannian manifolds are *geodesically complete.* In most of the applications in this article M is compact.

On a complete Riemannian manifold, a geodesic $\gamma(t) = \gamma(t; p, v)$, $t \geq 0$, in the direction v, is completely determined by an initial point $p = \gamma(0)$, and an initial velocity $v = (\mathrm{d}\gamma(t)/\mathrm{d}t)t = 0$. A *cut point* of p of the geodesic γ along v is $\gamma(r(v); p, v)$, where $r(v)$ is the supremum of all t_0 such that γ is distance minimizing between $p = \gamma(0)$ and $\gamma(t_0)$. The set of all cut points (along all v) is called the *cut locus* of p, denoted $\mathrm{Cut}(p)$. The geodesic distance $q \to \rho_g(p, q)$

may not be smooth at the cut locus Cut(p), as Example 11 below shows. Next, define the *exponential function* $\mathrm{Exp}_p : T_p(M) \to M : \mathrm{Exp}_p(v) = \gamma(1; p, v)$ the point in M reached by the geodesic in time $t = 1$, starting at p with an initial velocity v. It is known that Exp_p is a diffeomorphism on an open ball $B(0 : r_0)$ of $T_p(M)$, of radius $r_0 = r_0(p) < \infty$, onto $M\backslash \mathrm{Cut}(p)$ (Do Carmo 1992, p. 271). Here $r_0 = r_0(p)$ is the geodesic distance between p and Cut(p)). The inverse map $\mathrm{Exp}_p^{-1} : M\backslash \mathrm{Cut}(p) \to \mathrm{Exp}_p(B(0 : r_0)$ is called the *inverse exponential*, or the log *map*, $\log_p$, at p. The quantity $\mathrm{inj}(M) = \sup\{r_0(p); p \in M)\}$ is the *injectivity radius* of M. The $\log_p$ map also provides the so called *normal coordinates* for a neighborhood of p.

Example 11 (Exponential and Log Maps on the Sphere S^d). Consider the unit sphere $S^d = \{x \in \mathbb{R}^{d+1} : |x|^2 \sum_{j=1}^{d} (x^{(j)})^2 = 1\}$. Because $|\gamma(t)| = 1\ \forall t$ for a curve on S^d, the tangent space at p may be identified as the set of vectors in $\mathbb{R}^{d+1}$ orthogonal to p, $T_p(S^d) = \{v \in \mathbb{R}^{d+1} : pv' = 0\}$. Here we write p, v, etc. as row vectors. The geodesics are the big circles, so that the point reached at time one by the geodesic from p moving with an initial velocity v is the point on the big circle lying on the plane spanned by p and v at an arc distance $|v|$, i.e.,

$$\mathrm{Exp}_p(v) = \cos(|v|)p + \sin(|v|)v/|v|,\ v \neq 0, \mathrm{Exp}_p(0) = p\ (pv' = 0). \tag{A.1}$$

Also, the geodesic distance between p and q is the smaller of the lengths $|v|$ of the two arcs joining p and q on the big circle,

$$\rho_g(p, q) = \arccos pq' \in [0, \pi]. \tag{A.2}$$

Note that the cut locus of p is $\mathrm{Cut}(p) = \{-p\}$, and the distance between p and $-p$ is π, and $\mathrm{inj}(S^d) = \pi$. Hence the map $\log_p(q)$ is defined on $S^d\backslash\{-p\}$ and obtained by solving for v the equation $\exp_p(v) = q$. Now $|v| = \rho_g(p, q)$. Plugging this in (A.1) (and using (A.2)), one has

$$\sqrt{[1 - (pq')^2]}v = [q - (pq')p](\arccos pq'), (p \neq q),$$

which yields

$$\begin{aligned}\log_p(q) &= [q - (pq')p](\arccos p'q)/\sqrt{[1 - (pq')2]}\\ &= [\rho_g(p, q)/ \sin \rho_g(p, q)][q - (pq')p],\end{aligned} \tag{A.3}$$

for $q \neq p$, $q \neq -p$, $\log_p(p) = 0$. The map $\log_p(q)$ is a diffeomorphism on $S^d\backslash\{-p\}$ onto $\{v \in T_pS^d : |v| < \pi\}$. If one uses complex coordinates for p, q then pq' in the formula above are to be replaced by $\mathrm{Re}(pq^*)$, etc.

Most of the manifolds we consider in this article are of the form $M = N/\mathcal{G}$. Here N is a complete Riemannian manifold with a metric tensor $\rho_{g,N}$ and $\mathcal{G}$ is a compact Lie group of isometries *acting freely* on N, i.e., except for the identity map, no g in $\mathcal{G}$ has a fixed point. This means that the orbit O_p of a point p under $\mathcal{G}$ is in one-one correspondence with $\mathcal{G}$. As a subset of N, O_p is a submanifold of N of dimension that of $\mathcal{G}$. Its tangent space T_pO_p as a subspace of T_pN is called the *vertical subspace* of T_pN, denoted V_p. The subspace H_p of T_pN orthogonal to V_p is the *horizontal subspace.* M is then a Riemannian manifold with the metric tensor. The projection $\pi : N \longrightarrow M$ is a *Riemannian submersion.* The quotient $N/\mathcal{G}$ is then a Riemannian manifold.

The final important notion from geometry needed in this section is that of curvature. First, consider a smooth unit speed curve γ in $\mathbb{R}^2$: $1 = |\dot{\gamma}(t)|^2 = \langle\dot{\gamma}(t), \dot{\gamma}(t)\rangle$. Differentiation shows that $\ddot{\gamma}(t) = \mathrm{d}^2\gamma(t)/\mathrm{d}t^2$ is orthogonal to $\dot{\gamma}(t) : \ddot{\gamma}(t) = \kappa(t)N(t)$, where $N(t)$ is a unit vector orthogonal to $\dot{\gamma}(t)$ such that $(\dot{\gamma}(t), N(t))$ has the same orientation as $(\partial/\partial x_1, \partial/\partial x_2)$. Then $\kappa(t)$ is the curvature of γ at the point $\gamma(t)$. Next, at a point p on a regular surface S in $\mathbb{R}^3$, let $N = N(p)$ denote a unit normal to S at p. A plane π through $N(p)$ intersects S in a smooth curve. Let $\kappa(.; p, \pi)$ be the curvature of this curve. As π varies by degrees of rotation, the curvature varies. Let κ_1 be the maximum and κ_2 the minimum of these curvatures, and let $\kappa = \kappa_1\kappa_2$. The *Theorem Egregium* of Gauss says that $\kappa = \kappa(p)$ $(p \in S)$, the so-called *Gaussian curvature,* is intrinsic to the surface S, i.e., it is the same for all surfaces isometric to S (See, e.g., Boothby 1986, pp. 377–381). We now consider, somewhat informally, the case of a Riemannian manifold M. For $p \in M$ and $u, v \in T_p(M)$, consider the two dimensional subspace π spanned by u,v. Consider the two-dimensional submanifold swept out by geodesics in M with initial velocities lying in this subspace. The Gaussian curvature of this submanifold, thought of locally as a surface, is called the *sectional curvature* of M at p for the section π.

References

Afsari, B.: Riemannian l^p center of mass: existence, uniqueness, and convexity. Proc. Am. Math. Soc. **139**(2), 655–673 (2011)

Aggarwal, G., Chowdhury, A., Chellappa, R.: A system identification approach for video-based face recognition. In: Proceedings of the 17th International Conference on Pattern Recognition, ICPR 2004, vol. 4, pp. 175–178 (2004)

Alexandrov, A.D.: Uber eine verallgemeinerung der riemannschen geometrie. Schr. Forschungsinst. Math. 33–84 (1957)

Arsigny, A., Fillard, P., Pennec, X., Ayache, N.: Log-Euclidean metrics for fast and simple calculus on diffusion tensors. Magn. Reson. Med. **56**(2), 411–421 (2006)

Bandulasiri, A., Bhattacharya, R., Patrangenaru, V.: Nonparametric inference for extrinsic means on size-and-(reflection)-shape manifolds with applications in medical imaging. J. Multivar. Anal. **100**(9), 1867–1882 (2009)

Bandulasiri, A., Patrangenaru, V.: Algorithms for nonparametric inference on shape manifolds. In: Proceedings of JSM 2005, pp. 1617–1622 (2005)

Benjamini, Y., Hochberg, Y.: Controlling the false discovery rate: a practical and powerful approach to multiple testing. J. R. Stat. Soc. B. **57**(1), 289–300 (1995)

Bhattacharya, A.: Statistical analysis on manifolds: a nonparametric approach for inference on shape spaces. Sankhya **70**, 1–43 (2008)

Bhattacharya, A., Bhattacharya, R.: Statistics on Riemannian manifolds: asymptotic distribution and curvature. Proc. Am. Math. Soc. **136**, 2957–2967 (2008)

Bhattacharya, A., Bhattacharya, R.: Nonparametric Inference on Manifolds: With Applications to Shape Spaces. IMS Monograph, vol. 2. Cambridge University Press, Cambridge (2012)

Bhattacharya, A., Dunson, D.: Nonparametric Bayesian density estimation on manifolds with applications to planar shapes. Biometrika **97**, 851–865 (2010)
Bhattacharya, R.: On the uniqueness of intrinsic mean (2007, Unpublished manuscript)
Bhattacharya, R.: A nonparametric theory of statistics on manifolds. In: Eichelsbacher, P., Elsner, G., Kösters, H., Löwe, M., Merkl, F., Rolles, S. (eds.) Limit Theorems in Probability, Statistics and Number Theory. Springer Proceedings in Mathematics & Statistics, vol. 42, pp. 173–205. Springer, Heidelberg (2013)
Bhattacharya, R., Lin, L.: Omnibus CLTs for Fréchet means and nonparametric inference on non-Euclidean spaces. arXiv:1306.5806 (2013)
Bhattacharya, R., Lin, L.: Omnibus CLTs for Fréchet means and nonparametric inference on non-Euclidean spaces. Proc. Am. Math. Soc. **145**, 413–428 (2017)
Bhattacharya, R.N., Lin, L., Patrangenaru, V.: A Course in Mathematical Statistics and Large Sample Theory. Springer Texts in Statistics. Springer, New York (2016)
Bhattacharya, R., Oliver, R.: Nonparametric analysis of non-Euclidean data on shapes and images. Sankhya A **81**(1), 1–36 (2019)
Bhattacharya, R.N., Patrangenaru, V.: Nonparametric estimation of location and dispersion on Riemannian manifolds. J. Stat. Plan. Infer. **108**, 22–35 (2002). Volume in honor of the 80th birthday of Professor C.R. Rao
Bhattacharya, R.N., Patrangenaru, V.: Large sample theory of intrinsic and extrinsic sample means on manifolds. Ann. Stat. **31**, 1–29 (2003)
Bhattacharya, R.N., Patrangenaru, V.: Large sample theory of intrinsic and extrinsic sample means on manifolds-II. Ann. Stat. **33**, 1225–1259 (2005)
Bookstein, F.L.: Morphometric Tools for Landmark Data: Geometry and Biology. Cambridge University Press, Cambridge (1991)
Boothby, W.: An Introduction to Differentiable Manifolds and Riemannian Geometry, 2nd edn. Academic Press, New York (1986)
Carlsson, G.: Topology and data. Bull. Am. Math. Soc. **46**(2), 255–308 (2009)
Chavel, I.: Eigenvalues in Riemannian Geometry. Pure and Applied Mathematics. Elsevier Science, Amsterdam (1984)
Demmel, J., Eisenstat, S., Gilbert, J., Li, X., Liu, J.: A supernodal approach to sparse partial pivoting. SIAM J. Matrix Anal. Appl. **20**, 720–755 (1999)
Dey, T.K., Li, K.: Persistence-based handle and tunnel loops computation revisited for speed up. Comput. Graph. **33**(3), 351–358 (2009). IEEE International Conference on Shape Modelling and Applications
Dey, T.K., Li, K., Sun, J., Cohen-Steiner, D.: Computing geometry-aware handle and tunnel loops in 3D models. ACM Trans. Graph. **27**(3), 45:1–45:9 (2008)
Do Carmo, M.: Riemannian Geometry. Birkhäuser, Boston (1992)
Dryden, I., Kume, A., Le, H., Wood, A.T.: A multi-dimensional scaling approach to shape analysis. Biometrika **95**(4), 779–798 (2008)
Dryden, I.L., Mardia, K.V.: Statistical Shape Analysis. Wiley, New York (1998)
Ellingson, L., Patrangenaru, V., Ruymgaart, F.: Nonparametric estimation of means on Hilbert manifolds and extrinsic analysis of mean shapes of contours. J. Multivar. Anal. **122**, 317–333 (2013)
Ferguson, T.S.: A Bayesian analysis of some nonparametric problems. Ann. Stat. **1**, 209–230 (1973)
Ferguson, T.S.: Prior distributions on spaces of probability measures. Ann. Stat. **2**, 615–629 (1974)
Fisher, N., Lewis, T., Embleton, B.: Statistical Analysis of Spherical Data. Cambridge University Press, Cambridge (1987)
Fisher, R.: Dispersion on a sphere. Proc. R. Soc. Lond. Ser. A **217**, 295–305 (1953)

Fréchet, M.: Lés élements aléatoires de nature quelconque dans un espace distancié. Ann. Inst. H. Poincaré **10**, 215–310 (1948)
Gallot, S., Hulin, D., Lafontaine, J.: Riemannian Geometry. Universitext. Springer, Berlin (1990)
Ghosh, J., Ramamoorthi, R.: Bayesian Nonparametrics. Springer Series in Statistics. Springer, New York (2003)
Goodall, C.: Procrustes methods in the statistical analysis of shape. J. R. Stat. Soc. Ser. B (Methodol.) **53**(2), 285–339 (1991)
Goodlett, C., Davis, B., Jean, R., Gilmore, J., Gerig, G.: Improved correspondence for DTI population studies via unbiased atlas building. In: Medical Image Computing and Computer-Assisted Intervention (MICCAI), vol. 4191, pp. 260–267 (2006)
Gordon, C., Webb, D., Wolpert, S.: Isospectral plane domains and surfaces via Riemannian orbifolds. Inventiones mathematicae **110**(1), 1–22 (1992)
Gotsman, C., Gu, X., Sheffer, A.: Fundamentals of spherical parameterization for 3D meshes. ACM Trans. Graph. **22**(3), 358–363 (2003)
Gromov, M.: Structures métriques pour les variétés Rriemanniennes. Textes mathématiques. CEDIC/Fernand Nathan (1981)
Hotelling, H.: The generalization of student's ratio. Ann. Math. Stat. **2**(3), 360–378 (1931)
Hotz, T., Huckemann, S.: Intrinsic means on the circle: uniqueness, locus and asymptotics. Ann. Inst. Stat. Math. **67**(1), 177–193 (2015)
Hotz, T., Skwerer, S., Huckemann, S., Le, H., Marron, J. S., Mattingly, J. C., Miller, E., Nolen, J., Owen, M., Patrangenaru, V.: Sticky central limit theorems on open books. In: Annals of Applied Probability, pp. 2238–2258 (2013)
Huang, C., Styner, M., Zhu, H.: Penalized mixtures of offset-normal shape factor analyzers with application in clustering high-dimensional shape data. J. Am. Stat. Assoc. (2015)
Huckemann, S., Hotz, T., Munk, A.: Intrinsic shape analysis: geodesic PCA for Riemannian manifolds modulo isometric Lie group actions (with discussions). Statist. Sinica **20**, 1–100 (2010)
Irving, E.: Paleomagnetism and Its Application to Geological and Geographical Problems. Wiley, New York (1964)
Jain, V., Zhang, H.: A spectral approach to shape-based retrieval of articulated 3D models. Comput. Aided Des. **39**(5), 398–407 (2007)
Jones, P.W., Maggioni, M., Schul, R.: Manifold parametrizations by eigenfunctions of the Laplacian and heat kernels. Proc. Nat. Acad. Sci. **105**(6), 1803–1808 (2008)
Kac, M.: Can one hear the shape of a drum? Am. Math. Monthly **73**(4), 1–23 (1966)
Karcher, H.: Riemannian center of mas and mollifier smoothing. Commun. Pure Appl. Math. **30**, 509–554 (1977)
Kendall, D., Barden, D., Carne, T., Le, H.: Shape and Shape Theory. Wiley, New York (1999)
Kendall, D.G.: Shape manifolds, Procrustean metrics, and complex projective spaces. Bull. Lond. Math. Soc. **16**, 81–121 (1984)
Kendall, D.G.: A survey of the statistical theory of shape. Stat. Sci. **4**, 87–120 (1989)
Kendall, W.: Probability, convexity, and harmonic maps with small image I: uniqueness and fine existence. Proc. Lond. Math. Soc. **61**, 371–406 (1990)
Kendall, W., Le, H.: Limit theorems for empirical Fréchet means of independent and non-identically distributed manifold-valued random variables. Braz. J. Probab. Stat. **25**(3), 323–352 (2011)

Kindlmann, G., Tricoche, X., Westin, C.-F.: Delineating white matter structure in diffusion tensor MRI with anisotropy creases. Med. Image Anal. **11**(5), 492–502 (2007)

Le, H.: Locating Fréchet means with application to shape spaces. Adv. Appl. Prob. **33**, 324–338 (2001)

Lehmann, E.: Testing Statistical Hypotheses. Wiley Series in Probability and Mathematical Statistics. Probability and Mathematical Statistics. Wiley (1959)

Ma, Y., Soatto, S., Kosecká, J., Sastry, S.: An Invitation to 3-D Vision: From Images to Geometric Models. Interdisciplinary Applied Mathematics. Springer, New York (2005)

Mardia, K., Jupp, P.: Directional Statistics. Wiley, New York (2000)

Mardia, K., Patrangenaru, V.: Directions and projective shapes. Ann. Stat. **33**, 1666–1699 (2005)

McKilliam, R.G., Quinn, B.G., Clarkson, I.V.L.: Direction estimation by minimum squared arc length. IEEE Trans. Signal Process. **60**(5), 2115–2124 (2012)

Milnor, J.: Eigenvalues of the Laplace operator on certain manifolds. Proc. Nat. Acad. Sci. **51**(4), 542 (1964)

Milnor, J.: Topology from the Differentiable Viewpoint. University Press of Virginia, Charlottesville (1965)

Morra, J., Tu, Z., Apostolova, L., Green, A., Avedissian, C., Madsen, S., Parikshak, N., Toga, A., Jack Jr., C., Schuff, N., Weiner, M., Thompson, P.: Automated mappings of hippocampal atrophyin 1-year repeat MRI data from 490 subjects with alzheimers disease, mild cognitive impairment, and elderly controls. Neuroimage **S45**, 3–S15 (2000)

Patrangenaru, V., Liu, X., Sugathadasa, S.: A nonparametric approach to 3D shape analysis from digital camera images – I. J. Multivar. Anal. **101**(1), 11–31 (2010)

Pelletier, B.: Kernel density estimation on Riemannian manifolds. Stat. Prob. Lett. **73**, 297–304 (2005)

Ramsay, J.R.: Current status of cognitive-behavioral therapy as a psychosocial treatment for adult attention-deficit/hyperactivity disorder. Curr. Psychiatry Rep. **9**(5), 427–433 (2007)

Reshetnyak, Y.: Non-expansive maps in a space of curvature no greater than k. Sibirsk. Mat. Zh. **9**, 918–927 (1968)

Reuter, M.: Laplace Spectra for Shape Recognition. Books on Demand GmbH (2006)

Reuter, M., Biasotti, S., Giorgi, D., Patanè, G., Spagnuolo, M.: Discrete Laplace-Beltrami operators for shape analysis and segmentation. Comput. Graph. **33**(3), 381–390 (2009). IEEE International Conference on Shape Modelling and Applications

Rosenberg, S.: The Laplacian on a Riemannian Manifold: An Introduction to Analysis on Manifolds. EBSCO Ebook Academic Collection. Cambridge University Press, Cambridge (1997)

Sethuraman, J.: A constructive definition of Dirichlet priors. Statist. Sinica **4**, 639–50 (1994)

Shamir, A.: Segmentation and shape extraction of 3D boundary meshes. In: Wyvill, B., Wilkie, A. (eds.) Eurographics 2006 - State of the Art Reports. The Eurographics Association (2006)

Sparr, G.: Depth-computations from polyhedral images. In: Sandimi, G. (ed.) Proceedings of the 2nd European Conference on Computer Vision, pp. 378–386. Springer (1992). Also in Image and Vision Computing, **10**, 683–688

Sturm, K.: Probability Measures on Metric Spaces of Nonpositive Curvature. Preprint. SFB 611 (2003)

Sugathadasa, S.: Affine and Projective Shape Analysis with Applications. Ph.D. dissertation, Texas Tech University (2006)

Veeraraghavan, A., Roy-Chowdhury, A., Chellappa, R.: Matching shape sequences in video with applications in human movement analysis. IEEE Trans. Pattern Anal. Mach. Intell. **27**(12), 1896–1909 (2005)

Watson, G.S.: Statistics on Spheres. University Arkansas Lecture Notes in the Mathematical Sciences, vol. 6. Wiley, New York (1983)

Weyl, H.: Ueber die asymptotische verteilung der eigenwerte. Nachrichten von der Gesellschaft der Wissenschaften zu Göttingen, Mathematisch-Physikalische Klasse 110–117 (1911)

Yuan, Y., Zhu, H., Lin, W., Marron, J.S.: Local polynomial regression for symmetric positive definite matrices. J. R. Stat. Soc. Ser. B (Stat. Methodol.) 697–719 (2012)

Zelditch, S.: Spectral determination of analytic bi-axisymmetric plane domains. Geom. Funct. Anal. GAFA **10**(3), 628–677 (2000)

Ziezold, H.: On expected figures and a strong law of large numbers for random elements in quasi-metric spaces. In: Transactions of the Seventh Prague Conference on Information Theory, Statistical Functions, Random Processes and of the Eighth European Meeting of Statisticians, A, pp. 591–602 (1977). (Tech. Univ. Prague, Prague, 1974)

Conformal Measure Ensembles for Percolation and the FK–Ising Model

Federico Camia[1,2(✉)], René Conijn[3], and Demeter Kiss[4,5]

[1] NYU Abu Dhabi, Abu Dhabi, UAE
federico.camia@nyu.edu
[2] VU University Amsterdam, Amsterdam, The Netherlands
[3] Utrecht University, Utrecht, The Netherlands
reneconijn@gmail.com
[4] Statistical Laboratory, University of Cambridge, Cambridge, UK
kdemeter@gmail.com
[5] AIMR, Tohoku University, Sendai, Japan

Abstract. Under some general assumptions, we construct the scaling limit of open clusters and their associated counting measures in a class of two dimensional percolation models. Our results apply, in particular, to critical Bernoulli site percolation on the triangular lattice and to the critical FK–Ising model on the square lattice. Fundamental properties of the scaling limit, such as conformal covariance, are explored. As an application to Bernoulli percolation, we obtain the scaling limit of the largest cluster in a bounded domain. We also apply our results to the critical, two-dimensional Ising model, obtaining the existence and uniqueness of the scaling limit of the magnetization field, as well as a geometric representation for the continuum magnetization field which can be seen as a continuum analog of the FK representation of the Ising model.

Keywords: Percolation · Critical cluster · Scaling limit · Schramm–Smirnov topology · Ising model · Magnetization field

1 Dedication and Synopsis

This paper is dedicated to Chuck Newman on the occasion of his 70th birthday. The main goal of the paper is the study of the continuum scaling limit of counting measures for percolation and FK–Ising clusters in two dimensions, leading to the concept of *conformal measure ensemble*. This is a fitting topic for a paper dedicated to Chuck since the concept was originally discussed, in the context of the Ising model, about ten years ago, in New York, during a conversation by Chuck and the first author; it was one of the early discussions in the project that led to [8,9,11,12,15]. The concept first appeared in [15], although the term conformal measure ensemble was used for the first time in [8].

V. Sidoravicius (Ed.): *Sojourns in Probability Theory and Statistical Physics - II*, PROMS 299, pp. 44–89, 2019.
https://doi.org/10.1007/978-981-15-0298-9_2

Among other things, in this paper we construct the conformal measure ensemble for the critical, two-dimensional FK–Ising model, and use it to express the continuum scaling limit of the critical Ising magnetization field as a sum of the area measures of continuum FK–Ising clusters with fair, i.i.d. $\pm$ signs. This gives a geometric representation for the Ising magnetization which amounts to a continuum analog of the FK representation of the Ising model which is part of the Edwards–Sokal coupling. The key step consists in showing that the collection of appropriately rescaled counting measures of FK–Ising clusters has a scaling limit. With this result, we give an alternative proof of the uniqueness and conformal covariance of the scaling limit of the critical Ising magnetization, first proved in [9], and we show that the limiting magnetization is measurable with respect to the collection of loops that describe the full scaling scaling limit of the FK–Ising process [24]. Surprisingly, the FK–Ising conformal measure ensemble constructed in this paper plays a key role in the recent proof of exponential decay for the *near-critical* scaling limit of the planar Ising model [11].

In the case of critical site percolation on the triangular lattice, we show that the collection of appropriately rescaled counting measures of macroscopic clusters converges, in the scaling limit, to a collection of conformally covariant measures whose supports are the scaling limits of the macroscopic clusters themselves. As a consequence, we obtain that the largest percolation cluster in a bounded domain has a conformally covariant scaling limit. These results build on joint work of the first author with Chuck [13].

Chuck's papers mentioned above are only a tiny and biased sample of his impressive production. His many contributions to probability and statistical mechanics are both broad and profound. Through his work and teaching, Chuck has built an important legacy, and one can only wish that he will continue to guide and inspire younger researchers for many years to come.

2 Introduction

Several important models of statistical mechanics, such as percolation and the Ising and Potts models, can be described in terms of clusters. In the last twenty years, there has been tremendous progress in the study of the geometric properties of such models in the scaling limit. Much of that work has focused on interfaces, that is, cluster boundaries, taking advantage of the introduction of the Schramm–Loewner Evolution (SLE) by Schramm in [30]. In this paper, we are concerned with the scaling limit of the clusters themselves and their "areas." More precisely, we analyze the scaling limit of the collection of clusters and the associated counting measures (rescaled by an appropriate power of the lattice spacing).

Our main results are valid under some general assumptions, which can be verified for Bernoulli site percolation on the triangular lattice and for the FK–Ising model on the square lattice. Roughly speaking, our main results say that, under suitable assumptions, in a general two-dimensional percolation model, the collection of clusters together with their associated counting measures (appropriately rescaled) has a unique weak limit, in an appropriate topology, as the lattice spacing tends to zero. The collection of clusters converges to a collection of closed sets (the "continuum clusters"), while the collection of rescaled counting measures converges to a collection of continuum measures whose supports are the continuum clusters.

Our results are nontrivial at the critical point of the percolation model. For instance, in the case of critical site percolation on the triangular lattice, where a scaling limit in terms of cluster boundaries is known to exist and to be conformally invariant [13] (it can be described in terms of SLE_6 curves), we show that the continuum clusters are also conformally invariant, and that the associated measures are conformally covariant. The conformal covariance property of the collection of measures is a consequence of the conformal invariance of the critical scaling limit. Because of this property, we use the expression "conformal measure ensemble" to denote the collection of measures arising in the scaling limit of a critical percolation model.

The scaling limit of the rescaled counting measures is in the spirit of [19], and indeed we rely heavily on techniques and results from that paper. There is however a significant difference in that we distinguish between different clusters. In other words, we don't obtain a single measure that gives the combined size of all clusters inside a domain, but rather a collection of measures, one for each cluster. This is the main technical difficulty of the present paper. The reward is that handling individual clusters leads to new, interesting applications to Bernoulli percolation and the Ising model, which represent the main motivation of the paper and are discussed in detail in Sect. 3.

In the case of Bernoulli percolation, our main application is the scaling limit of the largest clusters in a bounded domain. The application to the Ising model requires a brief discussion. Consider a critical Ising model on the scaled lattice $\eta\mathbb{Z}^2$. Using the FK representation of the Ising model, one can write the total magnetization in a domain D as $\sum_i \sigma_i \nu_i^\eta(D)$, where the σ_i's are (± 1)-valued, symmetric random variables independent of each other and everything else, and $\nu_i^\eta = \sum_{u \in \mathcal{C}_i} \delta_u$ is the counting measure associated to the i-th FK cluster (δ_u denotes the Dirac measure concentrated at u and the order of the clusters is irrelevant) and $\nu_i^\eta(D) = |\mathcal{C}_i \cap D|$, where $\mathcal{C}_i$ is the i-th FK cluster. The first author and Newman [15] noticed that the power of η by which one should rescale the magnetization to obtain a limit, as $\eta \to 0$, is the same as the power that should ensure the existence of a limit for the rescaled counting measures. They then predicted that one should be able to give a meaning to the expression "$\Phi^0 = \sum_i \sigma_i \mu_i^0$," where Φ^0 is the limiting magnetization field, obtained from the scaling limit of the renormalized lattice magnetization, and $\{\mu_i^0\}$ is the collection of measures obtained from the scaling limit of the collection of rescaled versions of the counting measures $\{\nu_i^\eta\}$. The existence and uniqueness of the limiting magnetization field was proved in [9]; here we complete the program put forward in [15] for the two-dimensional critical Ising model by constructing the scaling limit of the collection of cluster measures and showing that the limiting magnetization field can indeed be expressed in terms of the limiting measures, thus providing a geometric representation (a sort of continuum FK representation based on continuum clusters) for the limiting magnetization field. As a byproduct, we obtain a proof of the existence and uniqueness of the limiting magnetization field alternative to [9] and independent of [18]. Our proof also shows that the limiting magnetization is measurable with respect to the collection of loops that describe the full scaling scaling limit of the FK–Ising process [24].

We conclude this introduction by pointing out that, despite coming from the scaling limit at the *critical* point, the FK–Ising conformal measure ensemble constructed in this paper plays a key role in the recent proof [11] of exponential decay for the *near-critical* scaling limit of the planar Ising model with an external magnetic field, a situation where conformal invariance is absent. Moreover, it is shown in [12] that the same conformal measure ensemble can be used to extend the geometric representation for the magnetization field established in this paper to the case of the *near-critical* magnetization field obtained in [10].

2.1 Definitions and Main Results

Let $\mathbb{L}$ denote a regular lattice with vertex set $V(\mathbb{L})$ and edge set $E(\mathbb{L})$. For u and v in $V(\mathbb{L})$, we write $u \sim v$ if $(u, v) \in E(\mathbb{L})$. We are interested in Bernoulli percolation and FK–Ising percolation in $\mathbb{L}$ with parameter p. When we talk about FK–Ising percolation, $\mathbb{L}$ will be the square lattice $\mathbb{Z}^2$. The FK clusters are defined as illustrated in Fig. 1, and we think of them as closed sets whose boundaries are the loops in the medial lattice shown in Fig. 1 (see [23] for an introduction to FK percolation).

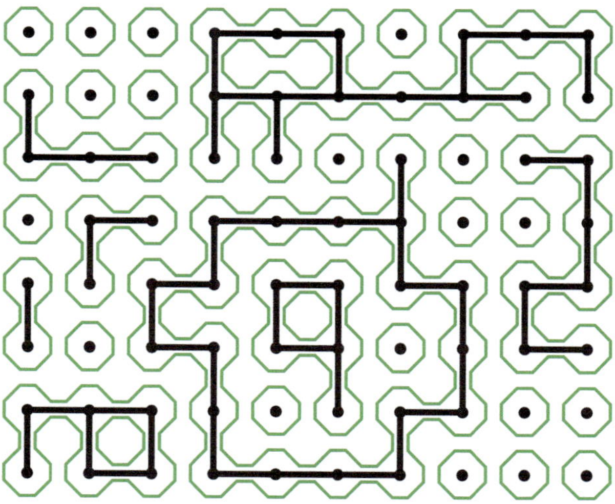

Fig. 1. Illustration of FK clusters. Black dots represent vertices of $\mathbb{Z}^2$, black horizontal and vertical edges represent FK bonds. The FK clusters are highlighted by lighter (green) loops on the medial lattice.

When dealing with Bernoulli percolation, $\mathbb{L}$ will be the triangular lattice $\mathbb{T}$, with vertex set

$$V(\mathbb{T}) := \{x + y\epsilon \in \mathbb{C} \mid x, y \in \mathbb{Z}\},$$

where $\epsilon = e^{\pi \mathbf{i}/3}$. The edge set $E(\mathbb{T})$ of $\mathbb{T}$ consists of the pairs $u, v \in V$ for which $\|u - v\|_2 = 1$. Further, let H_u denote the regular hexagon centered at $u \in V(\mathbb{T})$ with side length $1/\sqrt{3}$ with two of its sides parallel to the imaginary axis. Clusters are connected components of open or closed hexagons (see [22] for an introduction to Bernoulli percolation).

Let $\eta > 0$ and consider Bernoulli percolation on $\eta\mathbb{T}$ or the FK–Ising model on $\eta\mathbb{Z}^2$. We think of open and closed clusters as compact sets. To distinguish between them, we will call open clusters 'red' and closed clusters 'blue' (we deviate from the usual terminology of open and closed clusters on purpose: we reserve the words 'open' and 'closed' to describe the topological properties of sets). Let σ_η denote the union of the red clusters in $\eta\mathbb{L}$.

Further, let

$$\Lambda_r := \{z \in \mathbb{C} \mid |\Re z| \le r, |\Im z| \le r\}$$

denote the ball of radius r around the origin in the L^∞ norm. We set $\Lambda_r(u) = u + \Lambda_r$.

Our aim is to understand the limit of the set σ_η as η tends to 0. It is easy to see that the limit of σ_η in the Hausdorff topology as $\eta \to 0$ is trivial: it is the empty set when $p = 0$ and a.s. $\mathbb{C}$ for $p > 0$. Hence we concentrate on the connected components, i.e. clusters, of σ_η with diameter at least δ for some fixed $\delta > 0$. It is well-known (see for instance [1]) that, again, we get trivial limits unless $p = p_c$. (For $p < p_c$ the limit of each of the clusters is the empty set, while for $p > p_c$ the limit of the unique largest clusters is dense in $\mathbb{C}$, with the other clusters having the empty set as a limit.) Hence we consider $p = p_c$ in the following, and state informal versions of our main results after some additional definitions. The precise versions of our results are postponed to later sections.

For a set $A \subset \mathbb{C}$ and $u, v \in \mathbb{C}$ we write $u \overset{A}{\longleftrightarrow} v$ if there is a red path running in A which connects u to v. When A is omitted, it is assumed to be $\mathbb{C}$. Let $\mathrm{diam}(A)$ denote the L^∞ diameter of A. For $u \in \eta V$ denote by $\mathcal{C}^\eta(u)$ the connected component (i.e. cluster) of u in σ_η. If D is a simply connected domain with piecewise smooth boundary, we let $\mathscr{C}_D^\eta(\delta)$ denote the collection of connected components of σ_η, which are contained in D and have diameter larger than δ. That is,

$$\mathscr{C}_D^\eta(\delta) := \{\mathcal{C}^\eta(u) \mid u \in \eta V, \mathcal{C}^\eta(u) \subset D, \mathrm{diam}(\mathcal{C}^\eta(u)) \ge \delta\}. \tag{1}$$

On many places D is taken to be Λ_k, in that case we simplify notation by writing $\mathscr{C}_k^\eta(\delta) := \mathscr{C}_{\Lambda_k}^\eta(\delta)$. Finally let

$$\mathscr{C}^\eta(\delta) = \bigcup_{k \in \mathbb{N}} \mathscr{C}_k^\eta(\delta) \tag{2}$$

denote the collection of all connected components of σ_η with diameter at least δ.

In the following theorem, distances between subsets of $\mathbb{C}$ will be measured by the Hausdorff distance built on the L^∞ distance in $\mathbb{C}$: For $A, B \subseteq \mathbb{C}$,

$$d_H(A,B) := \inf\{\varepsilon > 0 \mid A + \Lambda_\varepsilon \supseteq B \text{ and } B + \Lambda_\varepsilon \supseteq A\}, \tag{3}$$

where $A + \Lambda_\varepsilon := \{x + y \in \mathbb{C} : x \in A, y \in \Lambda_\varepsilon\}$.

Let $\hat{\mathbb{C}}$ be the one-point (Alexandroff) compactification of $\mathbb{C}$, i.e. the Riemann sphere $\hat{\mathbb{C}} := \mathbb{C} \cup \{\infty\}$. A distance between subsets of $\hat{\mathbb{C}}$ which is equivalent to d_H on bounded sets is defined via the metric on $\mathbb{C}$ with distance function

$$\Delta(u,v) := \inf_{\varphi} \int \frac{1}{1 + |\varphi(s)|^2} ds,$$

where we take the infimum over all curves $\varphi(s)$ in $\mathbb{C}$ from u to v and $|\cdot|$ denotes the Euclidean norm.

The distance D_H between sets is then defined by

$$D_H(A,B) := \inf\{\varepsilon > 0 \mid \forall u \in A : \exists v \in B : \Delta(u,v) \leq \varepsilon \text{ and vice versa}\}. \tag{4}$$

The distance between finite collections i.e., sets of subsets of $\mathbb{C}$, denoted by $\mathscr{S}, \mathscr{S}'$, is defined as

$$\min_{\phi} \max_{S \in \mathscr{S}} d_H(S, \phi(S)) \tag{5}$$

where the infimum is taken over all bijections $\phi : \mathscr{S} \to \mathscr{S}'$. In case $|\mathscr{S}| \neq |\mathscr{S}'|$ we define the distance to be infinite. To account for possibly infinite collections, $\mathscr{S}$ and $\mathscr{S}'$, of subsets of $\hat{\mathbb{C}}$, we define

$$\begin{aligned} &dist(\mathscr{S}, \mathscr{S}') \\ &\quad := \inf\{\varepsilon > 0 \mid \forall A \in \mathscr{S}\ \exists B \in \mathscr{S}' : D_H(A,B) \leq \varepsilon \text{ and vice versa}\}. \end{aligned} \tag{6}$$

Convergence in the distance defined by (5) implies convergence in the distance *dist*, since the metrics d_H and D_H are equivalent on bounded domains.

Our first result is the following, see Theorem 11 for a slightly stronger version.

Theorem 1. *Let $k > \delta > 0$. Then, as $\eta \to 0$, $\mathscr{C}_k^\eta(\delta)$ converges in distribution, in the topology (5), to a collection of closed sets which we denote by $\mathscr{C}_k^0(\delta)$. Moreover, as $\delta \to 0$, $\mathscr{C}_k^0(\delta)$ has a limit in the metric (6), which we denote by $\mathscr{C}_k^0$.*

The next natural question is whether we can extract some more information from the scaling limit. In particular, can we count the number of vertices in each of the clusters in $\mathscr{C}^\eta(\delta)$ in the limit as η tends to 0? As we will see below, the number of vertices in the large clusters goes to infinity, hence we have to scale this number to get a non-trivial result. The correct factor is $\eta^{-2}\pi_1^\eta(\eta, 1)$, where $\pi_1^\eta(\eta, 1)$ denotes the probability that 0 is connected to $\partial\Lambda_1$ in σ_η. We arrive at the informal formulation of our next main result after some more notation.

For $S \subset \mathbb{C}$ let μ_S^η denote the normalized counting measure of its vertices, that is,

$$\mu_S^\eta := \frac{\eta^2}{\pi_1^\eta(\eta, 1)} \sum_{u \in S \cap \eta V} \delta_u, \tag{7}$$

where δ_u denotes the Dirac measure concentrated at u. Further, let $\mathscr{M}_k^\eta(\delta)$ denote the collection of normalized counting measures of the clusters in $\mathscr{C}_k^\eta(\delta)$. That is,

$$\mathscr{M}_k^\eta(\delta) := \{\mu_{\mathcal{C}}^\eta \,|\, \mathcal{C} \in \mathscr{C}_k^\eta(\delta)\}\,. \tag{8}$$

Similarly $\mathscr{M}^\eta(\delta) := \{\mu_{\mathcal{C}}^\eta \,|\, \mathcal{C} \in \mathscr{C}^\eta(\delta)\}$. We use the Prokhorov distance for the normalized counting measures. For finite Borel measures μ, ν on $\mathbb{C}$, it is defined as

$$d_P(\mu, \nu) \\ := \inf\{\varepsilon > 0 \,|\, \mu(S) \le \nu(S^\varepsilon) + \varepsilon,\, \nu(S) \le \mu(S^\varepsilon) + \varepsilon \text{ for all closed } S \subseteq \mathbb{C}\}\,.$$

where $S^\varepsilon = S + \Lambda_\varepsilon$. Then we construct a metric on collections of Borel measures from d_P similarly to (5). We also introduce a distance *Dist* between (infinite) collections of measures which is the same as (6) but with collections of sets replaced by collections of measures and with the distance D_H replaced by the Prokhorov distance d_P.

We arrive at the following result (see Theorem 13 for a slightly stronger version).

Theorem 2. *Let $k > \delta > 0$, then $\mathscr{M}_k^\eta(\delta)$ converges in distribution to a collection of finite measures which we denote by $\mathscr{M}_k^0(\delta)$. Moreover, as $\delta \to 0$, $\mathscr{M}_k^0(\delta)$ has a limit in the metric Dist, which we denote by $\mathscr{M}_k^0$.*

The next theorem is a full-plane analogue of Theorems 1 and 2.

Theorem 3. *Let $\mathbb{P}_k$ denote the joint distribution of $(\mathscr{C}_k^0, \mathscr{M}_k^0)$. There exists a probability measure $\mathbb{P}$ on the space of collections of subsets of $\hat{\mathbb{C}}$ and collections of measures, which is the full plane limit of the probability measures $\mathbb{P}_k$ in the sense that, for every bounded domain D, the restriction $\mathbb{P}_k|_D$ of $\mathbb{P}_k$ to $(\mathscr{C}_D^0, \mathscr{M}_D^0)$ converges to the restriction $\mathbb{P}|_D$ of $\mathbb{P}$ to $(\mathscr{C}_D^0, \mathscr{M}_D^0)$ as $k \to \infty$.*

The next theorem shows that the collections of clusters and measures from the previous theorem are invariant under rotations and translations, and transform covariantly under scale transformations. (The theorem could be extended to include more general fractal linear (Möbius) transformations by restricting to the Riemann sphere minus a neighborhood of infinity and its preimage. For simplicity, we restrict attention to linear transformations that map infinity to itself.) The random variables with distribution $\mathbb{P}$ introduced in the previous theorem are denoted by $(\mathscr{C}^0, \mathscr{M}^0)$.

Theorem 4. *Let f be a linear map from $\mathbb{C}$ to $\mathbb{C}$, that is $f(z) = rz + t$ with $r, t \in \mathbb{C}$. Assume that*

$$\lim_{\eta \to 0} \pi_1^\eta(a, b) = \left(\frac{a}{b}\right)^{\alpha_1 + o(1)}$$

for all $b > a > \eta$ and some $\alpha_1 \in [0,1]$, where $o(1)$ is understood as $b/a \to \infty$. We set

$$f(\mathscr{C}^0) := \{f(\mathcal{C}) : \mathcal{C} \in \mathscr{C}^0\}, \text{ and}$$
$$f(\mathscr{M}^0) := \{\mu^{0*} : \mu^0 \in \mathscr{M}^0\}$$

where μ^{0*} is the modification of push-forward measure of μ^0 along f defined as

$$\mu^{0*}(B) := |r|^{2-\alpha_1}\mu^0(f^{-1}(B))$$

for Borel sets B. Then the pairs $(f(\mathscr{C}^0), f(\mathscr{M}^0))$ and $(\mathscr{C}^0, \mathscr{M}^0)$ have the same distribution.

Remark 1. In the case of Bernoulli percolation, we will prove invariance/covariance under all conformal maps between any two bounded domains with piecewise smooth boundaries (see Theorems 6, 16 and 18).

Organization of the Paper

In the next section we discuss some applications of our results. First we consider applications to Bernoulli percolation on the triangular lattice. Secondly we provide a geometric representation for the magnetization field of the critical Ising model in terms of FK clusters.

In Sect. 4 we introduce the main tools and assumptions which we use throughout the paper, namely the loop process, the quad-crossing topology, arm events and the general assumptions under which we prove our main results. We finish Sect. 4 with checking that the assumptions hold for critical Bernoulli percolation on $\mathbb{T}$ and for the critical FK–Ising model on the square lattice. In Sects. 5, 6, 7 and 8 we give precise versions and proofs of Theorems 1, 2 and 3.

We investigate some fundamental properties of the continuum clusters and their normalized counting measures in Sect. 9. In particular, we also discuss the conformal invariance and covariance properties of the clusters in this section. We finish the paper with Sect. 10 where we prove the convergence of the largest clusters for Bernoulli percolation in a bounded domain.

3 Applications

3.1 Largest Bernoulli Percolation Clusters and Conformal Invariance/Covariance

Our first application concerns the scaling limit of the largest percolation clusters in a bounded domain with closed (blue) boundary condition. Denote by $\mathcal{M}^{\eta}_{(i)}$ the i-th largest cluster in $\Lambda_1 \cap \sigma_\eta$, where we measure clusters according to the number of vertices they contain.

In a sequence of papers, the behavior of the normalized number of vertices,

$$\frac{|\mathcal{M}^\eta_{(i)}|}{\eta^{-2}\pi^\eta_1(\eta,1)} = \mu^\eta_{\mathcal{M}^\eta_{(i)}}(\Lambda_1), \tag{9}$$

was investigated for $\eta > 0$ and $i \geq 1$. Probably the first such results appeared in [6] and [7]. Using Theorems 1 and 2 and results in Sect. 7 about convergence of clusters and portions of clusters in bounded domains, we deduce the following theorem.

Theorem 5. *For all $i \in \mathbb{N}$, the cluster $\mathcal{M}^\eta_{(i)}$ and its normalized counting measure $\mu^\eta_{\mathcal{M}^\eta_{(i)}}$ converge in distribution to a closed set $\mathcal{M}^0_{(i)}$ and a measure $\mu^0_{\mathcal{M}^0_{(i)}}$, respectively, as $\eta \to 0$.*

Proof. The result follows directly from Theorem 19 in Sect. 10. □

Recently some of the results from [6,7] where sharpened [4,5,25]. These sharpened results, in combination with Theorem 5, imply that the distribution of $\mu^0_{\mathcal{M}^0_{(i)}}(\Lambda_1)$ has no atoms [5], that its support is $(0,\infty)$ [4] and that it has a stretched exponential upper tail [25].

It is a celebrated result of Smirnov [32] that critical site percolation on the triangular lattice is conformally invariant in the limit as $\eta \to 0$. See also [13,14]. As we will show, under certain technical conditions, this implies that the collections of large clusters in the limit as $\eta \to 0$ are also conformally invariant, while their normalized counting measures are conformally covariant by the results in [19]. We denote by $\mathscr{B}^\eta_D(\delta)$ the collection of clusters, with diameter greater than $\delta > 0$, in a domain D with closed boundary condition. In Sect. 7 we will see that, as $\eta \to 0$, this collection converges in distribution to a limiting collection of clusters $\mathscr{B}^0_D(\delta)$. The latter converges as δ tends to 0 to the random collection $\mathscr{B}^0_D$. To indicate that we consider the measures of the clusters in $\mathscr{B}^0_D$ instead of the clusters in $\mathscr{C}^0_D$ we add a tilde, for example the collection of measures of the clusters in $\mathscr{B}^0_{\Lambda_1}$ is denoted by $\tilde{\mathscr{M}}^0_{\Lambda_1}$. We obtain the following result, which is stated in a slightly stronger form as Theorems 16 and 18.

Theorem 6. *Let f be a conformal map defined on an open neighbourhood of Λ_1, and $D = f(\Lambda_1)$. We set*

$$f(\mathscr{B}^0_{\Lambda_1}) := \{f(\mathcal{B}) : \mathcal{B} \in \mathscr{B}^0_{\Lambda_1}\}, \text{ and}$$
$$f(\tilde{\mathscr{M}}^0_{\Lambda_1}) := \{\mu^{0*} : \mu^0 \in \tilde{\mathscr{M}}^0_{\Lambda_1}\}$$

where μ^{0*} is the modification of the push-forward measure of μ^0 along f defined as

$$\mu^{0*}(B) := \int_{f^{-1}(B)} |f'(z)|^{91/48} d\mu^0(z)$$

for Borel sets B. Then the pairs $(f(\mathscr{B}^0_{\Lambda_1}), f(\tilde{\mathscr{M}}^0_{\Lambda_1}))$ and $(\mathscr{B}^0_D, \tilde{\mathscr{M}}^0_D)$ have the same distribution.

Proof. The result follows from a combination of Theorems 16 and 18, stated and proved in Sect. 9.2. □

3.2 Geometric Representation of the Critical Ising Magnetization Field

In this section we give an alternative proof of the existence and uniqueness of the limiting magnetization field obtained by taking the critical scaling limit of the magnetization in the two-dimensional Ising model, a result first proved in [9]. We also prove a geometric representation for the scaling limit of the critical Ising magnetization in two dimensions that was first conjecture in [15]. There, it was heuristically argued that the Ising magnetization field should be expressible in terms of the limiting cluster measures of the FK–Ising clusters, giving a sort of continuum FK representation based on continuum clusters; here, we provide the proof of a precise statement to that effect (see Theorem 7 below).

Consider a two-dimensional critical Ising model on $\eta\mathbb{Z}^2$ and its FK representation (see, e.g., [23]). In what follows, we will assume Wu's celebrated result on the power law decay of the critical Ising two-point function [33]. This assumption implies that, for critical FK–Ising percolation, $\eta^2/\pi^\eta(\eta,1)$ behaves like $\eta^{15/8}$ as $\eta \to 0$. (See Remark 1.5 of [9] for a discussion of this point.) We denote by Φ^η the lattice magnetization field, defined as

$$\Phi^\eta := \eta^{15/8} \sum_{x \in \eta\,\mathbb{Z}^2} S_x \delta_x,$$

where S_x is the spin at x and δ_x is the Dirac delta at x. We also introduce the ε-cutoff magnetization Φ^η_ε, define as

$$\Phi^\eta_\varepsilon := \sum_{\mathcal{C} \in \mathscr{C}^\eta(\varepsilon)} \mathcal{S}_{\mathcal{C}} \mu^\eta_{\mathcal{C}},$$

where the sum is over all FK–Ising clusters of diameter larger than ε (the order of the sum is irrelevant), the $\mathcal{S}_{\mathcal{C}}$'s are i.i.d. symmetric (± 1)-valued random variables, the $\mu^\eta_{\mathcal{C}}$'s are the FK–Ising normalized counting measures, and we think of Φ^η_ε as a random signed measure acting on the space C^∞_0 of infinitely differentiable functions with bounded support. Note that, if $\mathbf{1}_L$ denotes the indicator function of $[-L,L]^2$, $\langle \Phi^\eta_\varepsilon, \mathbf{1}_L \rangle$ is the magnetization in $[-L,L]^2$ produced by FK clusters of diameter at least ε.

Lemma 1. *For each $f \in C^\infty_0$, as $\eta \to 0$, $\langle \Phi^\eta_\varepsilon, f \rangle$ converges in distribution to the random variable*

$$\Phi^0_\varepsilon(f) := \sum_{\mathcal{C} \in \mathscr{C}^0(\varepsilon)} \mathcal{S}_{\mathcal{C}} \langle \mu^0_{\mathcal{C}}, f \rangle.$$

Proof. The statement follows from Theorem 3 by taking any k such that the domain of f is contained in Λ_k. □

Theorem 7. *If f is a bounded function of bounded support, as $\eta \to 0$, then $\langle \Phi^\eta, f \rangle$ converges in distribution to a random variable $\Phi^0(f)$ measurable with respect to the collection of loops and signs, and such that*

$$||\Phi^0(f) - \Phi^0_\varepsilon(f)||_2 \le C\, ||f||^2_\infty\, \varepsilon^{7/4}$$

for any $\varepsilon > 0$ and some positive constant $C < \infty$ independent of f. Moreover, if $(f_n)_{n\in\mathbb{N}}$ is a sequence of bounded functions of bounded support converging to f in the sup-norm as $n \to \infty$, then $\Phi(f_n) \to \Phi(f)$ in L_2 as $n \to \infty$.

Proof. We first identify a candidate for the limit $\Phi^0(f)$ of $\langle \Phi^\eta, f \rangle$. To that end, we consider $\Phi^0_\varepsilon(f)$ as an element of $L_2(\Omega, \mathbb{P})$ and let $\varepsilon \to 0$. The existence of the limit can be checked easily by considering sequences $\varepsilon_n \searrow 0$ and showing that $(\Phi^0_{\varepsilon_n}(f))_n$ is a Cauchy sequence. For any $m > n$, denoting by $||\cdot||_2$ the L_2-norm and using $\mathbb{E}$ for expectation with respect to $\mathbb{P}$, using the argument in the proof of Proposition 6.2 of [8], we have that

$$\begin{aligned}
||\Phi^0_{\varepsilon_n}(f) - \Phi^0_{\varepsilon_m}(f)||_2^2 &= \mathbb{E}\left(|\sum_{\mathcal{C}\in\mathscr{C}^0(\varepsilon_m)\setminus\mathscr{C}^0(\varepsilon_n)} \mathcal{S}_{\mathcal{C}}\, \mu^0_{\mathcal{C}}(f)|^2 \right) \\
&\le \limsup_{\eta\to 0} \mathbb{E}\left(\sum_{\mathcal{C}\in\mathscr{C}^\eta(\varepsilon_m)\setminus\mathscr{C}^\eta(\varepsilon_n)} (\mu^0_{\mathcal{C}}(f))^2 \right) \\
&\le \limsup_{\eta\to 0} \mathbb{E}\left[\sum_{\mathcal{C}\in\mathscr{C}^\eta\setminus\mathscr{C}^\eta(\varepsilon_n)} (\mu^\eta_{\mathcal{C}}(f))^2 \right] \\
&< C\left(\sup_{x\in D} |f(x)| \right)^2 \varepsilon_n^{7/4},
\end{aligned}$$

for some positive constant $C < \infty$ independent of f. Therefore, if f is a bounded function of bounded support, $\Phi^0_\varepsilon(f)$ converges, as $\varepsilon \to 0$ to an element $\Phi^0(f)$ of $L_2(\Omega, \mathbb{P})$; moreover, for any $\varepsilon > 0$,

$$||\Phi^0(f) - \Phi^0_\varepsilon(f)||_2 \le C\, ||f||^2_\infty\, \varepsilon^{7/4}. \tag{10}$$

Using the triangle inequality, for any $\eta > 0$, we can write

$$\begin{aligned}
||\Phi^0(f) - \langle \Phi^\eta, f \rangle\, ||_2 \le\, & ||\Phi^0(f) - \Phi^0_\varepsilon(f)||_2 \\
&+ ||\Phi^0_\varepsilon(f) - \Phi^\eta_\varepsilon(f)||_2 \\
&+ ||\Phi^\eta_\varepsilon(f) - \langle \Phi^\eta, f \rangle\, ||_2.
\end{aligned}$$

As $\eta \to 0$, the first term in the right hand side of the last inequality tends to zero because of (10). The third term can be made arbitrarily small by letting $\eta \to 0$ and taking ε small. Like (10), this follows from results and calculations in [15] and from the proof of Proposition 6.2 of [8]. For fixed $\varepsilon > 0$, the remaining term can be expressed as a finite sum, containing the normalized counting measures of clusters of diameter larger than ε that intersect the support of f. As $\eta \to 0$, this term tends to zero because of the convergence in probability of normalized counting measures proved in Theorem 7.2 under Assumption IV, and the L_3 bounds provided by Lemma 3.15.

The L_2-continuity of $\Phi^0(\cdot)$ follows from

$$\begin{aligned}
&||\Phi^0(f) - \Phi^0(f_n)||_2^2 \\
&\leq ||\Phi^0(f) - \Phi^0_\varepsilon(f)||_2^2 + ||\Phi^0_\varepsilon(f) - \Phi^0_\varepsilon(f_n)||_2^2 + ||\Phi^0_\varepsilon(f_n) - \Phi^0(f_n)||_2^2 \\
&\leq ||f - f_n||_\infty^2 \, \mathbb{E}\left(\sum_{\mathcal{C} \in \mathscr{C}^0(\varepsilon)} (\mu^0_{\mathcal{C}}(\mathbf{1}_L))^2 \right) + C(||f||_\infty^2 + ||f_n||_\infty^2)\varepsilon^{7/4},
\end{aligned}$$

where $\mathbf{1}_L$ denotes the indicator function of $[-L, L]^2$ and L is such that $\text{supp}(f)$, $\text{supp}(f_n) \subset [-L, L]^2$. The fact that the term $\mathbb{E}\left(\sum_{\mathcal{C} \in \mathscr{C}^0(\varepsilon)} (\mu^0_{\mathcal{C}}(\mathbf{1}_L))^2\right)$ is bounded follows, for instance, from Proposition B.2 of [9].

To conclude the proof of the theorem we note that, for every $\varepsilon > 0$, the sum defining $\Phi^0_\varepsilon(f)$ is a.s. finite; therefore $\Phi^0_\varepsilon(f)$ is measurable with respect to the collections of loops and signs. Since $\Phi^0(f)$ is the limit of $\Phi^0_\varepsilon(f)$ as $\varepsilon \to 0$, it is also measurable with respect to the collections of loops and signs. □

In the corollary below we consider the probability space Ω of continuum FK–Ising loops ($\text{CLE}_{16/3}$), clusters and area measures, together with the random signs assigned to the clusters. An element of that space is denoted ω and the joint probability distribution is denoted by $\mathbb{P}$. We let D be the space of infinitely differentiable functions with compact support equipped with the topology of uniform convergence of all derivatives, whose topological dual D' consists of all generalized functions. We remind the reader that, according to Theorem 7, the magnetization field Φ^0 is measurable with respect to ω.

Corollary 1. *There exists a random, continuous, linear functional $T \in D'$ with characteristic function $\chi_T(f) = \int_\Omega e^{i\Phi^0(f)} d\mathbb{P}(\omega)$, for all $f \in D$.*

Proof. Since D is a nuclear space, we can apply the Bochner–Minlos theorem (see, for example, Theorem 3.4.2 on p. 52 of [21]—a proof can be found in [20]). We define

$$\chi(f) := \int_\Omega e^{i\Phi^0(f)} d\mathbb{P}(\omega) \tag{11}$$

and check the following properties of χ.

1. Normalization: $\chi(0) = 1$.
2. Positivity: $\sum_{k,l=1}^m c_k \overline{c_l} \chi(f_k - f_l) \geq 0$ for every $m \in \mathbb{N}$, $f_1, \ldots, f_m \in D$ and $c_1, \ldots, c_m \in \mathbb{C}$.
3. Continuity: $\chi(f) \to 0$ as $f \to 0$ (in the topology of D).

The first property is clear. To establish the second property, let

$$F = \sum_{k=1}^m c_k e^{i\Phi^0(f_k)}$$

and note that the square of the $L_2(\Omega, \mathbb{P})$-norm of F is

$$0 \leq ||F||_2^2 = \int_\Omega \sum_{k,l=1}^{m} c_k \overline{c_l} e^{i\Phi^0(f_k - f_l)} d\mathbb{P}(\omega) \tag{12}$$

$$= \sum_{k,l=1}^{m} c_k \overline{c_l} \chi(f_k - f_l). \tag{13}$$

The remaining step is to establish continuity of χ. We think of $\Phi^0(f)$ as a sequence of random variables indexed by $f \in D$ as $f \to 0$ in the topology of D, which implies uniform convergence of f to zero. We have

$$||\Phi^0(f)||_2^2 \leq ||f||_\infty^2 \mathbb{E} \left(\sum_{\mathcal{C}} (\mu_{\mathcal{C}}^0(\mathbf{1}_L))^2 \right), \tag{14}$$

where $\mathbf{1}_L$ denotes the indicator function of $[-L, L]^2$ and L is such that $\text{supp}(f) \subset [-L, L]^2$. This implies convergence of $\Phi^0(f)$ to 0 in L_2 as $f \to 0$, which implies convergence in probability, which implies convergence in distribution, which is equivalent to pointwise convergence of characteristic functions, which gives us the type of continuity we need. Therefore, by an application of the Bochner–Minlos theorem, there exists a random, continuous, linear functional $T \in D'$ with characteristic function $\chi_T(f) = \chi(f)$. □

A result related to our Theorem 7 was recently proved by Miller, Sheffield and Werner [27]. They showed (see Theorem 7.5 of [27]) that forming clusters of $\text{CLE}_{16/3}$ loops by a percolation process with parameter $p = 1/2$ generates CLE_3, the Conformal Loop Ensemble with parameter 3. CLE_3 describes the full scaling limit of Ising spin-cluster boundaries [3] while $\text{CLE}_{16/3}$, as already mentioned, describes the full scaling limit of FK–Ising cluster boundaries [24]. We note that, although the magnetization can obviously be expressed using Ising spin clusters, as a sum of their signed areas, such a representation does not appear to be useful in the scaling limit because the area measures of spin clusters don't scale like the magnetization. The usefulness of the representation in terms of FK clusters is due to the fact that both the FK clusters and the magnetization need to be normalized by the same scale factor in the scaling limit. That is not true of the magnetization and the spin clusters.

4 Further Notation and Preliminaries

Above we interpreted the union of red hexagons in a percolation configuration σ_η, as a (random) subset of $\mathbb{C}$. In what follows, as an intermediate step, we will consider a percolation configuration as a (random) collection of loops. These loops form the boundaries of the clusters. We will describe this space in Subsect. 4.1. In order to define the clusters as subsets of the plane, we will also consider the (random) collection of quads ('topological squares' with two marked opposing sides) which are crossed horizontally. This leads us to the Schramm–Smirnov [31] topological space, which we briefly recall in the second subsection.

4.1 Space of Nonsimple Loops

The random collection of loops will be denoted by L_η for $\eta \geq 0$. The distance between two curves l, l' is defined as

$$d_c(l, l') := \inf \sup_{t \in [0,1]} \Delta(l(t), l'(t)), \tag{15}$$

where the infimum is over all parametrizations of the curves. The distance between closed sets of curves is defined similarly to the distance *dist* defined in (6) between collections of subsets of the Riemann sphere $\hat{\mathbb{C}}$. The space of closed sets of loops is a complete separable metric space.

For $\eta > 0$ the collection of (oriented) boundaries of the red clusters in σ_η is the closed set of loops, denoted by L_η. This set converges in distribution to L_0, called the *continuum nonsimple loop process* [13].

4.2 Space of Quad-Crossings

We borrow the notation and definitions from [19]. Let $D \subset \hat{\mathbb{C}}$ be open. A quad Q in D is a homeomorphism $Q : [0,1]^2 \to Q([0,1]^2) \subseteq D$. Let $\mathcal{Q}_D$ be the set of all quads, which we equip with the supremum metric

$$d(Q_1, Q_2) = \sup_{z \in [0,1]^2} |Q_1(z) - Q_2(z)|$$

for $Q_1, Q_2 \in \mathcal{Q}_D$.

A crossing of a quad Q is a closed connected subset of $Q\left([0,1]^2\right)$ which intersects $Q(\{0\} \times [0,1])$ as well as $Q(\{1\} \times [0,1])$. The crossings induce a natural partial order denoted by $\leq$ on $\mathcal{Q}_D$. We write $Q_1 \leq Q_2$ if all the crossings of Q_2 contain a crossing of Q_1. For technical reasons, we also introduce a slightly less natural partial order on $\mathcal{Q}_D$: we write $Q_1 < Q_2$ if there are open neighbourhoods $\mathcal{N}_i$ of Q_i such that for all $N_i \in \mathcal{N}_i$, $i \in \{1, 2\}$, $N_1 \leq N_2$. We consider the collection of all closed hereditary subsets of $\mathcal{Q}_D$ with respect to $<$ and denote it by $\mathscr{H}_D$. It is the collection of the closed sets $\mathcal{S} \subset \mathcal{Q}_D$ such that if $Q \in \mathcal{S}$ and $Q' \in \mathcal{Q}_D$ with $Q' < Q$ then $Q' \in \mathcal{S}$.

For a quad $Q \in \mathcal{Q}_D$ let $\boxminus_Q$ denote the set

$$\boxminus_Q := \{\mathcal{S} \in \mathscr{H}_D \mid Q \in \mathcal{S}\},$$

which corresponds with the configurations where Q is crossed. For an open subset $\mathcal{U} \subset \mathcal{Q}_D$ let $\boxdot_\mathcal{U}$ denote the set

$$\boxdot_\mathcal{U} := \{\mathcal{S} \in \mathscr{H}_D \mid \mathcal{U} \cap \mathcal{S} = \emptyset\},$$

which corresponds with the configurations where none of the quads of $\mathcal{U}$ is crossed. We endow $\mathscr{H}_D$ with the topology $\mathscr{T}_D$ which is the minimal topology containing the sets $\boxminus_Q^c$ and $\boxdot_\mathcal{U}^c$ as open sets for all $Q \in \mathcal{Q}_D$ and $\mathcal{U} \subset \mathcal{Q}_D$ open. We have:

Theorem 8. (Theorem 1.13 of [31]). *Let D be an open subset of $\hat{\mathbb{C}}$. Then the topological space $(\mathscr{H}_D, \mathscr{T}_D)$ is a compact metrizable Hausdorff space.*

Using this topological structure, we construct the Borel σ-algebra on $\mathscr{H}_D$. We get:

Corollary 2. (Corollary 1.15 of [31]). Prob$(\mathscr{H}_D)$, *the space of Borel probability measures of $(\mathscr{H}_D, \mathscr{T}_D)$, equipped with the weak* topology is a compact metrizable Hausdorff space.*

Notational Remarks. (i) In the following we abuse the notation of a quad Q. When we refer to Q as a subset of $\hat{\mathbb{C}}$, we consider its range $Q([0,1]^2) \subset \hat{\mathbb{C}}$.

(ii) Note that a percolation configuration σ_η, as defined in the introduction, naturally induces a quad-crossing configuration $\omega_\eta \in \mathscr{H}_{\hat{\mathbb{C}}}$, namely

$$\omega_\eta := \left\{ Q \in \mathcal{Q}_{\hat{\mathbb{C}}} \,|\, \sigma_\eta \text{ contains a crossing of } Q \right\}. \tag{16}$$

Furthermore, $\mathbb{P}_\eta$ will denote the law governing $(\omega_\eta \times L_\eta)$.

Further we will need the following definitions for restrictions of the configuration to a subset of the Riemann Sphere.

Definition 1. *Let $D \subseteq \hat{\mathbb{C}}$ be an open set and $\omega \in \mathscr{H}_{\hat{\mathbb{C}}}$. Then $\omega|_D$, the restriction of ω to D, is defined as*

$$\omega|_D := \{Q \in \omega : Q \subset D\}.$$

The image of $\omega|_D$ under a conformal map $f : D \to \hat{\mathbb{C}}$ is defined as

$$f(\omega|_D) := \{f(Q) : Q \in \omega|_D\} \in \mathscr{H}_{f(D)}.$$

The restriction of the loop process to D is defined as

$$L|_D := \{l : \exists \tilde{l} \in L \ s.t. \ l \text{ is an excursion of } \tilde{l} \text{ in } D\}.$$

The image of $L|_D$ under a conformal map $f : D \to \hat{\mathbb{C}}$ is defined as

$$f(L|_D) := \{f(l) : l \in L|_D\}.$$

Furthermore, $\mathbb{P}_{\eta,D}$ denotes the law of $(\omega_{\eta,D}, L_{\eta,D}) := (\omega_\eta|_D, L_\eta|_D)$ for $\eta \geq 0$.

4.3 Assumptions

Below we list the assumptions which are used throughout the article.

The edge set in the sublattice on $D \subset \mathbb{C}$ of $\eta\mathbb{L}$ is $(\eta E(\mathbb{L}))|_D := \{(u, v) \in \eta E(\mathbb{L}) : u, v \in \eta V(\mathbb{L}) \cap D\}$. The discrete boundary of $D \subset \mathbb{C}$ of the lattice $\eta\mathbb{L}$ is defined by:

$$\partial_\eta D := \{u \in \eta V(\mathbb{L}) \cap D : \exists v \in \eta\mathbb{L} : u \sim v \text{ and } v \in \eta\mathbb{L} \cap (\mathbb{C} \setminus D)\}.$$

A boundary condition ξ is a partition of the discrete boundary of D. A set in this partition denotes the vertices which are connected via red hexagons or edges (depending on the model) in $\mathbb{C} \setminus D$. When ξ is omitted, it means we are considering the full plane model and are not specifying any boundary conditions on the discrete boundary of D.

Assumption 1 (Domain Markov Property). *Let $D \subset E \subset \mathbb{C}$ be open sets. Further let $S \subset \overline{E \setminus D}$ and $T \subset \overline{D}$ closed sets. Then*

$$\mathbb{P}_\eta(\sigma_D = T \cap D \,|\, \sigma_{\overline{E\setminus D}} = S) = \mathbb{P}_\eta(\sigma_D = T \,|\, \xi) =: \mathbb{P}^\xi_\eta(\sigma_D = T)$$

where $\sigma_D = \sigma_\eta \cap D$ and ξ is the discrete boundary condition on D induced by $\sigma_{\overline{E\setminus D}} = S$.

For some models the randomness is on the vertices (e.g. Bernoulli site percolation) and for others on the edges (e.g. FK–Ising percolation). For the models of the first form we define $\Omega_{\eta,D} := \eta V(\mathbb{L}) \cap D$ and for models of the second form $\Omega_{\eta,D} := (\eta E(\mathbb{L}))|_D$.

Assumption 2 (Strong positive association/FKG). *The finite measures are strongly positively-associated. More precisely, let $D \subset \mathbb{C}$ be a bounded closed set. For every boundary condition ξ on $\partial_\eta D$ and increasing functions $f, g : \{red, blue\}^{\Omega_{\eta,D}} \to \mathbb{R}$, we have*

$$\mathbb{E}^\xi_\eta[f \cdot g] \geq \mathbb{E}^\xi_\eta[f] \cdot \mathbb{E}^\xi_\eta[g].$$

Hence for increasing events A, B and boundary condition ξ on $\partial_\eta D$:

$$\mathbb{P}^\xi_\eta(A \cap B) \geq \mathbb{P}^\xi_\eta(A)\mathbb{P}^\xi_\eta(B).$$

It is well known that monotonicity in the boundary condition is equivalent to strongly positively-association, if the measure is strictly positive (has the finite energy property), i.e. every configuration has strictly positive probability. (See e.g. [23, Theorem 2.24].) Furthermore it is well known that positive association survives the limit as the lattice grows towards infinity. See for example [23, Proposition 4.10].

In the following assumption $l(Q)$ denotes the extremal length of Q, that is, let $\phi : Q \to [0, a] \times [0, 1]$ conformal such that $\phi(Q(\{0\} \times [0, 1])) = \{0\} \times [0, 1]$ and $\phi(Q(\{1\} \times [0, 1])) = \{a\} \times [0, 1]$, then $l(Q) = a$.

Assumption 3 (RSW). *Let $M > 0$. There exist $\delta > 0$ such that, for every quad Q with $l(Q) \leq M$ and every boundary condition ξ on the discrete boundary of $Q([0, 1]^2)$:*

$$\mathbb{P}^\xi_\eta(\omega_\eta \in \boxminus_Q) \geq \delta$$

and for every quad Q with $l(Q) \geq M$ and every boundary condition ξ on the discrete boundary of $Q([0, 1]^2)$:

$$\mathbb{P}^\xi_\eta(\omega_\eta \notin \boxminus_Q) \leq 1 - \delta.$$

Assumption 4 (Full Scaling Limit). *As $\eta \to 0$, the law of L_η converges weakly to a random infinite collection of loops L_0 in the induced Hausdorff metric on collections of loops induced by the distance* (15) *(similar to the metric dist defined in* (6)*). Moreover, the limiting law is conformally invariant.*

4.4 Arm Events

For $S \subset \hat{\mathbb{C}}$, let $\partial S, int\,(S)$, $\bar{S}$ denote the boundary, interior and the closure of S, respectively. We call the elements of $\{0,1\}^k$, $k \geq 0$ as colour-sequences. For ease of notation, we omit the commas in the notation of the colour sequences, e.g. we write (101) for $(1,0,1)$.

Definition 2. *Let $l \in \mathbb{N}$, $\kappa \in \{0,1\}^l$, $S \subseteq \hat{\mathbb{C}}$ and D, E be two disjoint open, simply connected subsets of $\hat{\mathbb{C}}$ with piecewise smooth boundary. Let $D \overset{\kappa,S}{\longleftrightarrow} E$ denote the event that there are $\delta > 0$ and quads $Q_i \in \mathcal{Q}_S$, $i = 1, 2, \ldots, l$ which satisfy the following conditions.*

1. *$\omega \in \boxminus_{Q_i}$ for $i \in \{1, 2, \ldots, l\}$ with $\kappa_i = 1$ and $\omega \in \boxminus^c_{Q_i}$ for $i \in \{1, 2, \ldots, l\}$ with $\kappa_i = 0$.*
2. *For all $i \neq j \in \{1, 2, \ldots, l\}$ with $\kappa_i = \kappa_j$, the quads Q_i and Q_j, viewed as subsets of $\hat{\mathbb{C}}$, are disjoint, and are at distance at least δ from each other and from the boundary of S;*
3. *$\Lambda_\delta + Q_i(\{0\} \times [0,1]) \subset D$ and $\Lambda_\delta + Q_i(\{1\} \times [0,1]) \subset E$ for $i \in \{1, 2, \ldots, l\}$ with $\kappa_i = 1$;*
4. *$\Lambda_\delta + Q_i([0,1] \times \{0\}) \subset D$ and $\Lambda_\delta + Q_i([0,1] \times \{1\}) \subset E$ for $i \in \{1, 2, \ldots, l\}$ with $\kappa_i = 0$;*
5. *The intersections $Q_i \cap D$, for $i = 1, 2, \ldots, l$, are at distance at least δ from each other, the same holds for $Q_i \cap E$;*
6. *A counterclockwise order of the quads Q_i $i = 1, 2, \ldots, l$ is given by ordering counterclockwise the connected components of $Q_i \cap D$ containing $Q_i(0,0)$.*

When the subscript S is omitted, it is assumed to be $\hat{\mathbb{C}}$.

Remark 2. It is a simple exercise to show that the events $D \overset{\kappa,S}{\longleftrightarrow} E$ are *Borel*$(\mathscr{T}_{\hat{\mathbb{C}}})$-measurable. See [19, Lemma 2.9] for more details.

In what follows we consider some special arm events. For $z \in \mathbb{C}, a > 0$ let $H_1(z,a)$, $H_2(z,a)$, $H_3(z,a)$, $H_4(z,a)$ denote the left, lower, right, and upper half planes which have the right, top, left and bottom sides of $\Lambda_a(z)$ on their boundary, respectively. For $z \in \mathbb{C}$, $0 < a < b$ we set

$$A(z; a, b) := \Lambda_b(z) \setminus \Lambda_a(z).$$

Furthermore, for $i = 1, 2, 3, 4$, $\kappa \in \{0,1\}^l$ and $\kappa' \in \{0,1\}^{l'}$ with $l, l' \geq 0$ we define the event where there are $l + l'$ disjoint arms with colour-sequence $\kappa \vee \kappa' := (\kappa_1, \ldots, \kappa_l, \kappa'_1, \ldots, \kappa'_{l'})$ in $A(z; a, b)$ so that the l' arms, with colour-sequence κ', are in the half-plane $H_i(z,a)$. That is,

$$\mathcal{A}^i_{\kappa,\kappa'}(z; a, b) :=$$
$$\left\{\Lambda_a(z) \overset{\kappa\vee\kappa'}{\longleftrightarrow} \left(\hat{\mathbb{C}} \setminus \Lambda_b(z)\right)\right\} \cap \left\{\Lambda_a(z) \overset{\kappa', H_i(z,a)}{\longleftrightarrow} \left(\hat{\mathbb{C}} \setminus \Lambda_b(z)\right)\right\} \quad (17)$$

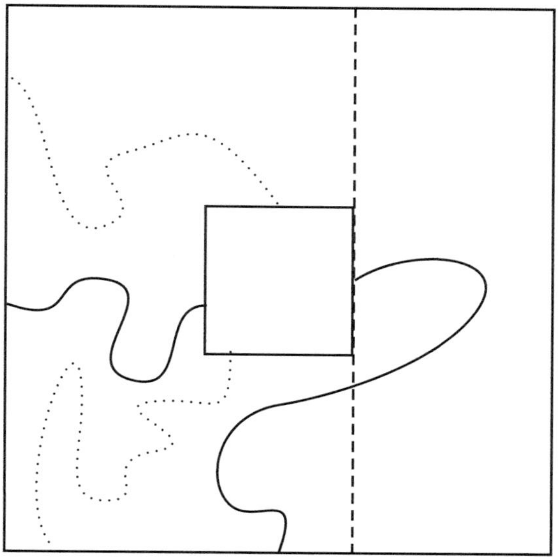

Fig. 2. Illustration of the event $\mathcal{A}^1_{(1),(010)}(a,b)$.

In the notation above, when z is omitted, it is assumed to be 0. When $\kappa' = \emptyset$, both the subscript κ' and the superscript i will typically be omitted. See Fig. 2 for an illustration of an arm event.

Finally, for $0 < a < b$ and boundary condition ξ on $\partial_\eta \Lambda_b$ we set

$$\pi_1^{\eta,\xi}(a,b) := \mathbb{P}_\eta^\xi(\mathcal{A}_{(1)}(a,b)), \qquad \pi_4^{\eta,\xi}(a,b) := \mathbb{P}_\eta^\xi(\mathcal{A}^1_{(1010),\emptyset}(a,b)),$$
$$\pi_6^{\eta,\xi}(a,b) := \mathbb{P}_\eta^\xi(\mathcal{A}^1_{(010101),\emptyset}(a,b)), \qquad \pi_{0,3}^{\eta,\xi}(a,b) := \mathbb{P}_\eta^\xi(\mathcal{A}^1_{\emptyset,(010)}(a,b)),$$
$$\pi_{1,3}^{\eta,\xi}(a,b) := \mathbb{P}_\eta^\xi(\mathcal{A}^1_{(1),(010)}(a,b)).$$

Remark 3. The (technical) reason to define $H_i(z,a)$ in this slightly unnatural way will become clear in the proof of Lemma 12.

4.5 Consequences of RSW

Lemma 2 (Quasi multiplicativity). *Suppose that Assumptions 1–3 hold. There is a constant $C > 0$ such that*

$$\mathbb{P}_\eta^\xi(\mathcal{A}_{(1)}(a,b)) \leq C \frac{\pi_1^{\eta,\xi}(a,c)}{\pi_1^{\eta,\xi}(b,c)}$$

for all $a,b,c,\eta > 0$ with $\eta < a < b < c$ and boundary condition ξ on $\partial_\eta \Lambda_c$.

Lemma 3. *Suppose that Assumptions 1–3 hold. There are constants $\lambda_1 \in (0,1)$ and $C > 0$ such that*

$$\pi_1^{\eta,\xi}(\eta,b) \geq C \left(\frac{a}{b}\right)^{\lambda_1} \mathbb{P}_\eta^\xi(\mathcal{A}_{(1)}(\eta,a))$$

for all $b > a > \eta$ *and boundary condition* ξ *on* $\partial_\eta \Lambda_b$.

Lemma 4. *Suppose that Assumptions 1–3 hold. There are positive constants* C, λ_6 *such that*

$$\pi_6^{\eta,\xi}(a,b) \le C\left(\frac{a}{b}\right)^{2+\lambda_6}, \qquad \pi_{0,3}^{\eta,\xi}(a,b) \le C\left(\frac{a}{b}\right)^2 \tag{18}$$

for all $0 < \eta < a < b$ *and boundary condition* ξ *on* $\partial_\eta \Lambda_b$.

Lemma 5. *Suppose that Assumptions 1–3 hold. There are positive constants* $C, \lambda_{1,3}$ *such that*

$$\pi_{1,3}^{\eta,\xi}(a,b) \le C\left(\frac{a}{b}\right)^{2+\lambda_{1,3}} \tag{19}$$

for all $0 < \eta < a < b$ *and boundary condition* ξ *on* $\partial_\eta \Lambda_b$.

Lemma 6. *Suppose that Assumptions 1–3 hold. There are constants* $C, \lambda > 0$ *such that*

$$\frac{\pi_1^{\eta,\xi}(a,b)}{\pi_4^{\eta,\xi}(a,b)} \ge C\left(\frac{b}{a}\right)^{\lambda}$$

for all $b > a > \eta$ *and boundary condition* ξ *on* $\partial_\eta \Lambda_b$.

For the sake of generality, we have stated the bounds in the previous lemmas in the presence of boundary conditions. However, in the rest of the paper only the full-plane versions of the bounds will appear, so the superscript ξ will be dropped. (The versions with boundary conditions are necessary to obtain results that we use in this paper, but whose proofs we do not reproduce.) For the next lemma we need some additional notation.

Definition 3. *For* $\eta, a > 0$ *let*

$$\mathcal{V}_a^\eta := \{v \in \Lambda_{a/2} \cap \eta V \,|\, v \overset{1}{\longleftrightarrow} \partial\Lambda_a \text{ in } \omega_\eta\}$$

denote the number of vertices in $\Lambda_{a/2}$ *connected to* $\partial\Lambda_a$ *in* σ_η.

Lemma 7. *Suppose that Assumptions 1–3 hold. Then there are positive constants* c, C *such that*

$$\mathbb{P}_\eta(|\mathcal{V}_a^\eta| \ge x(a/\eta)^2 \pi_1^\eta(\eta,a)) \le Ce^{-cx}$$

for all $a > \eta$ *and* $x \ge 0$.

Lemma 8. *Suppose that Assumptions 1–3 hold. Then there is a constant* $C > 0$ *such that*

$$\mathbb{E}_\eta[|\mathcal{W}_a^\eta|^3] \le C\eta^{-6}\pi_1^\eta(\eta,a)^3$$

for all $0 < \eta < a < 1/2$, *where*

$$\mathcal{W}_a^\eta := \{v \in \Lambda_1 \cap \eta V \,|\, v \overset{1}{\longleftrightarrow} \partial\Lambda_a(v) \text{ in } \omega_\eta\}.$$

Proof of Lemmas 2–8. Lemmas 4 and 5 follow from Assumptions 1–3, as explained in e.g. [22,29] for the case of Bernoulli percolation and in [17, Corollary 1.5 and Remark 1.6] for the case of FK–Ising percolation. (The additional boundary conditions, which are not present in the above mentioned corollary and remark in [17], do not affect the results. This can easily be deduced from Equation (5.1) in [17].)

Also Lemmas 3 and 6 follow from standard RSW, FKG arguments.

Lemma 2 is similar to [17, Theorem 1.3], which is shown to follow from our assumptions 1–3. The boundary condition on $\partial_\eta \Lambda_c$ has no effect on the proof, because the RSW result is uniform in the boundary conditions. (Furthermore there is no need to "make" the arms well separated on $\partial_\eta \Lambda_c$.)

An easy proof of Lemma 7 for critical percolation can be found in [28]. It is easy to see that the same proof can be modified in such a way that the result follows from Lemmas 2–6, and hence from Assumptions 1–3. For percolation, Lemma 7 can also be found in [6, Lemma 6.1], and for FK–Ising percolation in [9, Lemma 3.10].

Finally Lemma 8 can be proved easily using Lemma 2. See for example [19, Lemma 4.5] or the proof of Lemma 7. □

4.6 Additional Preliminaries

Lemma 9. *Suppose that Assumptions 1–4 hold. The set of crossed quads is, almost surely, measurable with respect to the collection of loops.*

Proof of Lemma 9. A proof of this can be found in [19, Section 2.3] and follows almost immediately from arguments given in [13, Section 5.2]. The proof of the measurability of quad crossings with respect to the collection of loops makes use of three properties of the loop process, which all follow from RSW techniques (see the first three items of Theorem 3 in [13, Section 5.2]). Because of this, the measurability is a simple consequence of our Assumptions 1–4. □

Remark 4. Assumption 4, together with the separability of $\mathscr{H}_{\hat{\mathbb{C}}}$, implies that there is a coupling $\mathbb{P}$ so that $\omega_\eta \to \omega_0$ a.s. as $\eta \to 0$.

Before we proceed to the next lemma, we recall the following result on the scaling limits of arm events. A slightly weaker version of the following lemma appeared as [19, Lemma 2.9]. Its proof extends immediately to the more general case.

Lemma 10 (Lemma 2.9 of [19]**).** *Suppose that Assumptions 1–4 hold. Then, under a coupling $\mathbb{P}$ of $(\mathbb{P}_\eta)_{\eta \ge 0}$ such that $\omega_\eta \to \omega_0$ almost surely, we have for events $\mathcal{D} \in \{\{A \overset{(1),S}{\longleftrightarrow} B\}, \{A \overset{(010),S}{\longleftrightarrow} B\}, \mathcal{A}^i_{\kappa,\kappa'}(z; a, b)\}$,*

$$\mathbf{1}_\mathcal{D}(\omega_\eta) \to \mathbf{1}_\mathcal{D}(\omega_0) \qquad \textit{in } \mathbb{P}\textit{-probability},$$

for $(\kappa, \kappa') \in \{((1), \emptyset), ((1010), \emptyset), ((010101), \emptyset), (\emptyset, (010)), ((1), (010))\}$, rectangle $S \subseteq \mathbb{C}$, $i \in \{1, 2, 3, 4\}$, $0 < a < b$ and A, B disjoint open subsets of $\mathbb{C}$ with piecewise smooth boundary.

The lemma above implies that for all $a, b > 0$ with $a < b$ the probability $\pi_1^\eta(a,b)$ converges as $\eta \to 0$. We write $\pi_1^0(a,b)$ for the limit. General arguments [2, Section 4] using Lemma 2 above show that

$$\pi_1^0(a,b) = \left(\frac{a}{b}\right)^{\alpha_1 + o(1)} \tag{20}$$

for some $\alpha_1 \geq 0$ where $o(1)$ is understood as $b/a \to \infty$. Lemma 3 shows that $\alpha_1 < 1$.

We need some additional notation for the next theorems. For $z \in \mathbb{C}$ and $a > 0$ let $\Lambda'_a(z) := \{u \in \mathbb{C} \,|\, \Re(u-z), \Im(u-z) \in [-a,a)\}$. Note that $\Lambda_a(z)$ and $\Lambda'_a(z)$ differ only on their boundary. For an annulus $A = A(z; a, b)$ let

$$\mu_{1,A}^\eta := \frac{\eta^2}{\pi_1^\eta(\eta,1)} \sum_{v \in \Lambda'_a(z) \cap \eta V} \delta_v \mathbf{1}_{\{v \leftrightarrow \partial\Lambda_b(z) \text{ in } \omega_\eta\}} \tag{21}$$

denote the counting measure of the vertices in $\Lambda'_a(z)$ with an arm to $\partial\Lambda_b(z)$ at scale η.

Theorem 9. *Suppose that Assumptions 1–4 hold. Let $A = A(z; a, b)$ be an annulus, and $\mathbb{P}$ be a coupling such that $\omega_\eta \to \omega_0$ a.s. as $\eta \to 0$. Then the measures $\mu_{1,A}^\eta$ converge weakly to $\mu_{1,A}^0$ in probability under the coupling $\mathbb{P}$ as η tends to 0. Furthermore, $\mu_{1,A}^0$ is a measurable function of ω_0. In particular, the pair $(\omega_\eta, \mu_{1,A}^\eta)$ converges to $(\omega_0, \mu_{1,A}^0)$ in distribution as $\eta \to 0$.*

Theorem 9 is proved for site percolation on the triangular lattice in [19] where it is Theorem 5.1. Namely, it is easy to check that the proof of [19, Theorem 5.1] shows that the measures $\mu_{1,A}^\eta \xrightarrow{P} \mu_{1,A}^0$ under the coupling $\mathbb{P}$ converge weakly in probability as $\eta \to 0$. For FK–Ising, a sketch proof for a theorem similar to this was given in [9]. Unfortunately the proof contains a mistake, but luckily the mistake can be easily fixed. Below we give an informal sketch of the proof of Theorem 9, following the proof in [9] and briefly explaining how to fix it.

The strategy is to approximate, in the L^2-sense, the one-arm measure by the number of mesoscopic boxes connected to $\partial\Lambda_b(z)$, multiplied by a constant depending on the size of the boxes. Here mesoscopic means much larger than the mesh size η but much smaller than a.

In order to get L^2-bounds on the error terms, first we use a coupling argument to argue that the boxes which are far away from each other are almost independent. Namely, with high probability one can draw a red circuit around one of the boxes, which is also conditioned on having a long red arm (because of positive association, that event can only increase the probability of a red circuit). This red circuit makes, via the Domain Markov Property, the contribution of the surrounded box independent of that of the other boxes. The total contribution of the boxes which are close to each other is negligible. Secondly we use a ratio limit argument, based on the existence of the one-arm exponent α_1 from (20), to show that the contribution of a single box is approximately a constant, which only depends on the size of the mesoscopic box.

The small mistake in [9] mentioned above is in the assumption that the convergence in Lemma 10 is almost sure, as claimed in an earlier version of [19]. However, as noted in the final version of [19], one can only prove convergence in probability. Luckily, arguments in [19] show that convergence in probability, together with L^3 bounds from Lemma 8, is sufficient to prove convergence in L^2 of the number of mesoscopic boxes connected to $\partial\Lambda_b(z)$ times a constant depending on the size of these boxes.

4.7 Validity of the Assumptions

The Case of Critical Percolation. Now we check that the Assumptions above hold for critical site percolation on the triangular lattice.

Theorem 10. *For critical site percolation on the triangular lattice, the Assumptions 1–4 hold.*

Proof of Theorem 10. The Domain Markov Property, Assumption 1, is trivial for Bernoulli percolation. Assumption 2 is well known (see, e.g., [23, Theorem 3.8]). RSW, Assumption 3, is also well known (see, for example, [22, 29]). The existence of the full scaling limit in Assumption 4 is proved by the first author and Newman in [13]. We note that the value of α_1 for Bernoulli percolation is 5/48, as proved in [26]. □

The Case of the Critical FK–Ising Model. The Domain Markov Property and strong positive association are standard and well known (see, e.g., [23]). The recent development of the RSW theory for the FK–Ising model proves Assumption 3. Namely, Assumption 3 follows from Theorem 1.1 in [17] combined with the fact that the discrete extremal length, used in [17], is comparable to its continuous counterpart, used here (see [16, Proposition 6.2]). Recently, a proof of the uniqueness of the full scaling limit for the critical FK–Ising model has been completed by Kemppainen and Smirnov [24]. Theorem 1.1 in their paper implies Assumption 4. We note that the value of α_1 for the Ising model is 1/8. As shown in [15], this can be seen from the behavior of the Ising two-point function at criticality [33].

5 Approximations of Large Clusters

In what follows we give two approximations of open clusters with diameter at least $\delta > 0$, which are completely contained in Λ_k. The first one relies solely on the arm events described in the previous section, while the other is 'the natural' one, namely it is simply the union of ε-boxes which intersect the cluster. The advantage of the first approximation is that it can also be defined in the limit as the mesh size goes to 0. First we prove Proposition 1 below, which shows that, on a certain event, these two approximations coincide. Then in Sect. 5.1 we give a lower bound for the probability of that event.

For simplicity, we set $k = 1$ from now on. The constructions and proofs for different values of k are analogous. Let $\mathbb{Z}[\mathbf{i}] = \{a + b\mathbf{i} \mid a, b \in \mathbb{Z}\}$. For $\varepsilon > 0$, let B_ε be the following collection of squares of side length ε:

$$B_\varepsilon := \left\{\Lambda_{\varepsilon/2}(\varepsilon z) \mid z \in \Lambda_{\lceil 1/\varepsilon \rceil} \cap \mathbb{Z}[\mathbf{i}]\right\}.$$

Fix $\omega \in \mathscr{H}_{\hat{\mathbb{C}}}$. We define the graph $G_\varepsilon = G_\varepsilon(\omega)$ as follows. Its vertex set is B_ε. The boxes $\Lambda_{\varepsilon/2}(\varepsilon z), \Lambda_{\varepsilon/2}(\varepsilon z') \in B_\varepsilon$ are connected by an edge if $||z - z'||_\infty = 1$ or if $\omega \in \{\Lambda_{\varepsilon/2}(\varepsilon z) \overset{(1)}{\longleftrightarrow} \Lambda_{\varepsilon/2}(\varepsilon z')\}$. For a graph H with $V(H) \subseteq B_\varepsilon$ we set

$$U(H) := \bigcup_{\Lambda \in V(H)} \Lambda \subseteq \Lambda_{1+2\varepsilon}. \tag{22}$$

Let $L(H)$ denote the set of leftmost vertices of H. That is,

$$L(H) := \{\Lambda_{\varepsilon/2}(\varepsilon z) \in V(H) \mid \forall z' \in \mathbb{Z}[\mathbf{i}] \text{ with } \Lambda_{\varepsilon/2}(\varepsilon z') \in V(H), \Re z \le \Re z'\}.$$

Similarly, we define $R(H), T(H), B(H)$ as the rightmost, top and bottom sets of vertices of H, respectively. Let $SH(H)$ (resp. $SV(H)$) denote the narrowest double-infinite horizontal (resp. vertical) strip containing $U(H)$. Finally, let $SR(H)$ denote the smallest rectangle containing $U(H)$ with sides parallel to one of the axes. Thus $SR(H) = SH(H) \cap SV(H)$.

Definition 4. *For $z, z' \in \mathbb{C}$, we set $\text{dist}_1(z, z') = |\Re(z - z')|$ and $\text{dist}_2(z, z') = |\Im(z - z')|$. We call dist_1 (resp. dist_2) the distance in the horizontal (resp. vertical) direction. We also use the notation $d_\infty(z, z') := ||z - z'||_\infty = \text{dist}_1(z, z') \vee \text{dist}_2(z, z')$ for the L^∞ distance.*

For disjoint sets $A, B \subset \hat{\mathbb{C}}$ we set $\text{dist}_i(A, B) := \inf\{\text{dist}_i(z, z') : z \in A, z' \in B\}$ for $i = 1, 2$.

Let $\eta > 0$, $\Lambda = \Lambda_{\varepsilon/2}(z) \in B_\varepsilon$ and $\Lambda' = \Lambda_{\varepsilon/2}(z') \in B_\varepsilon$. Suppose there is a cluster which is completely contained in Λ_1, such that Λ contains a leftmost vertex of this cluster and Λ' a rightmost vertex. Then Λ and Λ' are connected by 2 blue arms and one red arm in between them.

This leads us to the following definition, which gives us a way to characterize the clusters using only arm events.

Definition 5. *Let $\omega \in \mathscr{H}_{\hat{\mathbb{C}}}$ and $G_\varepsilon = G_\varepsilon(\omega)$ the graph defined above. Let H be a subgraph of $G_\varepsilon(\omega)$. We say that H is good, if it satisfies the following conditions:*

1. *H is complete,*
2. *$U(H) \subseteq \Lambda_1$,*
3. *H is maximal, that is, if $\Lambda \in V(G_\varepsilon)$ and $(\Lambda, \Lambda') \in E(G_\varepsilon)$ for all $\Lambda' \in V(H)$, then $\Lambda \in V(H)$,*
4. *$\text{diam}(U(H)) \ge \delta$,*
5. *for all $\Lambda \in L(H)$ and $\Lambda' \in R(H)$ we have $\omega \in \{\Lambda \overset{(010),SV(H)}{\longleftrightarrow} \Lambda'\}$, a similar condition holds for $\Lambda \in T(H)$ and $\Lambda' \in B(H)$, with $SV(H)$ replaced by $SH(H)$.*

For a set $S \subseteq \mathbb{C}$ and $\varepsilon > 0$ let $K_\varepsilon(S)$ denote the complete graph on the vertex set

$$\{\Lambda_{\varepsilon/2}(\varepsilon z) \,|\, z \in \mathbb{Z}[\mathbf{i}] \text{ and } \Lambda_{\varepsilon/2}(\varepsilon z) \cap S \neq \emptyset\}.$$

Further, we introduce the shorthand notation

$$U_\varepsilon(S) := U(K_\varepsilon(S)) = \bigcup_{z \in \mathbb{Z}[\mathbf{i}] : \Lambda_{\varepsilon/2}(\varepsilon z) \cap S \neq \emptyset} \Lambda_{\varepsilon/2}(\varepsilon z).$$

For $\mathcal{C}_\eta \in \mathscr{C}_1^\eta(\delta)$, the graph $K_\varepsilon(\mathcal{C}_\eta)$ approximates $\mathcal{C}_\eta$ in the sense that

$$d_H(\mathcal{C}_\eta, U_\varepsilon(\mathcal{C}_\eta)) < \varepsilon.$$

This is the second approximation of large clusters we referred to in the beginning of this section. Our next aim is to find an event where the two approximations coincide.

In what follows we use the quantities defined above in the case where $\omega = \omega_\eta$ for some $\eta \geq 0$. We denote the particular choice of η in the superscript, for example $G_\varepsilon^\eta := G_\varepsilon(\omega_\eta)$. We shall prove:

Proposition 1. *Let $\eta, \varepsilon, \delta > 0$ with $1/10 > \delta > 10\varepsilon$. Suppose that $\omega_\eta \in \mathcal{E}(\varepsilon, \delta)$, where $\mathcal{E}(\varepsilon, \delta)$ is defined in (23) below.*

(i) Then for each good subgraph H of G_ε^η there is a unique cluster $\mathcal{C}^\eta \in \mathscr{C}_1^\eta(\delta)$ such that $H = K_\varepsilon(\mathcal{C}^\eta)$.
(ii) Conversely, if $\mathcal{C}^\eta \in \mathscr{C}_1^\eta(\delta)$, then $K_\varepsilon(\mathcal{C}^\eta)$ is a good subgraph of G_ε^η.

Proof of Proposition 1. Proposition 1 follows from the combination of Lemmas 11 and 12 with the definition (23) below. □

For $\varepsilon, \delta > 0$ we define the event

$$\mathcal{E}(\varepsilon, \delta) := \mathcal{NA}(\varepsilon, \delta) \cap \mathcal{NC}(\varepsilon, \delta). \tag{23}$$

First we define the event $\mathcal{NC}(\varepsilon, \delta)$ below, then we introduce $\mathcal{NA}(\varepsilon, \delta)$ in Definition 7.

Definition 6. *Let $0 < 10\varepsilon < \delta < 1$. We write $\mathcal{NC}(\varepsilon, \delta)^c$ for the union of events*

$$\mathcal{A}^j_{\emptyset,(010)}(z; \varepsilon/2, \delta/2 - 3\varepsilon) \cap \mathcal{A}^{j+2}_{\emptyset,(010)}(z'; \varepsilon/2, \delta/2 - 3\varepsilon) \tag{24}$$

for $j = 1, 2$, and $z, z' \in \Lambda_{\lceil 1/\varepsilon \rceil} \cap \mathbb{Z}[\mathbf{i}]$ with $\mathrm{dist}_j(z, z') \in (\delta - 3\varepsilon, \delta + 3\varepsilon)$.

Definition 6 implies the following lemma, which explains the choice of the event $\mathcal{NC}(\varepsilon, \delta)$.

Lemma 11. *Let $0 < 10\varepsilon < \delta < 1$. On $\omega_\eta \in \mathcal{NC}(\varepsilon, \delta)$ there is no cluster $\mathcal{C}^\eta$, which is completely contained in Λ_1 with diameter between $\delta - 2\varepsilon$ and δ.*

We define the event $\mathcal{NA}(\varepsilon, \delta)$ which will be crucial in what follows.

Definition 7. *Let ε, δ with $0 < 10\varepsilon < \delta < 1$. We set $\mathcal{NA}_1(\varepsilon, \delta)$ for the complement of the event*

$$\bigcup_{z \in \Lambda_{\lceil 1/\varepsilon \rceil} \cap \mathbb{Z}[\mathbf{i}]} \bigcup_{j=1}^{4} \mathcal{A}^j_{1,(010)}(\varepsilon z; \varepsilon/2, \delta/2 - 3\varepsilon).$$

We write $\mathcal{NA}_2(\varepsilon, \delta)^c$ for the union of events

$$\mathcal{A}^j_{\emptyset,(010)}(z; \varepsilon/2, \delta/2 - 3\varepsilon) \tag{25}$$

for $j = 1, 2, 3, 4$, and $z \in \Lambda_{\lceil 1/\varepsilon \rceil} \cap \mathbb{Z}[\mathbf{i}]$ with $\min_{i \in \{1,2\}} \mathrm{dist}_i(\Lambda_{\varepsilon/2}(z), \partial \Lambda_1) \le \varepsilon$. We define $\mathcal{NA}(\varepsilon, \delta) := \mathcal{NA}_1(\varepsilon, \delta) \cap \mathcal{NA}_2(\varepsilon, \delta)$.

Lemma 12. *Let $\eta, \varepsilon, \delta > 0$ with $0 < 10\varepsilon < \delta < 1$ and suppose that $\omega_\eta \in \mathcal{NA}(\varepsilon, \delta)$.*

(i) If $\mathcal{C}^\eta \in \mathscr{C}^\eta_1(\delta)$, then $K_\varepsilon(\mathcal{C}^\eta)$ is a good subgraph of G^η_ε.
(ii) Conversely, for any good subgraph H of G^η_ε, there is a unique cluster $\mathcal{C}^\eta \in \mathscr{C}^\eta_1(\delta - 2\varepsilon)$ such that $H = K_\varepsilon(\mathcal{C}^\eta)$.

Proof of Lemma 12. Let ε, δ as in the lemma, and $\omega_\eta \in \mathcal{NA}(\varepsilon, \delta)$. First we prove part (i) above. Apart from conditions (2) and (3), the conditions in Definition 5 are trivially satisfied. The fact that $\omega_\eta \in \mathcal{NA}_2(\varepsilon, \delta)$ implies that condition (2) is satisfied. We prove condition (3) by contradiction.

Suppose that condition (3) is violated. Then there is $\Lambda \in V(G^\eta_\varepsilon) \setminus V(K_\varepsilon(\mathcal{C}^\eta))$ such that $(\Lambda, \Lambda') \in E(G^\eta_\varepsilon)$ for all $\Lambda' \in V(K_\varepsilon(\mathcal{C}^\eta))$.

We can assume that the diameter of $\mathcal{C}^\eta$ is realized in the horizontal direction. Take $L \in L(K_\varepsilon(\mathcal{C}^\eta))$ and $R \in R(K_\varepsilon(\mathcal{C}^\eta))$. Let γ denote a path in $\mathcal{C}^\eta$ connecting L and R. We can further assume that $\mathrm{dist}_1(\Lambda, L) > \delta/2 - \varepsilon$. Note that γ is not connected to Λ. However, Λ is connected to L. Hence the blue boundary of $\mathcal{C}^\eta$ separates γ from the connection between Λ and L. We get, from L to distance $\delta/2 - \varepsilon$, three half plane arms with colour sequence (010), and a fourth red arm from the connection between Λ and L. In particular, $\omega_\eta \in \mathcal{NA}_1(\varepsilon, \delta)^c$, giving a contradiction and proving part (i) of Lemma 12.

Now we proceed to the proof of part (ii). We may assume that the diameter of $U(H)$ is realized between a leftmost and a rightmost point of it. Let $L \in L(H)$, $R \in R(H)$ and γ be a path in $SR(H)$ connecting L and R. Furthermore, let $\Lambda' \in V(G^\eta_\varepsilon)$ be such that γ is connected to Λ' by a path in $\sigma_\eta \cap \Lambda_1$.

We show that $(\Lambda, \Lambda') \in E(G^\eta_\varepsilon)$ for all $\Lambda \in V(H)$. Suppose the contrary, i.e. there is $\Lambda \in V(H)$ such that $(\Lambda, \Lambda') \notin E(G^\eta_\varepsilon)$. Then Λ is not connected to γ. Furthermore, we may assume that $\mathrm{dist}_1(\Lambda, L) > \delta/2 - \varepsilon$. Then as above, we find three half plane arms with colour sequence (010) and a fourth red arm starting at L to distance $\delta/2 - \varepsilon$. In particular, $\omega_\eta \in \mathcal{NA}_1(\varepsilon, \delta)^c$, which contradicts the assumption on ω_η above.

Hence $\Lambda' \in V(H)$ since H is maximal. Thus $K_\varepsilon(\mathcal{C}^\eta(\gamma))$ is a subgraph of H, where $\mathcal{C}^\eta(\gamma)$ denotes the connected component of γ in σ_η. Note that $K_\varepsilon(\mathcal{C}^\eta(\gamma))$

is a good subgraph because it satisfies condition (4), since $\text{dist}_1(L, R) > \delta$, and condition (3), by part (i) of Lemma 12. This completes the proof of part (ii) and that of Lemma 12. □

The proof above implies the following useful property of the event $\mathcal{NA}(\varepsilon, \delta)$.

Lemma 13. *Let $\eta, \varepsilon, \delta > 0$ with $0 < 10\varepsilon < \delta < 1$. If $\omega_\eta \in \mathcal{NA}(\varepsilon, \delta)$, then we have $|\mathscr{C}_1^\eta(\delta)| \leq 32\varepsilon^{-2}$.*

Proof of Lemma 13. Let $\mathcal{C}, \mathcal{C}' \in \mathscr{C}_1^\eta(\delta)$ be clusters with diameter at least δ in the horizontal direction. The proof of Lemma 12 shows that on the event $\mathcal{NA}(\varepsilon, \delta)$, $L(K_\varepsilon(\mathcal{C}))$ and $L(K_\varepsilon(\mathcal{C}'))$ are disjoint. The same holds for pairs of clusters with vertical diameter at least δ. Thus $|\mathscr{C}_1^\eta(\delta)| \leq 2(2\lceil 1/\varepsilon \rceil)^2) \leq 32\varepsilon^{-2}$. □

5.1 Bounds on the Probability of the Events $\mathcal{NC}(\varepsilon, \delta)$ and $\mathcal{NA}(\varepsilon, \delta)$

Our aim in this section is to prove the following bound on the probability of the complement of $\mathcal{E}(\varepsilon, \delta)$, defined in (23).

Proposition 2. *Let ε, δ with $0 < 10\varepsilon < \delta < 1$. Suppose that Assumptions 1–3 hold. Then there are positive constants $C = C(\delta), \lambda$ such that for all $\eta \in (0, \varepsilon)$ we have*

$$\mathbb{P}_\eta\left(\mathcal{E}(\varepsilon, \delta)^c\right) \leq C\varepsilon^\lambda.$$

The proof of the proposition above follows from Lemmas 14 and 15 below. We start with an upper bound on the probability of the complement of $\mathcal{NA}(\varepsilon, \delta)$.

Lemma 14. *Suppose that Assumptions 1–3 hold. Let ε, δ with $0 < 10\varepsilon < \delta < 1$. Then there are constants $C = C(\delta), \lambda > 0$ such that*

$$\mathbb{P}_\eta(\mathcal{NA}\,(\varepsilon, \delta)^c) \leq C\varepsilon^\lambda \tag{26}$$

for all $\eta < \varepsilon$. In particular, $|\mathscr{C}_1^\eta(\delta)|$ is tight in η for all fixed $\delta > 0$.

Proof of Lemma 14. For ε, δ with $0 < 10\varepsilon < \delta < 1$ simple union bounds together with Lemmas 4 and 5 give

$$\mathbb{P}_\eta(\mathcal{NA}_1\,(\varepsilon, \delta)^c) \leq 10\varepsilon^{-2}\left(\frac{\varepsilon}{\delta}\right)^{2+\lambda_{1,3}} = 10\frac{\varepsilon^{\lambda_{1,3}}}{\delta^{2+\lambda_{1,3}}},$$
$$\mathbb{P}_\eta(\mathcal{NA}_2\,(\varepsilon, \delta)^c) \leq 40\varepsilon^{-1}\left(\frac{\varepsilon}{\delta}\right)^2 = 40\frac{\varepsilon}{\delta^2}.$$

This, combined with the definition of the event $\mathcal{NA}(\varepsilon, \delta)$, provides the desired upper bound. The tightness of $|\mathscr{C}_1^\eta(\delta)|$ follows from the combination of Lemma 13 and (26). □

Lemma 15. *Suppose that Assumptions 1–3 hold. Let ε, δ with $0 < 10\varepsilon < \delta < 1$. Then there is a constant $C > 0$ such that for all $\eta \in (0, \varepsilon)$ we have*

$$\mathbb{P}_\eta(\mathcal{NC}(\varepsilon, \delta)^c) \leq C\frac{\varepsilon}{\delta^2}.$$

Proof of Lemma 15. A simple union bound combined with Lemma 4 provides the desired result. □

6 Construction of the Set of Large Clusters in the Scaling Limit

Now we are ready to construct the limiting object from Theorem 1. Before we do so, we note that Corollary 2, combined with Assumption 4 and Lemma 9, implies that there is a coupling of ω_η's for $\eta \geq 0$, denoted by $\mathbb{P}$, such that

$$\mathbb{P}(\omega_\eta \to \omega_0 \text{ as } \eta \to 0) = 1,$$

where ω_0 has law $\mathbb{P}_0$.

Fix some $\delta > 0$. Let $\omega \in \mathscr{H}$ be a quad-crossing configuration. We define

$$n_0(\omega) := \inf\left\{n \geq 0 \,|\, \omega \in \mathcal{E}(3^{-n'}, \delta) \text{ for all } n' \geq n\right\},$$

where we use the convention that the infimum of the empty set is ∞ and the event $\mathcal{E}(\varepsilon, \delta)$ is defined in (23). It is clear that $\mathcal{E}(3^{-n}, \delta) \in Borel(\mathscr{T}_{\hat{\mathbb{C}}})$, hence the function n_0 is $Borel(\mathscr{T}_{\hat{\mathbb{C}}})$ measurable. Note that $\omega_\eta \in \mathcal{E}(\eta/10, \delta)$ for $0 < \eta < 10\delta$. Hence $n_0(\omega_\eta) < \infty$ for all $0 < \eta < 10\delta$. Furthermore, we write $g_n(\omega, \delta)$ for the number of good subgraphs in $G_{3^{-n}}(\omega)$.

Let $\eta > 0$, $n \geq n_0(\omega_\eta)$, and H^η be a good subgraph in $G^\eta_{3^{-n}} = G_{3^{-n}}(\omega_\eta)$. Proposition 1 shows that for all $n' \geq n$, there is a unique good subgraph H'^η of $G^\eta_{3^{-n'}}$ such that $U(H^\eta) \supseteq U(H'^\eta)$.

Let $g^\eta_n = g_n(\omega_\eta, \delta)$. For each $n \geq 0$, we fix an ordering of the graphs with vertex sets in $B_{3^{-n}}$. For $j = 1, 2, \ldots, g^\eta_{n_0}$, let $H^\eta_{j,n_0} := H_{j,n_0(\omega_\eta)}(\omega_\eta)$ denote the jth good subgraph of $G^\eta_{3^{-n_0}}$. Then for $n \geq n_0(\omega_\eta)$, let $H^\eta_{j,n}$ denote the unique good subgraph of $G^\eta_{3^{-n}}$ such that $U(H^\eta_{j,n_0}) \supseteq U(H^\eta_{j,n})$.

For $\eta \geq 0$ and $j = 1, 2, \ldots, g^\eta_{n_0}$ we set

$$\mathcal{C}^\eta_j(\delta) := \bigcap_{n \geq n_0(\omega_\eta)} U(H^\eta_{j,n}) \tag{27}$$

on the event $n_0(\omega_\eta) < \infty$, while on the event $n_0(\omega_\eta) = \infty$ we set $\mathcal{C}^\eta_j(\delta) = \{-1/2, 1/2\}$ for all $j \geq 1$. (Note that we can replace $\{-1/2, 1/2\}$ by any disconnected subset of Λ_1.) Since the sequence of compact sets $U(H^\eta_{j,n})$ is decreasing, the intersection in (27) is non-empty on the event $n_0(\omega_\eta) < \infty$. Proposition 1 shows that for $\eta > 0$, we get the collection of clusters $\mathscr{C}^\eta_1(\delta)$, that is,

$$\mathscr{C}^\eta_1(\delta) = \{\mathcal{C}^\eta_j(\delta) \,:\, 1 \leq j \leq g^\eta_{n_0}\}.$$

Before we state and prove the following precise version of Theorem 1, let us comment on the topology used there. We employ a slightly different topology than the one in (5), defined as follows.

Let $\mathfrak{C}$ denote the set of non-empty closed subsets of Λ_1 endowed with the Hausdorff distance d_H as defined in (3). Let $l(\mathfrak{C})$ denote the space of sequences in $\mathfrak{C}$. We endow it with the metric d_l defined as

$$d_l(\underline{C}, \underline{C}') := \sum_{j=1}^{\infty} d_H(C_j, C'_j) 2^{-j} \tag{28}$$

for $\underline{C} = (C_j)_{j\geq 1}, \underline{C}' = (C'_j)_{j\geq 1}$. Note that convergence in d_l is equivalent with coordinate-wise convergence. Furthermore, $l^\infty(\mathfrak{C})$ inherits the compactness from $\mathfrak{C}$.

For $\eta \geq 0$, we extend the definition (27) by setting $\mathcal{C}_j^\eta(\delta) := \{-1/2, 1/2\}$ for $j > g_{n_0}^\eta$. We write $\underline{\mathcal{C}}_1^\eta(\omega_\eta, \delta) := (\mathcal{C}_j^\eta(\delta))_{j\geq 1}$.

For a quad-crossing configuration ω, $\underline{\mathcal{C}}_1^\eta = \underline{\mathcal{C}}_1^\eta(\omega)$ denotes the vector of all (macroscopic) clusters in ω defined as follows. The first $g_{n_0}(\omega, 3^{-1})$ entries of $\underline{\mathcal{C}}_1^\eta(\omega)$ coincide with those of $\underline{\mathcal{C}}_1^\eta(\omega, 3^{-1})$. For $m \geq 4$, the next $g_{n_0}(\omega, m^{-1}) - g_{n_0}(\omega, (m-1)^{-1})$ entries coincide with those elements in $\underline{\mathcal{C}}_1^\eta(\omega, m^{-1})$ which are not listed earlier in $\underline{\mathcal{C}}_1^\eta(\omega)$, with their relative order.

Now we are ready to state the following precise and slightly stronger version of Theorem 1.

Theorem 11. *Suppose that Assumptions 1–4 hold. Let $\delta > 0$ and let $\mathbb{P}$ be a coupling such that $\omega_\eta \to \omega_0$ a.s. as $\eta \to 0$. Then $\underline{\mathcal{C}}_1^\eta(\delta) \to \underline{\mathcal{C}}_1^0(\delta)$ in probability in the metric d_l as $\eta \to 0$. In particular, the pair $(\omega_\eta, \underline{\mathcal{C}}_1^\eta(\delta))$ converges in distribution to $(\omega_0, \underline{\mathcal{C}}_1^0(\delta))$ as $\eta \to 0$. Moreover, the same convergence result holds for $\underline{\mathcal{C}}_1^\eta$. Furthermore, $\underline{\mathcal{C}}_1^0(\delta)$ and $\underline{\mathcal{C}}_1^0$ are measurable functions of ω_0.*

Remark 5. Note that the connected sets of Λ_1 form a compact subspace of $\mathfrak{C}$. Hence $\{-1/2, 1/2\}$ is separated from the clusters $\mathcal{C}_j^\eta$ for $j = 1, \ldots, g_{n_0}^\eta$. Thus the convergence of the vectors $\underline{\mathcal{C}}_1^\eta(\delta)$ in the metric d_l implies the convergence of $\mathscr{C}_1^\eta(\delta)$ in the topology (5). Namely, the bijection is given by the ordering of the entries in the corresponding vectors, while the proof of Lemma 13 implies that, in the sequence, there is no pair of clusters converging to the same closed set. The convergence in the metric (6) follows from the equivalence of the metrics d_H and D_H.

Before we turn to the proof of Theorem 11, we prove the following lemma.

Lemma 16. *Suppose that Assumptions 1–4 hold. Let $\mathbb{P}$ be a coupling such that $\omega_\eta \to \omega_0$ $\mathbb{P}$-a.s. as $\eta \to 0$. Then*

$$\mathbb{P}(n_0(\omega_0) = \infty) = 0.$$

Moreover, $n_0(\omega_\eta) \to n_0(\omega_0)$ in probability under $\mathbb{P}$ as $\eta \to 0$.

Proof of Lemma 16. For each fixed $\varepsilon, \delta > 0$ the event $\mathcal{E}(\varepsilon, \delta)$ can be written as a finite union of intersections of some events appearing in Lemma 10. Thus

$$\mathbb{P}_0(\mathcal{E}(\varepsilon, \delta)^c) = \lim_{\eta\to 0} \mathbb{P}_\eta(\mathcal{E}(\varepsilon, \delta)^c) \leq C\varepsilon^\lambda$$

with C and λ as in Proposition 2. Hence

$$\sum_{n=1}^{\infty} \mathbb{P}_0(\mathcal{E}(3^{-n}, \delta)^c) < \infty.$$

Thus the Borel–Cantelli lemma shows that $\mathbb{P}(n_0(\omega_0) = \infty) = 0$.

Let $k \geq 1$. Lemma 10 and Proposition 2 imply that

$$\begin{aligned}
\mathbb{P}(|n_0(\omega_\eta) - n_0(\omega_0)| \geq 1) \\
&\leq \mathbb{P}(n_0(\omega_\eta) > k) + \mathbb{P}(n_0(\omega_0) > k) \\
&\quad + \mathbb{P}(|n_0(\omega_\eta) - n_0(\omega_0)| \geq 1, n_0(\omega_0) \vee n_0(\omega_\eta) \leq k) \\
&\leq \sum_{l \geq k+1} (\mathbb{P}_\eta(\mathcal{E}(3^{-l}, \delta)^c) + \mathbb{P}_0(\mathcal{E}(3^{-l}, \delta)^c)) \qquad (29) \\
&\quad + \mathbb{P}\big(\exists l \leq k \text{ s.t. } \mathbf{1}_{\{\omega_\eta \in \mathcal{E}(3^{-l},\delta)\}} \neq \mathbf{1}_{\{\omega_0 \in \mathcal{E}(3^{-l},\delta)\}}\big) \\
&\leq C \sum_{l \geq k+1} 3^{-\lambda l} + \sum_{l=1}^{k} \mathbb{P}\big(\mathbf{1}_{\{\omega_\eta \in \mathcal{E}(3^{-l},\delta)\}} \neq \mathbf{1}_{\{\omega_0 \in \mathcal{E}(3^{-l},\delta)\}}\big)
\end{aligned}$$

with some constant $C > 0$. Taking $\eta \to 0$ in (29) with a suitable constant C' we get

$$\lim_{\eta \to 0} \mathbb{P}(|n_0(\omega_\eta) - n_0(\omega_0)| \geq 1) \leq C' 3^{-\lambda k}$$

for all $k > 0$. This shows that $n_0(\omega_\eta) \to n_0(\omega_0)$ in probability as $\eta \to 0$, and concludes the proof of Lemma 16. □

Proof of Theorem 11. Let $\delta > 0$ and let $\mathbb{P}$ be a coupling such that $\omega_\eta \to \omega_0$ a.s. We will work under $\mathbb{P}$ in what follows. Note that for each $n \in \mathbb{N}$, the event $\mathcal{E}(3^{-n}, \delta)$, the graph $G_{3^{-n}}(\omega)$ and the good subgraphs of $G_{3^{-n}}(\omega)$ are functions of the outcomes of finitely many arm events appearing in Lemma 10. Thus, as $\eta \to 0$, each of

- $\mathbf{1}_{\{\omega_\eta \in \mathcal{E}(3^{-n},\delta)\}}$,
- $G_{3^{-n}}(\omega_\eta)$, and
- the ordered set of good subgraphs of $G_{3^{-n}}(\omega_\eta)$

converges in probability to the same quantity with ω_η replaced by ω_0. This implies the following convergence statements in probability as $\eta \to 0$:

(1) by Lemma 16, $n_0(\omega_\eta) \to n_0(\omega_0) < \infty$,
(2) $g_n^\eta \to g_n^0$ for all $n \geq 1$, in particular, $g^\eta_{n_0(\omega_\eta)} \to g^0_{n_0(\omega_0)}$,
(3) $H^\eta_{j,n} \to H^0_{j,n}$ for $j = 1, 2, \ldots, g_{n_0(\omega_0)}$ and $n \geq n_0(\omega_0)$.

Let $n \geq n_0(\omega_\eta) \vee n_0(\omega_0)$, then

$$\begin{aligned}
d_H(\mathcal{C}_j^\eta, \mathcal{C}_j^0) &\leq d_H(\mathcal{C}_j^\eta, U(H^\eta_{j,n})) + d_H(U(H^\eta_{j,n}), U(H^0_{j,n})) + d_H(U(H^0_{j,n}), \mathcal{C}_j^0) \\
&\leq 3^{-n} + d_H(U(H^\eta_{j,n}), U(H^0_{j,n})) + 3^{-n} \qquad (30)
\end{aligned}$$

for $j = 1, 2, \ldots, g^\eta_{n_0} \wedge g^0_{n_0}$. Thus taking the limit $\eta \to 0$ in (30), by (1–3) above, we get

$$\lim_{\eta \to 0} \mathbb{P}(d_H(\mathcal{C}_j^\eta, \mathcal{C}_j^0) > 3 \cdot 3^{-n}, n \geq n_0(\omega_0) \vee n_0(\omega_\eta)) = 0 \qquad (31)$$

for $j \geq 1$. Then taking the limit $n \to \infty$, Lemma 16 shows that $\mathcal{C}_j^\eta \to \mathcal{C}_j^0$ in probability in the Hausdorff metric as $\eta \to 0$ for all $j \geq 1$. Since convergence in $l^\infty(\mathfrak{C})$ coincides with coordinate-wise convergence, we get that $\lim_{\eta\to 0} \underline{\mathcal{C}}_1^\eta(\delta) = \underline{\mathcal{C}}_1^0(\delta)$ in probability, as required.

The proof of the claims of Theorem 11 for $\underline{\mathcal{C}}_1^\eta$ is analogous. It follows from the convergence of $\underline{\mathcal{C}}_1^\eta(\delta)$ with $\delta = 3^{-m}$ for $m \geq 1$. The measurability of $\underline{\mathcal{C}}_1^0(\delta)$ and $\underline{\mathcal{C}}_1^0$ with respect to ω_0 follows easily from their definition involving arm events (see Remark 2). Thus the proof of Theorem 11 is complete. □

7 Scaling Limit in a Bounded Domain

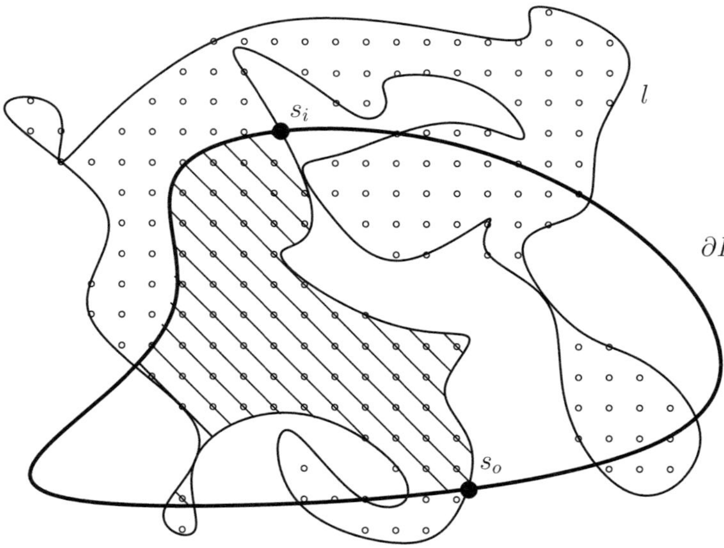

Fig. 3. Illustration of a cluster in D. The small open circles denote the interior of the loop l. The shaded area intersected with the cluster of the loop is equal to $\mathcal{B}(\mathcal{E})$.

In this section we will deduce the convergence of all clusters and "pieces" of clusters contained in a bounded domain D from the convergence of clusters and loops completely contained in $\Lambda_k \supset D$, for some k sufficiently large. We call $\mathscr{B}_D^\eta(\delta)$ the collection of all clusters or portions of clusters of diameter at least δ contained in D^η, where D^η denotes an appropriate discretization of D. In the case of $\mathbb{Z}^2$, the boundary of D^η is a circuit in the medial lattice that surrounds all the vertices of $\mathbb{Z}^2$ contained in D and minimizes the distance to ∂D. Analogously, in the case of the triangular lattice, $\mathbb{T}$, the boundary of D^η is a circuit in the dual (hexagonal) lattice that surrounds all the vertices of $\mathbb{T}$ contained in D and minimizes the distance to ∂D. More precisely, for every cluster $\mathcal{C} \in \mathscr{C}^\eta(\delta)$ that intersect D^η, consider the set of all connected components $\mathcal{B}$ of $\mathcal{C} \cap D^\eta$ with diameter at least $\delta > 0$. For every $\eta, \delta > 0$, we let $\mathscr{B}_D^\eta(\delta)$ denote the union of $\mathscr{C}_D^\eta(\delta)$ with the set of all such connected components $\mathcal{B}$. (Note that clusters contained in Λ_k but not completely contained in D^η are split into different elements of $\mathscr{B}_D^\eta(\delta)$ (see Fig. 3). For the case of Bernoulli percolation, the collection $\mathscr{B}_D^\eta(\delta)$ is precisely the set of all clusters in D^η with closed boundary condition.

As in Sect. 6, instead of the collection $\mathscr{B}_D^\eta(\delta)$, we consider the sequence $\underline{\mathcal{B}}_D^\eta(\delta)$ of clusters with diameter at least δ, with the metric d_l. Now we are ready to state the theorem on the convergence of all portions of clusters in $\sigma_\eta \cap D$ for a bounded domain D.

Theorem 12. *Suppose that Assumptions 1–4 hold. Let D be a simply connected bounded domain with piecewise smooth boundary. Let $\mathbb{P}$ be a coupling where $(\omega_\eta, L_\eta) \to (\omega_0, L_0)$ a.s. as $\eta \to 0$. Then, for any $\delta > 0$, $\underline{\mathcal{B}}_D^\eta(\delta) \to \underline{\mathcal{B}}_D^0(\delta)$ in probability in the metric d_l as $\eta \to 0$. In particular, the triple $(\omega_\eta, L_\eta, \underline{\mathcal{B}}_D^\eta(\delta))$ converges in distribution to $(\omega_0, L_0, \underline{\mathcal{B}}_D^0(\delta))$ as $\eta \to 0$. Moreover, the same convergence result holds for $\underline{\mathcal{B}}_D^\eta$. Furthermore, $\underline{\mathcal{B}}_D^0(\delta)$ and $\mathscr{B}_D^0$ are measurable functions of the pair (ω_0, L_0).*

Proof of Theorem 12. Let (ω_η, L_η) and (ω_0, L_0) be as in the statement of Theorem 12. The probability that all the clusters that intersect D are completely contained in Λ_k is at least one minus the probability of having a red arm from the boundary of D to $\partial\Lambda_k$. The latter probability goes to zero as $k \to \infty$, hence there is a finite $k \in \mathbb{N}$ such that there is no red arm from D to $\partial\Lambda_{k-1}$ in ω_0. We take the smallest such k. With this choice, all clusters in $\mathscr{C}^\eta$ that intersect D are contained in Λ_k.

We first give an orientation to the loops contained in Λ_k in such a way that clockwise loops are the outer boundaries of red clusters and counterclockwise loops are the outer boundaries of blue clusters. For each clockwise loop ℓ intersecting ∂D, we consider all excursions $\mathcal{E}$ inside D of diameter at least δ. Each excursion $\mathcal{E}$ runs from a point s_{in} on ∂D to a point s_{out} on ∂D. We call the counterclockwise segment of ∂D from s_{in} to s_{out} the base of $\mathcal{E}$. We call $\overline{\mathcal{E}}$ the concatenation of $\mathcal{E}$ with its base. We define the interior $\mathrm{I}(\overline{\mathcal{E}})$ of $\overline{\mathcal{E}}$ to be the closure of the set of points with nonzero winding number for the curve $\overline{\mathcal{E}}$.

We call $\mathscr{E}_\mathcal{E}$ the collection of all clockwise excursions in D of the same loop ℓ with base contained inside the base of $\mathcal{E}$. If $\mathcal{C}$ is the cluster whose outer boundary is the loop ℓ, we define $\mathcal{B}(\mathcal{E})$ as follows:

$$\mathcal{B}(\mathcal{E}) := \overline{\mathrm{I}(\overline{\mathcal{E}}) \setminus \{\cup_{\mathcal{E}' \in \mathscr{E}_\mathcal{E}} \mathrm{I}(\overline{\mathcal{E}'})\}} \cap \mathcal{C},$$

where by $\cup_{\mathcal{E}' \in \mathscr{E}_\mathcal{E}} \mathrm{I}(\overline{\mathcal{E}'})$ we mean $\lim_{\xi \to 0} \cup_{\mathcal{E}' \in \mathscr{E}_\mathcal{E}, \mathrm{diam}\mathcal{E}' > \xi} \mathrm{I}(\overline{\mathcal{E}'})$, and the limit exists because it is the limit of an increasing sequence of closed sets.

For any $\delta > 0$, $\mathscr{B}_D^0(\delta)$ is the collection of all sets $\mathcal{B}(\mathcal{E})$ defined above, for all clockwise excursions $\mathcal{E}$ in D of diameter at least δ.

For any $\eta > 0$, the collection $\mathscr{B}_D^\eta(\delta)$ contains all clusters completely contained in D plus all the connected components of the intersections of clusters in

Λ_k with D. $\mathscr{B}^{\eta}_D(\delta)$ can be obtained with the following construction which mimics the continuum construction given earlier. We first give an orientation to the loops contained in Λ_k in such a way that loops that have red in their immediate interior are oriented clockwise and loops that have blue in their immediate interior are oriented counterclockwise. For each clockwise loop ℓ^η intersecting ∂D^η, we consider all excursions $\mathcal{E}^\eta$ inside D^η of diameter at least δ. Each excursion $\mathcal{E}^\eta$ runs from a point s^η_{in} on ∂D^η to a point s^η_{out} on ∂D^η. We call the counterclockwise segment of ∂D^η from s^η_{in} to s^η_{out} the base of $\mathcal{E}^\eta$. We call $\overline{\mathcal{E}^\eta}$ the concatenation of $\mathcal{E}^\eta$ with its base. We define the interior $\mathrm{I}\left(\overline{\mathcal{E}^\eta}\right)$ of $\overline{\mathcal{E}^\eta}$ to be the set of hexagons contained inside $\overline{\mathcal{E}^\eta}$.

We call $\mathscr{E}^\eta_{\mathcal{E}^\eta}$ the collection of all clockwise excursions in D^η of the same loop ℓ^η with base contained inside the base of $\mathcal{E}^\eta$. If $\mathcal{C}^\eta$ is the cluster whose outer boundary is the loop ℓ^η, we define $\mathcal{B}^\eta(\mathcal{E}^\eta)$ as follows:

$$\mathcal{B}^\eta(\mathcal{E}^\eta) := \overline{\mathrm{I}\left(\overline{\mathcal{E}^\eta}\right) \setminus \left\{ \cup_{(\mathcal{E}^\eta)' \in \mathscr{E}^\eta_{\mathcal{E}^\eta}} \mathrm{I}\left(\overline{(\mathcal{E}^\eta)'}\right) \right\}} \cap \mathcal{C}^\eta.$$

We now note that the almost sure convergence $(\omega_\eta, L_\eta) \to (\omega_0, L_0)$, combined with Lemma 4, implies the same for the excursions in D. (Lemma 4 insures, via standard arguments, that an excursion cannot come close to the boundary of D without touching it, so that large lattice and continuum excursions will match with high probability for η sufficiently small. For more details on how to use Lemma 4, the interested reader is referred to Lemma 6.1 of [13].) Together with the convergence of the clusters, this implies that $(\omega_\eta, L_\eta, \underline{\mathcal{B}}^\eta_D(\delta))$ converges in distribution to $(\omega_0, L_0, \underline{\mathcal{B}}^0_D(\delta))$ as $\eta \to 0$, the ordering is simply given by the ordering of the clusters completely contained in D and a clockwise ordering of the points s_{in} (s^η_{in}). The above result is valid for any $\delta > 0$, so letting $\delta \to 0$ gives the second part of the theorem. □

8 Limits of Counting Measures of Clusters

In this section we state and prove Theorem 13, a precise and slightly stronger version of Theorem 2. We do this for the more general case of (portions of) clusters $\mathscr{B}^\eta_D(\delta)$ in a domain with piecewise smooth boundary D. The convergence of measures of the clusters which are completely contained in Λ_k follows immediately. For ease of notation we assume D to be Λ_1.

Let $\mathfrak{M}$ denote the set of finite Borel measures on Λ_1 endowed with the Prokhorov metric. Recall that $\mathfrak{M}$ is a separable metric space.

For $\eta \geq 0$, $n \in \mathbb{N}$ and $S \subseteq \Lambda_1$, we define

$$\mu^\eta_{S,n} := \sum_{z \in \mathbb{Z}[\mathbf{i}] : \Lambda_{3 \cdot 3^{-n}/2}(3^{-n}z) \cap S \neq \emptyset} \mu^\eta_{1, A(3^{-n}z; 3^{-n}/2, \delta/2 - 3^{-n})}. \tag{32}$$

This is the sum of counting measures $\mu^\eta_{1,A(z;3^{-n}/2,\delta/2-3^{-n})}$ such that $z \in 3^{-n}\mathbb{Z}[\mathbf{i}]$ and the inner box $\Lambda_{3^{-n}/2}(z)$ or one of its neighbors has nonempty intersection with S.

Simple arguments show the following:

Observation 1. *Let B be a Borel subset of $\mathbb{C}$ and $S \subseteq \Lambda_1$. Then, for fixed $\eta > 0$, $\mu^\eta_{S,n}(B) \geq \mu^\eta_{S,n'}(B)$ for $n' \geq n$ with probability* 1.

It is easy to check that, for all fixed $\eta > 0$ and $\mathcal{B} \in \mathscr{B}^\eta_{\Lambda_1}(\delta)$, the following limit exists

$$\lim_{n\to\infty} \mu^\eta_{\mathcal{B},n} \tag{33}$$

and is actually equal to $\mu^\eta_{\mathcal{B}}$ as defined in (7). This motivates us to define, for any cluster $\mathcal{B} \in \mathscr{B}^0_{\Lambda_1}(\delta)$, $\mu^0_{\mathcal{B}}$ by (33) with $\eta = 0$ if the limit exists, and set $\mu^0_{\mathcal{B}} = 0$ when it does not.

Let $l(\mathfrak{M})$ denote the set of infinite sequences in $\mathfrak{M}$ with bounded distance from the empty measure. Similarly to (28), we set

$$d_l(\underline{\nu}, \underline{\phi}) := \sum_{j=1}^{\infty} \frac{d_P(\nu_j, \phi_j)}{1 + d_P(\nu_j, \phi_j)} 2^{-j}$$

for $\underline{\nu}, \underline{\phi} \in l(\mathfrak{M})$. It is easy to check that $l(\mathfrak{M})$ is separable, but not compact. Let $h^\eta(\delta) := |\mathscr{B}^\eta_{\Lambda_1}(\delta)|$, for $\eta \geq 0$. It follows from Lemma 16, together with the tightness of the number of excursions of diameter at least δ in Λ_1, that $h^0(\delta)$ is a.s. finite. For $\eta \geq 0$, we define $\underline{\mu}^\eta = (\mu^\eta_j)_{j\geq 1}$, the vector of measures $\mu^\eta_j := \mu^\eta_{\mathcal{B}_j}$ for $\mathcal{B}_j \in \mathscr{B}^\eta_{\Lambda_1}(\delta)$ and $j = 1, 2, \ldots, h^\eta(\delta)$, and we set $\mu^\eta_j = 0$ for $j > h^\eta(\delta)$. We define $\underline{\mu}^\eta$ similarly to $\underline{\mathcal{C}}^\eta$.

Now we are ready to state the main result from this section.

Theorem 13. *Suppose that Assumptions 1–4 hold. Let D be a simply connected bounded domain with piecewise smooth boundary. Let $\mathbb{P}$ be a coupling such that $(\omega_\eta, L_\eta) \to (\omega_0, L_0)$ a.s. as $\eta \to 0$. Then $\underline{\mu}^\eta_D(\delta) \to \underline{\mu}^0_D(\delta)$ in probability as $\eta \to 0$, where $\underline{\mu}^0_D(\delta)$ is a measurable function of the pair (ω_0, L_0). In particular, the triple $(\omega_\eta, L_\eta, \underline{\mu}^\eta_D(\delta))$ converges in distribution to $(\omega_0, L_0, \underline{\mu}^0_D(\delta))$ as $\eta \to 0$. The same convergence result holds when $\underline{\mu}^\eta_D(\delta)$ is replaced by $\underline{\mu}^\eta_D$.*

The same conclusions hold for the measures of the clusters in $\hat{\mathbb{C}}$ which intersect a bounded domain D, that is, keeping the information of connections outside D.

Remark 6. Lemma 18 below shows that clusters whose diameter is at least $\delta > 0$ have nonzero mass. Thus the convergence in Theorem 13 implies convergence in the metric analogous to (5) based on the Prokhorov metric d_P, and so Theorem 2 is proved.

Let us first show that Theorem 3 follows easily from Theorems 11 and 13.

Proof of Theorem 3. The proof is analogous to the proof of Theorem 6 of [13], so we only give a sketch. Let D be any bounded subset of $\mathbb{C}$ and $k_1 > k_2$ be such that $D \subset \Lambda_{k_2}$. The measures $\mathbb{P}_{k_1}$ and $\mathbb{P}_{k_2}$ can be coupled in such a way that they coincide inside D, in the sense that they induced the same marginal distribution on $(\mathscr{C}_D^0, \mathscr{M}_D^0)$. This is because they are obtained from the scaling limit of the same full-plane lattice measure $\mathbb{P}_\eta$. The consistency relations needed to apply Kolmogorov's extension theorem are then satisfied, which insures the existence of a limit $\mathbb{P}$. □

The following lemma plays an important role in the proof of Theorem 13. Let $||\nu||_{TV}$ denote the total variation of a signed measure ν.

Lemma 17. *Suppose that Assumptions 1–3 hold. Let $\delta > 0$. Then there are positive constants $C = C(\delta), \varphi$ such that, for $n \in \mathbb{N}$ and $\eta > 0$ with $0 < 10\eta < 3^{-n} < \delta/10$,*

$$\mathbb{P}_\eta(\exists \mathcal{B} \in \mathscr{B}^\eta_{\Lambda_1}(\delta), S \subseteq \Lambda_1 \ s.t. \ d_H(\mathcal{B}, S) < \varepsilon/2, ||\mu^\eta_{\mathcal{B}} - \mu^\eta_{S,n}||_{TV} \geq \varepsilon^\varphi) \leq C \cdot \varepsilon^\varphi$$

where $\varepsilon = 3^{-n}$.

Proof of Theorem 13 *given Lemma* 17. Let $\mathbb{P}$ be as in Theorem 13, $\delta > 0$. It follows from Theorem 12 that the clusters in $\mathscr{B}^\eta_{\Lambda_1}(\delta)$ converge in probability as $\eta \to 0$.

Moreover, Theorem 9 shows that each of the measures

$$\mu^\eta_{1,A(3^{-n}z;3^{-n}/2,\delta/2-3^{-n})} \quad \text{for } n \geq 1 \text{ and } z \in \mathbb{Z}[\mathbf{i}] \text{ with } 3^{-n}z \in \Lambda_1.$$

converges in probability in the Prokhorov metric, as $\eta \to 0$, to the analogous measure where η is replaced by 0.

This implies that, for all fixed n and $S \subset \Lambda_1$, $\mu^\eta_{S,n} \to \mu^0_{S,n}$ weakly in probability as $\eta \to 0$. The monotonicity of the measures $\mu^\eta_{S,n}$ in n for a fixed subset S and fixed η (Observation 1) carries through to the limit as $\eta \to 0$, thus the weak limit $\mu^0_S = \lim_{n\to\infty} \mu^0_{S,n}$ exists almost surely. Furthermore, since each of the measures $\mu^0_{S,n}$ is a function of (ω_0, L_0) and is a.s. finite, we conclude that μ^0_S is a.s. finite and is a function of (ω_0, L_0).

Let $\mathcal{B}$ be the j-th element of $\underline{\mathcal{B}}^0_{\Lambda_1}(\delta)$ and let $\mathcal{B}^\eta_j$ be the j-th element of $\underline{\mathcal{B}}^\eta_{\Lambda_1}(\delta)$, where $\underline{\mathcal{B}}^0_{\Lambda_1}(\delta)$ and $\underline{\mathcal{B}}^0_{\Lambda_1}$ are the sequences of clusters that appear in Theorem 12. Fix $\kappa > 0$. Lemma 17 implies that, for some constants $\varphi, C = C(\delta)$, for $\kappa > \varepsilon^\varphi$, $\eta < \varepsilon/10$ and $3^{-n} = \varepsilon$, we have

$$\begin{aligned}
&\mathbb{P}(d_P(\mu^0_{\mathcal{B}}, \mu^\eta_{\mathcal{B}^\eta_j}) > 3\kappa) \\
&\leq \mathbb{P}(d_P(\mu^0_{\mathcal{B}}, \mu^0_{\mathcal{B},n}) > \kappa) + \mathbb{P}(d_P(\mu^0_{\mathcal{B},n}, \mu^\eta_{\mathcal{B},n}) > \kappa) \\
&\quad + \mathbb{P}(||\mu^\eta_{\mathcal{B},n} - \mu^\eta_{\mathcal{B}^\eta_j}||_{TV} > \kappa, d_H(\mathcal{B}, \mathcal{B}^\eta_j) < \varepsilon/2) + \mathbb{P}(d_H(\mathcal{B}, \mathcal{B}^\eta_j) \geq \varepsilon/2) \\
&\leq \mathbb{P}(d_P(\mu^0_{\mathcal{B}}, \mu^0_{\mathcal{B},n}) > \kappa) + \mathbb{P}(d_P(\mu^0_{\mathcal{B},n}, \mu^\eta_{\mathcal{B},n}) > \kappa) \\
&\quad + C\kappa + \mathbb{P}(d_H(\mathcal{B}, \mathcal{B}^\eta_j) \geq \varepsilon/2)
\end{aligned} \tag{34}$$

where d_P denotes the Prokhorov distance of Borel measures.

Now we take the limit first as $\eta \to 0$ then as $n \to \infty$ in (34). From the arguments above and Theorem 12 we deduce that

$$\lim_{\eta \to 0} \mathbb{P}(d_P(\mu^0_{\mathcal{B}}, \mu^{\eta}_{\mathcal{B}^{\eta}_j}) > 3\kappa) \leq C\kappa$$

for all $\kappa > 0$. Thus the measures $\mu^{\eta}_{\mathcal{B}^{\eta}_j}$ tend to $\mu^0_{\mathcal{B}}$ weakly in probability as $\eta \to 0$.

Recall that the convergence in $l^{\infty}(\mathfrak{M})$ is equivalent to coordinate-wise convergence. Thus $\underline{\mu}^{\eta}(\delta) \to \underline{\mu}^0(\delta)$ in probability as $\eta \to 0$. We have already proved in the lines above that $\underline{\mu}^0(\delta)$ is a measurable function of (ω_0, L_0), thus we deduced the results in Theorem 13 for $\underline{\mu}^{\eta}(\delta)$.

The results for $\underline{\mu}^{\eta}$ follow from the lines above by arguments similar to those at the end of the proof of Theorem 11. This concludes the proof of Theorem 13. □

We finish this section by proving Lemma 17 above. Its proof relies on Lemma 7.

Proof of Lemma 17. Let η, n, δ as in Lemma 17. To simplify the notation, we set $\varepsilon := 3^{-n}$, $\delta' := \delta/2 - 3\varepsilon$ and $\beta := \frac{\lambda}{2(\lambda+\lambda_1)}$, with λ_1 as in Lemma 3 and λ as in Lemma 6.

We define the following collection of 'pivotal' boxes:

$$\mathrm{Piv}^{\eta}(\varepsilon, \varepsilon^{\beta}) := \{\Lambda_{\varepsilon/2}(\varepsilon z) \mid z \in \mathbb{Z}[\mathbf{i}] \cap \Lambda_{\varepsilon^{-1}+1}; \omega_{\eta} \in \mathcal{A}_{(1010),\emptyset}(\varepsilon z; 3\varepsilon/2, \varepsilon^{\beta})\}.$$

Furthermore, we let $\nu^{\eta}_{\varepsilon^{\beta}}$ denote the normalized counting measure of the vertices close to the boundary of Λ_1 which have an open arm to distance $5\varepsilon^{\beta}$:

$$\nu^{\eta}_{\varepsilon^{\beta}} := \frac{\eta^2}{\pi^{\eta}_1(\eta, 1)} \sum_{v \in A(0; 1-\varepsilon^{\beta}, 1) \cap \eta V} \delta_v \mathbf{1}\{v \overset{(1)}{\longleftrightarrow} \partial \Lambda_{5\varepsilon^{\beta}}(v)\}. \tag{35}$$

Roughly speaking, $\nu^{\eta}_{\varepsilon^{\beta}}$ is introduced to account for boxes near $\partial\Lambda_1$ where two large pieces of a cluster come close to each other. Such boxes are not necessarily 'pivotal' since the two large pieces may connect just outside Λ_1, in which case the boxes are not counted in Piv^{η}.

Take $\mathcal{B} \in \mathscr{B}^{\eta}_{\Lambda_1}(\delta)$ and $S \subseteq \Lambda_1$ such that $d_H(S, \mathcal{B}) < \varepsilon/2$. Note that $d_H(S, \mathcal{B}) < \varepsilon/2$ implies that the counting measure $\mu^{\eta}_{S,n}$ is larger than or equal to the counting measure $\mu^{\eta}_{\mathcal{B}}$. As a consequence it is easy to check that, for these $\mathcal{B}$ and S, we have

$$\begin{aligned} &||\mu^{\eta}_{S,n} - \mu^{\eta}_{\mathcal{B}}||_{TV} \\ &\quad \leq ||\nu^{\eta}_{\varepsilon^{\beta}}||_{TV} + \sum_{z \in \mathbb{Z}[\mathbf{i}] \,:\, \Lambda_{\varepsilon/2}(\varepsilon z) \in \mathrm{Piv}^{\eta}(\varepsilon, \varepsilon^{\beta})} ||\mu^{\eta}_{1, A(\varepsilon z; 3\varepsilon/2, \delta')}||_{TV} \qquad (36) \\ &\quad \leq ||\nu^{\eta}_{\varepsilon^{\beta}}||_{TV} + |\mathrm{Piv}^{\eta}(\varepsilon, \varepsilon^{\beta})| \sup_{z \in \mathbb{Z}[\mathbf{i}] \cap \Lambda_{\varepsilon^{-1}+1}} ||\mu^{\eta}_{1, A(\varepsilon z; 3\varepsilon/2, 3\varepsilon)}||_{TV}. \end{aligned}$$

Letting $a_\varepsilon^\eta := \varepsilon^{-(2+\varphi)}\pi_4^\eta(3\varepsilon/2, \varepsilon^\beta)$, from (36) we deduce that

$$\begin{aligned}
&\mathbb{P}_\eta(\exists \mathcal{B} \in \mathscr{B}^\eta_{\Lambda_1}(\delta), S \subseteq \Lambda_1 \text{ s.t. } d_H(S,\mathcal{B}) < \varepsilon/2, ||\mu^\eta_{\mathcal{B}} - \mu^\eta_{S,n}||_{TV} \geq \varepsilon^\varphi) \\
&\qquad \leq \mathbb{P}_\eta(||\nu^\eta_{\varepsilon^\beta}||_{TV} \geq \frac{1}{2}\varepsilon^\varphi) + \mathbb{P}_\eta(|\mathrm{Piv}^\eta(\varepsilon,\varepsilon^\beta)| \geq a^\eta_\varepsilon) \\
&\qquad + \ \mathbb{P}_\eta(\sup_{z\in\Lambda_{\varepsilon^{-1}+1}\cap\mathbb{Z}[\mathbf{i}]} ||\mu^\eta_{1,A(\varepsilon z;3\varepsilon/2,3\varepsilon)}||_{TV} > \varepsilon^\varphi/2a^\eta_\varepsilon), \qquad (37)
\end{aligned}$$

for some φ to be fixed later. By the Markov inequality, we have

$$\mathbb{P}_\eta(|\mathrm{Piv}^\eta(\varepsilon,\varepsilon^\beta)| \geq a^\eta_\varepsilon) \leq C_1\varepsilon^\varphi \qquad (38)$$

for some positive constant $C_1 = C_1(\delta)$ for all $\varphi > 0$.

Now we bound the third term in (37). With some positive constants C_2, C_3, C_4 depending on δ, and recalling Definition 3, we have that

$$\begin{aligned}
&\mathbb{P}_\eta(\sup_{z\in\Lambda_{\varepsilon^{-1}+1}\cap\mathbb{Z}[\mathbf{i}]} ||\mu^\eta_{1,A(\varepsilon z;3\varepsilon/2,3\varepsilon)}||_{TV} > \varepsilon^\varphi/2a^\eta_\varepsilon) \\
&\qquad \leq C_2\varepsilon^{-2}\mathbb{P}_\eta(||\mu^\eta_{1,A(3\varepsilon/2,3\varepsilon)}||_{TV} > \varepsilon^\varphi/2a^\eta_\varepsilon) \\
&\qquad = C_2\varepsilon^{-2}\mathbb{P}_\eta(|\mathcal{V}^\eta_{3\varepsilon}| \geq \varepsilon^\varphi\eta^{-2}\pi_1^\eta(\eta,1)/2a^\eta_\varepsilon) \\
&\qquad \leq C_2\varepsilon^{-2}\exp\left(-C_3\varepsilon^{2\varphi}\frac{\pi_1^\eta(\eta,1)}{\pi_1^\eta(\eta,3\varepsilon)\pi_4^\eta(3\varepsilon/2,\varepsilon^\beta)}\right) \qquad (39) \\
&\qquad \leq C_2\varepsilon^{-2}\exp\left(-C_4\varepsilon^{2\varphi}\frac{\pi_1^\eta(3\varepsilon,\varepsilon^\beta)}{\pi_4^\eta(3\varepsilon/2,\varepsilon^\beta)}\pi_1^\eta(\varepsilon^\beta,1)\right),
\end{aligned}$$

where, in the second inequality, we used Lemma 7 and, in the last line, we used Lemma 2 twice. Lemmas 3 and 6, (39) and the choice of β give that

$$\begin{aligned}
&\mathbb{P}_\eta(\sup_{z\in\Lambda_{\varepsilon^{-1}+1}\cap\mathbb{Z}[\mathbf{i}]} ||\mu^\eta_{1,A(\varepsilon z;3\varepsilon/2,3\varepsilon)}||_{TV} > \varepsilon^\varphi/2a^\eta_\varepsilon) \\
&\leq C_2\varepsilon^{-2}\exp(-C_5\varepsilon^{2\varphi+\lambda(\beta-1)+\lambda_1\beta}) \\
&= C_2\varepsilon^{-2}\exp(-C_5\varepsilon^{2\varphi-\lambda/2}) \qquad (40)
\end{aligned}$$

with $C_5 > 0$. Computations similar to those above give the following upper bound for the second term in (37):

$$\begin{aligned}
\mathbb{P}_\eta(||\nu^\eta_{\varepsilon^\beta}||_{TV} \geq \frac{1}{2}\varepsilon^\varphi) &\leq C_6\varepsilon^{-\beta}\exp\left(-C_7\varepsilon^{\varphi-\beta}\frac{\pi_1^\eta(\eta,1)}{\pi_1^\eta(\eta,\varepsilon^\beta)}\right) \\
&\leq C_6\varepsilon^{-\beta}\exp\left(-C_8\varepsilon^{\varphi-\beta+\beta\lambda_1}\right) \qquad (41)
\end{aligned}$$

for suitable constants C_6, C_7, C_8. We set $\varphi = \frac{\lambda\wedge(\beta(1-\lambda_1))}{4} > 0$. A combination of (37), (38), (40) and (41) finishes the proof of Lemma 17. □

9 Properties of the Continuum Clusters and Their Normalized Counting Measures

We start with the connections between the clusters and their counting measures. The first result of the section shows, roughly speaking, that the scaling limit of the clusters as closed sets contains the same information as their normalized counting measures. Then we show conformal invariance of the clusters and conformal covariance of their normalized counting measures.

9.1 Basic Properties

Recall the notation $\mathscr{C}^\eta(\delta)$ from (2). We set $\mathscr{C}^0 = \bigcup_{n=1}^{\infty} \mathscr{C}^0(3^{-n})$. For $\mathcal{C} \in \mathscr{C}^0$ and $0 < \psi < 1/2$ we write

$$\tilde{\mu}^0_{\mathcal{C},\psi} := \frac{4\psi^2}{\pi_1^0(2\psi,1)} \sum_{z\in\mathbb{Z}[\mathbf{i}]:\Lambda_{\psi/2}(\psi z)\cap\mathcal{C}\neq\emptyset} \delta_{\psi z}. \tag{42}$$

Theorem 14. *Suppose that Assumptions 1–4 hold. Then* $\operatorname{supp}(\mu^0_{\mathcal{C}}) = \mathcal{C}$ *for all* $\mathcal{C} \in \mathscr{C}^0$. *Moreover,*

$$\tilde{\mu}^0_{\mathcal{C},\psi} \to \mu^0_{\mathcal{C}} \text{ weakly in probability as } \psi \to 0 \tag{43}$$

for all $\mathcal{C} \in \mathscr{C}^0$.

The proof of the theorem above relies on the following two lemmas.

Lemma 18. *Suppose that Assumptions 1–3 hold. Let* $k, \delta > 0$. *Then for all* $\varphi > 0$ *there is* $x_\varphi = x_\varphi(k,\delta) > 0$ *such that*

$$\mathbb{P}_\eta(\exists \mathcal{C} \in \mathscr{B}^\eta_k(\delta) \text{ with } ||\mu^\eta_{\mathcal{C}}||_{TV} < x_\varphi) < \varphi \tag{44}$$

for all $\eta \in (0,\delta)$.

Proof of Lemma 18. For critical percolation the proof of Lemma 18 follows from the proof of [5, Theorem 1.2]: (3.18) of [5] with $x = 0$ can be shown in the same manner as for $x > 0$. Alternatively, Lemma 18 can be deduced from a combination of [7, Theorem 3.1 (i), Theorem 3.3 (i) and Lemma 4.4], using tightness of the number of clusters of diameter at least δ.

It is easy to verify that actually all these arguments just need Assumptions 1–3. □

The second is essentially [19, Proposition 4.13] see also [19, Eqn. (4.39)]. Let A be the annulus $A = A(a,b)$ with $0 < a < b$ and $\mathcal{C} \in \mathscr{C}^0$. For $\eta \geq 0$ and $0 < \psi < 1/2$ we set

$$\tilde{\mu}^\eta_{A,\psi} := \frac{4\psi^2}{\pi_1^\eta(2\psi,1)} \sum_{z\in\mathbb{Z}[\mathbf{i}]\cap\Lambda_{\psi^{-1}a}} \mathbf{1}\{\Lambda_{\psi/2}(\psi z) \overset{1}{\longleftrightarrow} \partial\Lambda_b\}\delta_{\psi z}.$$

Lemma 19. (Proposition 4.13 of [19]). *Suppose that Assumptions 1–4 hold. Let $f : \mathbb{C} \to \mathbb{R}$ be a continuous function with compact support, and let $A = A(a, b)$ be an annulus with $0 < a < b$. Then*

$$\tilde{\mu}^0_{A,\psi}(f) \to \mu^0_A(f) \text{ in } L^2 \text{ as } \psi \to 0. \tag{45}$$

Remark 7. For the proof of Theorem 14, convergence in probability is enough in (45).

Proof of Theorem 14. Since $\mathscr{C}^0 = \bigcup_{n=1}^{\infty} \mathscr{C}^0(3^{-n})$ and $\mathscr{C}^0(3^{-n}) = \bigcup_{k\in\mathbb{N}} \mathscr{C}^0_k(3^{-n})$, to prove the first part of the theorem, it suffices to show that $\text{supp}(\mu^0_{\mathcal{C}}) = \mathcal{C}$ with probability 1 for all $\mathcal{C} \in \mathscr{C}^0_k(\delta)$ for any fixed $\delta > 0$ and $k \in \mathbb{N}$. We will work under a coupling $\mathbb{P}$ such that $\omega_\eta \to \omega_0$ a.s.

Equations (32) and (33) show that, for all $\mathcal{C} \in \mathscr{C}^0(\delta)$, $\text{supp}(\mu^0_{\mathcal{C}})$ is contained in the (3^{-n})-neighborhood of $\mathcal{C}$ for every n, with probability 1. Hence, $\text{supp}(\mu^0_{\mathcal{C}}) \subseteq \mathcal{C}$ for all $\mathcal{C} \in \mathscr{C}^0(\delta)$ with probability 1.

We turn to the proof of $\text{supp}(\mu^0_{\mathcal{C}}) \supseteq \mathcal{C}$. Take $\varphi > 0$ and x_φ as in Lemma 18. By covering Λ_k with at most $4(k/\varepsilon)^2$ squares with side length ε, we get

$$\begin{aligned}
&\mathbb{P}_\eta(\exists z \in \mathbb{Z}[\mathbf{i}], \exists \mathcal{C} \in \mathscr{C}^\eta(\delta) \text{ s.t. } \Lambda_{\varepsilon/2}(\varepsilon z) \cap \mathcal{C} \neq \emptyset \text{ and } \mu^\eta_{\mathcal{C}}(\Lambda_\varepsilon(\varepsilon z)) < x_\varphi) \\
&\qquad \leq 4(k/\varepsilon)^2 \mathbb{P}_\eta(\exists \mathcal{B} \in \mathscr{B}^\eta_{\Lambda_\varepsilon}(\varepsilon/2) \text{ with } ||\mu^\eta_{\mathcal{B}}||_{TV} < x_\varphi) \\
&\qquad \leq 4(k/\varepsilon)^2 \varphi.
\end{aligned} \tag{46}$$

By Theorem 13 we have that $\underline{\mu}^\eta(\delta) \to \underline{\mu}^0(\delta)$ in probability in the metric d_l for all $\delta > 0$ as $\eta \to 0$. This, combined with the tightness of $|\mathscr{C}^0_k(\delta)|$, (46) and the Portmanteau theorem, gives that

$$\begin{aligned}
&\mathbb{P}_0(\exists z \in \mathbb{Z}[\mathbf{i}], \exists \mathcal{C} \in \mathscr{C}^0_k(\delta) \text{ s.t. } \Lambda_{\varepsilon/2}(\varepsilon z) \cap \mathcal{C} \neq \emptyset \text{ and } \mu^0_{\mathcal{C}}(\Lambda_\varepsilon(\varepsilon z)) < x_\varphi) \\
&\qquad \leq 4(k/\varepsilon)^2 \varphi
\end{aligned} \tag{47}$$

for all $\varepsilon \in (0, \delta/10)$. We take the limit $\varphi \to 0$ in (47) and get

$$\mathbb{P}_0(\exists z \in \mathbb{Z}[\mathbf{i}], \exists \mathcal{C} \in \mathscr{C}^0_k(\delta) \text{ s.t. } \Lambda_{\varepsilon/2}(\varepsilon z) \cap \mathcal{C} \neq \emptyset \text{ and } \mu^0_{\mathcal{C}}(\Lambda_\varepsilon(\varepsilon z)) = 0) = 0, \tag{48}$$

which shows that $\text{supp}(\mu^0_{\mathcal{C}}) + \Lambda_\varepsilon \supseteq \mathcal{C}$ for all $\mathcal{C} \in \mathscr{C}^0_k(\delta)$ with probability 1 for each fixed $\varepsilon > 0$. Thus $\text{supp}(\mu^0_{\mathcal{C}}) \supseteq \mathcal{C}$ for all $\mathcal{C} \in \mathscr{C}^0$ with probability 1, and finishes the proof of the first statement of Theorem 14.

Since the proof of (43) is analogous to that of Lemma 17, we only give a sketch. Let $\mathcal{C} \in \mathscr{C}^0(\delta)$ with $\delta > 0$, and let $f : \mathbb{C} \to \mathbb{R}$ be a continuous function with compact support. Recall the definition (32) of $\mu^0_{\mathcal{C},n}$. We set

$$\bar{\mu}^0_{\mathcal{C},n,\psi} := \sum_{z\in\mathbb{Z}[\mathbf{i}]:\Lambda_{3\varepsilon/2}(\varepsilon z)\cap\mathcal{C}\neq\emptyset} \tilde{\mu}^0_{A(\varepsilon z,\varepsilon/2,\delta/2-\varepsilon),\psi}, \quad \text{with } \varepsilon = 3^{-n}.$$

Note that when we replace $\mu^0_{A(\varepsilon z,\varepsilon/2,\delta/2-\varepsilon)}$ by $\tilde{\mu}^0_{A(\varepsilon z,\varepsilon/2,\delta/2-\varepsilon),\psi}$ in the definition of $\mu^0_{\mathcal{C},n}$, we arrive at the measure $\bar{\mu}^0_{\mathcal{C},n,\psi}$. Thus, for any fixed $\varepsilon > 0$, Lemma 19 shows that $\bar{\mu}^0_{\mathcal{C},n,\psi}(f)$ and $\mu^0_{\mathcal{C},n}(f)$ are close to each other in L^2 when ψ is small. In particular, $\bar{\mu}^0_{\mathcal{C},n,\psi} \to \mu^0_{\mathcal{C},n}$ weakly in probability as $\psi \to 0$.

Arguments similar to those in the proof of Lemma 17 give that $\tilde{\mu}^0_{\mathcal{C},\psi}$ and $\bar{\mu}^0_{\mathcal{C},n,\psi}$ are close to each other in total variation distance (hence in Prokhorov distance as well) with high probability when ψ and $\varepsilon = 3^{-n}$ are both small.

By the proof of Theorem 13, $\mu^0_{\mathcal{C},n}$ is close to $\mu^0_{\mathcal{C}}$ in Prokhorov distance with high probability when n is large. Thus

$$\tilde{\mu}^0_{\mathcal{C},\psi} \approx \bar{\mu}^0_{\mathcal{C},n,\psi} \xrightarrow{\psi \to 0} \mu^0_{\mathcal{C},n} \xrightarrow{n \to \infty} \mu^0_{\mathcal{C}},$$

where the limits are in Prokhorov metric in probability, and $\tilde{\mu}^0_{\mathcal{C},\psi} \approx \bar{\mu}^0_{\mathcal{C},n,\psi}$ means that the Prokhorov distance between these measures is small with high probability when $\varepsilon = 3^{-n}$ and ψ are both small. Thus (43) follows, and Theorem 14 is proved. □

9.2 Conformal Invariance and Covariance

In this section we prove Theorem 4 and the stronger conformal covariance of Bernoulli percolation clusters as stated in Theorem 6.

Let us first restrict ourselves to critical site percolation on the triangular lattice. At the end of this section we will show how to obtain the weaker invariance of Theorem 4 from our general assumptions.

Recall Definition 1 of the restriction of a configuration to a bounded domain D.

Theorem 15. *For $\eta \geq 0$, let $\mathbb{P}_\eta$ denote the measure for critical site percolation on the triangular lattice. Let $D \subseteq \mathbb{C}$ be a domain and $f : D \to \mathbb{C}$ be a conformal map. The laws of $(f(\omega_{0,D}), f(L_{0,D}))$ and $(\omega_{0,f(D)}, L_{0,f(D)})$ coincide.*

The conformal invariance of the continuum loop process was proved in [13, Theorem 3, item 4]. The conformal invariance of the quad crossings follows immediately because of the measurability with respect to the loop process [19,31].

The construction of the continuum clusters and their measures was obtained in Sects. 5, 6, 7 and 8 by approximating the cluster by boxes $\Lambda_{\varepsilon/2}(z)$. In order to prove conformal invariance/covariance we would like to approximate the clusters by conformally transformed boxes $f(\Lambda_{\varepsilon/2}(z))$. More precisely, let $\phi > 0$ and $f : \Lambda_{1+\phi} \to \hat{\mathbb{C}}$ be a conformal map. We set $D = f(\Lambda_1)$ and $D' = f(\Lambda_{1+\phi})$. Let d_f denote the push-forward of the L^∞ metric on $\Lambda_{1+\phi}$. That is,

$$d_f(x, y) := ||f^{-1}(x) - f^{-1}(y)||_\infty$$

for $x, y \in D'$. Note that f is defined in an open neighborhood of Λ_1 because when we approximate the cluster measures using one arm measures, we need to consider annuli whose inner square is contained in Λ_1 but which are not completely contained in Λ_1.

Clearly, $(\Lambda_{1+\phi}, d_\infty)$ and (D', d_f) are isomorphic as metric spaces. Thus all the geometric constructions in Sect. 5 can be repeated for the clusters in D just by applying the map f. We denote these analogues of the objects by an additional 'f' subscript. Thus all the statements, apart from those in Sect. 5.1, remain valid if we keep the constants such as ε, δ unchanged, but add an additional subscript f in the objects appearing in the claims. Moreover, the bounds in Sect. 5.1 remain valid asymptotically, as $\eta \to 0$, if we use the transformed boxes $f(\Lambda_{\varepsilon/2}(z))$ to define the relevant events because of the conformal invariance of the scaling limit.

Next note that there is a positive constant $K = K(f)$ such that $|f'(u)| \in [1/K, K]$ for $u \in \Lambda_{1+\phi/2}$. Thus d_f and the L^∞-metric are equivalent on D. As above, we add a subscript 'f' for the metrics built from d_f. Thus $d_{H,f}$ and $d_{P,f}$ are equivalent to d_H and d_P respectively, where $d_{H,f}$ and $d_{P,f}$ are built on d_f.

We can obtain the clusters in D in two ways: via the square boxes $\Lambda_{\varepsilon/2}(z)$, that is, using the metric L^∞ in D, or via the transformed boxes $f(\Lambda_{\varepsilon/2}(z))$, that is, using the metric d_f. The equivalence of the metrics implies that these two approximations provide the same continuum clusters in the scaling limit.

Now notice that the scaling limit in D in terms of quad crossings is distributed like the image under f of the scaling limit in Λ_1, because of the conformal invariance of quad crossing configurations. This implies that the construction in D, using the transformed boxes $f(\Lambda_{\varepsilon/2}(z))$, gives clusters that have the same distribution as the images of the continuum clusters in Λ_1. This proves the following theorem.

Theorem 16. *For $\eta \geq 0$, let $\mathbb{P}_\eta$ denote the measure for critical site percolation on the triangular lattice. Let $\phi > 0$, $f : \Lambda_{1+\phi} \to \hat{\mathbb{C}}$ be a conformal map, and $D := f(\Lambda_1)$.*

Then the laws of $\mathscr{B}^0_D$ and $f(\mathscr{B}^0_{\Lambda_1})$ are identical, where

$$f(\mathscr{B}^0_{\Lambda_1}) := \{f(\mathcal{B}) \,:\, \mathcal{B} \in \mathscr{B}^0_{\Lambda_1}\}.$$

In addition to the convergence of arm measures, [19] contains a proof of the conformal covariance of these measures. The relevant result is Theorem 6.7 in [19], stated below.

Theorem 17. *For $\eta \geq 0$, let $\mathbb{P}_\eta$ denote the measure for critical site percolation on the triangular lattice. Let $D \subseteq \mathbb{C}$ be a domain and $f : D \to \mathbb{C}$ be a conformal map. Let $A \subset \mathbb{C}$ be a proper annulus with piecewise smooth boundary with $\overline{A} \subset D$. For a Borel set $B \subseteq f(D)$, let*

$$\mu^{0*}_{1,A}(B) := \int_{f^{-1}(B)} |f'(z)|^{2-\alpha_1} d\mu^0_{1,A}(z).$$

Then the laws of $\mu^0_{1,f(A)} = \mu^0_{1,f(A)}(\omega_{0,f(D)})$ and $\mu^{0*}_{1,A} = \mu^{0*}_{1,A}(\omega_{0,D})$ coincide.

The boundedness of f' discussed earlier implies that approximating the cluster measures in D by one-arm measures of annuli of the form $f(\Lambda_{\delta/2} \setminus \Lambda_{\varepsilon/2})$ provides the same limit as approximating the same measures by one-arm measures of annuli of the form $\Lambda_{\delta/2} \setminus \Lambda_{\varepsilon/2}$. Hence, one can carry out the arguments in the proof of Lemma 17 using one-arm measures of annuli of the form $f(\Lambda_{\delta/2} \setminus \Lambda_{\varepsilon/2})$. This observation and Theorem 17 imply the following result, where $\tilde{\mathscr{M}}^0_D$ denotes the collection of measures of all clusters in $\mathscr{B}^0_D$, and μ^{0*} is defined in Theorem 6.

Theorem 18. *For $\eta \geq 0$, let $\mathbb{P}_\eta$ denote the measure for critical site percolation on the triangular lattice. Let $\phi > 0$, $f : \Lambda_{1+\phi} \to \hat{\mathbb{C}}$ be a conformal map, and $D := f(\Lambda_1)$. Then the laws of $\tilde{\mathscr{M}}^0_D$ and $f(\tilde{\mathscr{M}}^0_{\Lambda_1})$ are identical, where*

$$f(\tilde{\mathscr{M}}^0_{\Lambda_1}) := \{\mu^{0*} : \mu^0 \in \tilde{\mathscr{M}}^0_{\Lambda_1}\}.$$

We conclude this section with a brief discussion of the proof of Theorem 4.

Proof of Theorem 4. The theorem follows from a straightforward modification of the arguments above, using the rotation and translation invariance and scaling covariance of the 1-arm measures under Assumptions 1–4, which follow easily from the proof of Theorem 9 (see also [19, Equation (6.1) and Proposition 6.4]). □

10 Proof of the Convergence of the Largest Bernoulli Percolation Clusters

Now we turn to the precise version and to the proof of Theorem 5.

Theorem 19. *Let $\mathbb{P}$ be a coupling such that $(\omega_\eta, L_\eta) \to (\omega_0, L_0)$ a.s. as $\eta \to 0$. Then for all $i \in \mathbb{N}$ the i-th largest cluster $\mathcal{M}^\eta_{(i)}$ converges in $\mathbb{P}$-probability to $\mathcal{M}^0_{(i)}$ as $\eta \to 0$, where $\mathcal{M}^0_{(i)}$ is a measurable function of (ω_0, L_0). In particular, $(\omega_\eta, L_\eta, \mathcal{M}^\eta_{(i)}) \to (\omega_0, L_0, \mathcal{M}^0_{(i)})$ in distribution. The same convergence holds for the measures $\mu^\eta_{\mathcal{M}^\eta_{(i)}}$.*

Let us start with some preliminary results. Recall the definition of collections of (portions of) clusters $\mathscr{B}^\eta_{\Lambda_1}(\delta)$ in Sect. 7.

Proposition 3 ([5, Proposition 3.2])**.** *Let $\delta \in (0, 1)$. For all $\varphi > 0$ there exist $\eta_0, \alpha > 0$ such that, for all $\eta < \eta_0$,*

$$\mathbb{P}_\eta(\exists \mathcal{B}, \mathcal{B}' \in \mathscr{B}^\eta_{\Lambda_1}(\delta) : \mathcal{B} \neq \mathcal{B}' : |\mu^\eta_{\mathcal{B}}(\Lambda_1) - \mu^\eta_{\mathcal{B}'}(\Lambda_1)| < \alpha) < \varphi.$$

Proof of Proposition 3. In [5] a proof for Proposition 3 was given for bond percolation on the square lattice, however the proof also works for other models, like site percolation on the triangular lattice, as noted in Remark (i) after Theorem 1.1 in [5].

□

Lemma 20 (Lemma 4.4 of [7]). *There are positive constants c, C such that for all $x, y > 0$*

$$\mathbb{P}_\eta(\exists \mathcal{B} \in \mathscr{B}^\eta_{\Lambda_1} : \mu^\eta_{\mathcal{B}}(\Lambda_1) > x \text{ and } \mathrm{diam}(\mathcal{B}) < y) < Cy^{-1} \exp(-cx/\sqrt{y})$$

for all $\eta < \eta_0 = \eta_0(x, y)$.

The next proposition follows easily from a combination of Lemma 20 and [7, Theorems 3.1, 3.3 and 3.6] (see also [5]).

Proposition 4. *Let $i \in \mathbb{N}$ be fixed. For all $\varphi > 0$ there exist $\delta > 0, \eta_0 > 0$ such that, for all $\eta < \eta_0$,*

$$\mathbb{P}_\eta(\exists j \leq i : \mathcal{M}^\eta_{(j)} \notin \mathscr{B}^\eta_{\Lambda_1}(\delta)) < \varphi.$$

Proof of Theorem 19. Let $i \in \mathbb{N}$ be fixed and $\mathbb{P}$ be a coupling such that $(\omega_\eta, L_\eta) \to (\omega_0, L_0)$ a.s. as $\eta \to 0$. First we show that the i-th largest clusters in the scaling limit can almost surely be defined as a function of the pair (ω_0, L_0). Then we show that the i-th largest cluster $\mathcal{M}^\eta_{(i)}$ in the discrete configuration ω_η converges to the i-th largest continuum cluster.

Let $m \in \mathbb{N}$. Theorems 12 and 13 show that the sequence of clusters $\underline{\mathcal{B}}^0_{\Lambda_1}(3^{-m})$ and their corresponding measures $\underline{\mu}^0(3^{-m})$ are a.s. well defined.

We define the *volume* of a continuum cluster $\mathcal{B} \in \mathscr{B}^0_{\Lambda_1}$ as $\mu^0_{\mathcal{B}}(\Lambda_1)$. Lemma 7 shows that the volumes of the clusters $\mathcal{B} \in \mathscr{B}^0_{\Lambda_1}(3^{-m})$ are a.s. finite. Moreover, Lemma 16, together with the tightness of the number of excursions in Λ_1 of diameter at least 3^{-m}, gives that $h^0(3^{-m}) := |\mathscr{B}^0_{\Lambda_1}(3^{-m})|$ is a.s. finite. Thus we can reorder the sequence of clusters $\underline{\mathcal{B}}^0_{\Lambda_1}(3^{-m})$ in decreasing order by their volume. We break ties in some deterministic way. However, we will see below that ties occur with probability 0. Let $\mathcal{M}^0_{(j)}(3^{-m})$ denote the j-th cluster in this new ordering.

Let $\varphi > 0$ be arbitrary and take α and η_0 as in Proposition 3. Then, for $\eta < \eta_0$,

$$\begin{aligned}
&\mathbb{P}(\exists \mathcal{B}, \mathcal{B}' \in \mathscr{B}^0_{\Lambda_1}(3^{-m}) : \mathcal{B} \neq \mathcal{B}', |\mu^0_{\mathcal{B}}(\Lambda_1) - \mu^0_{\mathcal{B}'}(\Lambda_1)| < \alpha/2) \\
&\quad \leq \mathbb{P}(\exists \mathcal{B}, \mathcal{B}' \in \mathscr{B}^\eta_{\Lambda_1}(3^{-m}) : \mathcal{B} \neq \mathcal{B}', |\mu^\eta_{\mathcal{B}}(\Lambda_1) - \mu^\eta_{\mathcal{B}'}(\Lambda_1)| < \alpha) \\
&\qquad + \mathbb{P}(\exists j \leq h^0(3^{-m}) : |\mu^\eta_{\mathcal{B}^\eta_j}(\Lambda_1) - \mu^0_{\mathcal{B}^0_j}(\Lambda_1)| > \alpha/4) \qquad (49) \\
&\quad \leq \varphi + \mathbb{P}(\exists j \leq h^0(3^{-m}) : |\mu^\eta_{\mathcal{B}^\eta_j}(\Lambda_1) - \mu^0_{\mathcal{B}^0_j}(\Lambda_1)| > \alpha/4).
\end{aligned}$$

The second term in the last line of (49) tends to 0 as $\eta \to 0$, since $h^0(3^{-m})$ is a.s. finite and $\underline{\mu}^\eta(3^{-m}) \to \underline{\mu}^0(3^{-m})$ in probability by Theorem 13. Since $\varphi > 0$ was arbitrary, this shows that

$$\mathbb{P}(\exists \mathcal{B}, \mathcal{B}' \in \mathscr{B}^0_{\Lambda_1}(3^{-m}) : \mathcal{B} \neq \mathcal{B}', |\mu^0_{\mathcal{B}}(\Lambda_1) - \mu^0_{\mathcal{B}'}(\Lambda_1)| = 0) = 0.$$

That is, with probability 1, there are no ties in the ordering of continuum clusters described above.

Now we show that, for all $j \leq i$,

$$\mathbb{P}(\exists m_0 \in \mathbb{N} \text{ s.t. } \mathcal{M}^0_{(j)}(3^{-m_0}) = \mathcal{M}^0_{(j)}(3^{-m}) \text{ for all } m \geq m_0) = 1. \qquad (50)$$

Consider the event

$$E = \{\exists j_0 \leq i : \nexists m_0 \in \mathbb{N} \text{ s.t. } \mathcal{M}^0_{(j_0)}(3^{-m_0}) = \mathcal{M}^0_{(j_0)}(3^{-m}) \text{ for all } m \geq m_0\}$$

and the events

$$E_n^m = \{\exists \mathcal{B} \in \mathscr{B}^0_{\Lambda_1}(3^{-m}) \text{ s.t. } \operatorname{diam}(\mathcal{B}) < 3^{-m+1} \text{ and } \mu^0_{\mathcal{B}}(\Lambda_1) > 1/n\}.$$

Note that

$$E \subset \bigcup_{n=1}^{\infty} \{E_n^m \text{ for infinitely many } m \in \mathbb{N}\}.$$

Theorems 11 and 13 and Lemma 20 imply that, for each $m, n \geq 1$, there is $\eta_0 = \eta_0(m, n)$ such that

$$\begin{aligned}\mathbb{P}(E_n^m) &\leq 2\mathbb{P}(\exists \mathcal{B} \in \mathscr{B}^\eta_{\Lambda_1} : \mu^\eta_{\mathcal{B}}(\Lambda_1) > 1/(2n) \text{ and } \operatorname{diam}(\mathcal{B}) < 6 \times 3^{-m}) \\ &\leq C 3^{m-1} \exp\Big(-\frac{c}{2\sqrt{6}n} 3^{m/2}\Big)\end{aligned}$$

for all $\eta \leq \eta_0$. Since $\sum_{m=1}^\infty 3^m \exp\Big(-\frac{c}{2\sqrt{6}n} 3^{m/2}\Big) < \infty$, it follows from the Borel–Cantelli Lemma that $\mathbb{P}(E_n^m$ for infinitely many $m \in \mathbb{N}) = 0$ for every $n \geq 1$. Hence $\mathbb{P}(E) = 0$, which proves (50).

For each $j \leq i$, we set $\mathcal{M}^0_{(j)} := \mathcal{M}^0_{(j)}(3^{-m_0})$, where m_0 is as in the event on the left hand side of (50). It remains to show that $\mathcal{M}^\eta_{(i)}$ converges in probability to $\mathcal{M}^0_{(i)}$, as well as the analogous statement for their measures. Let $\varepsilon, \alpha > 0$ and $m > 0$. First we check that

$$\begin{aligned}&\mathbb{P}(d_H(\mathcal{M}^\eta_{(i)}, \mathcal{M}^0_{(i)}) > \varepsilon) \\ &\qquad \leq \mathbb{P}(\exists j \leq i : \mathcal{M}^0_{(j)} \neq \mathcal{M}^0_{(j)}(3^{-m})) \\ &\qquad\quad + \mathbb{P}(\exists j \leq i : \mathcal{M}^\eta_{(j)} \neq \mathcal{M}^\eta_{(j)}(3^{-m})) \\ &\qquad\quad + \mathbb{P}(\exists \mathcal{B}, \mathcal{B}' \in \mathscr{B}^0_{\Lambda_1}(3^{-m}) : \mathcal{B} \neq \mathcal{B}', |\mu^0_{\mathcal{B}}(\Lambda_1) - \mu^0_{\mathcal{B}'}(\Lambda_1)| < \alpha) \quad (51) \\ &\qquad\quad + \mathbb{P}(\exists \mathcal{B}, \mathcal{B}' \in \mathscr{B}^\eta_{\Lambda_1}(3^{-m}) : \mathcal{B} \neq \mathcal{B}', |\mu^\eta_{\mathcal{B}}(\Lambda_1) - \mu^\eta_{\mathcal{B}'}(\Lambda_1)| < \alpha) \\ &\qquad\quad + \mathbb{P}(\exists k \leq h^0(3^{-m}) : |\mu^\eta_{\mathcal{B}^\eta_k}(\Lambda_1) - \mu^0_{\mathcal{B}^0_k}(\Lambda_1)| > \alpha/3) \\ &\qquad\quad + \mathbb{P}(\exists k \leq h^0(3^{-m}) : d_H(\mathcal{B}^\eta_k, \mathcal{B}^0_k) > \varepsilon),\end{aligned}$$

where $\mathcal{B}^\eta_k$ (resp. $\mathcal{B}^0_k$) is the k-th cluster of $\mathcal{B}^\eta_{\Lambda_1}(3^{-m})$ (resp. $\mathscr{B}^0_{\Lambda_1}(3^{-m})$) in the order used in the proofs of Theorems 11 and 13.

We justify (51) as follows. On the complement of the first two events on the right hand side of (51), the i-th largest clusters at scale η and 0 (i.e., in the scaling limit) have diameter at least 3^{-m}. On the complement of the third and fourth event on the right hand side of (51), the normalized volumes of the different clusters with diameter at least 3^{-m} are at least α apart at both scales η and 0. Thus, on the complement of the first five events on the right hand side of (51), the ordering according to their volume of the k largest clusters at scale η and 0 coincide; that is, for all $j \leq i$, there is a unique $k_j \leq h^0(3^{-m})$ such that $\mathcal{M}^\eta_{(j)} = \mathcal{B}^\eta_{k_j}$ and $\mathcal{M}^0_{(j)} = \mathcal{B}^0_{k_j}$. This, together with the last term in the right hand side of (51), proves (51).

Let $\varphi > 0$ be arbitrary. By (50) and Proposition 4, we find m and $\eta_0 > 0$ such that the first and second term on the right hand side of (51) are less than $\varphi/6$ for all $\eta < \eta_0$. Then we use the bounds in (49) and Proposition 3 and find $\alpha, \eta_1 > 0$ so that the third and fourth term on the right hand side of (51) are less than $\varphi/6$ for all $\eta < \eta_1$. Finally, we apply Theorem 13 to control the fifth term and Theorem 11 to control the sixth term, and deduce that $\limsup_{\eta\to 0} \mathbb{P}(d_H(\mathcal{M}^{\eta}_{(i)}, \mathcal{M}^{0}_{(i)}) > \varepsilon) < \varphi$. Since φ and ε were arbitrary, this shows that $\mathcal{M}^{\eta}_{(i)} \to \mathcal{M}^{0}_{(i)}$ in probability as $\eta \to 0$.

The proof for the convergence of normalized counting measures goes in a similar way: notice that if we replace the fifth term on the right hand side of (51) with

$$\mathbb{P}(\exists j \leq h^0(3^{-m}) : d_p(\mu^{\eta}_{\mathcal{B}^{\eta}_j}, \mu^{0}_{\mathcal{B}^{0}_j}) > \alpha/3),$$

then we get an upper bound for the probability $\mathbb{P}(\exists j \leq i : d_P(\mu^{\eta}_{\mathcal{M}^{\eta}_{(j)}}, \mu^{0}_{\mathcal{M}^{0}_{(j)}}) > \alpha/3)$. This completes the proof of Theorem 19. □

Acknowledgements. The work of the first author was supported in part by the Netherlands Organization for Scientific Research (NWO) through grant Vidi 639.032.916. The work of the second author was partly supported by NWO Top grant 613.001.403. The second author was at VU University Amsterdam while most of the research was carried out. The third author thanks NWO for its financial support and Centrum Wiskunde & Informatica (CWI) for its hospitality during the time when he was a PhD student, when the project was initiated. All three authors thank Rob van den Berg for fruitful discussions. The first author thanks Chuck Newman for his friendship and invaluable guidance during many years, and for being a constant inspiration.

References

1. Aizenmann, M., Barsky, D.J.: Sharpness of the phase transition in percolation models. Commun. Math. Phys. **108**(3), 489–526 (1987)
2. Beffara, V., Nolin, P.: On monochromatic arm exponents for 2D critical percolation. Ann. Probab. **39**(4), 1286–1304 (2011)
3. Benoist, S., Hongler, C.: The scaling limit of critical Ising interfaces is CLE(3). Ann. Probab. **47**(4), 2049–2086 (2019)
4. van den Berg, J., Conijn, R.: On the size of the largest cluster in $2D$ critical percolation. Electron. Commun. Probab. **17**(58), 1–13 (2012)
5. van den Berg, J., Conijn, R.: The gaps between the sizes of large clusters in 2D critical percolation. Electron. Commun. Probab. **18**(92), 1–9 (2013)
6. Borgs, C., Chayes, J.T., Kesten, H., Spencer, J.: Uniform boundedness of critical crossing probabilities implies hyperscaling. Random Struct. Alg. **15**(3–4), 368–413 (1999)
7. Borgs, C., Chayes, J.T., Kesten, H., Spencer, J.: The birth of the infinite cluster: finite-size scaling in percolation. Comm. Math. Phys. **224**(1), 153–204 (2001)

8. Camia, F.: Towards conformal invariance and a geometric representation of the 2D Ising magnetization field. Markov Process. Relat. Fields **18**, 89–110 (2012)
9. Camia, F., Garban, C., Newman, C.M.: Planar Ising magnetization field I. Uniqueness of the critical scaling limit. Ann. Probab. **43**(2), 528–571 (2015)
10. Camia, F., Garban, C., Newman, C.M.: Planar Ising magnetization field II. Properties of the critical and near-critical scaling limits. Ann. Inst. H. Poincaré Probab. Stat. **52**(1), 146–161 (2016)
11. Camia, F., Jiang, J., Newman, C.M.: Exponential decay for the near-critical scaling limit of the planar Ising model. In: Communications on Pure and Applied Mathematics (2017)
12. Camia, F., Jiang, J., Newman, C.M.: FK–Ising coupling applied to near-critical planar models. Stoch. Process. Appl., To appear
13. Camia, F., Newman, C.M.: Two-dimensional critical percolation: the full scaling limit. Commun. Math. Phys. **268**(1), 1–38 (2006)
14. Camia, F., Newman, C.M.: Critical percolation exploration path and SLE_6: a proof of convergence. Probab. Theory Relat. Fields **139**(3–4), 473–519 (2007)
15. Camia, F., Newman, C.M.: Ising (conformal) fields and cluster area measures. Proc. Nat. Acad. Sci. **106**(14), 5457–5463 (2009)
16. Chelkak, D.: Robust discrete complex analysis: a toolbox. Ann. Probab. **44**(1), 628–683 (2016)
17. Chelkak, D., Duminil-Copin, H., Hongler, C.: Crossing probabilities in topological rectangles for critical planar FK-Ising model. Electron. J. Probab. **21**, 1–28 (2016). Paper no. 5
18. Chelkak, D., Hongler, C., Izyurov, K.: Conformal invariance of spin correlations in the planar Ising model. Annal. Math. **181**, 1087–1138 (2015)
19. Garban, C., Pete, G., Schramm, O.: Pivotal, cluster and interface measures for critical planar percolation. J. Am. Math. Soc. **26**, 939–1024 (2013)
20. Gelfand, I.M., Vilenkin, N.Ya.: Generalized Functions, vol. 4. Academic Press, New York (1964)
21. Glimm, J., Jaffe, A.: Quantum Physics. Springer, New York (1981)
22. Grimmett, G.: Percolation, 2nd edn. Springer, Heidelberg (1999)
23. Grimmett, G.: The Random-Cluster Model. Springer, Heidelberg (2006)
24. Kemppainen, A., Smirnov, S.: Conformal invariance in random cluster models. II. Full scaling limit as a branching SLE. arXiv:1609.08527 (2016)
25. Kiss, D.: Large deviation bounds for the volume of the largest cluster in 2D critical percolation. Electron. Commun. Probab. **19**(32), 1–11 (2014)
26. Lawler, G.F., Schramm, O., Werner, W.: One-arm exponent for critical 2D percolation. Electron. J. Probab. **7**, 13 p. (2002). Paper no. 2
27. Miller, J., Sheffield, S., Werner, W.: CLE percolations. Forum Math. Pi **7**, E4, 102 p. (2017)
28. Nguyen, B.G.: Typical cluster size for two-dimensional percolation processes. J. Stat. Phys. **50**(3–4), 715–726 (1988)
29. Nolin, P.: Near-critical percolation in two dimensions. Electron. J. Probab. **13**, 1562–1623 (2008)
30. Schramm, O.: Scaling limits of loop-erased random walks and uniform spanning trees. Israel J. Math. **118**, 221–288 (2000)
31. Schramm, O., Smirnov, S.: On the scaling limits of planar percolation. Ann. Probab. **39** (2011). With an appendix by Christophe Garban

32. Smirnov, S.: Critical percolation in the plane: conformal invariance, Cardy's formula, scaling limits. Comptes Rendus de l'Académie des Sciences - Series I - Mathematics **333**(3), 239–244 (2001)
33. Wu, T.T.: Theory of Toeplitz determinants and the spin correlations of the two-dimensional Ising model. I. Phys. Rev. **149**, 380–401 (1966)

Stochastic Hydrogeology: Chuck Newman Had a Good Idea About Where to Start

Colin L. Clark[1] and Larry Winter[2(✉)]

[1] Program in Applied Mathematics, Department of Mathematics, University of Arizona, 617 N. Santa Rita Ave., Tucson, AZ 85721, USA
clark@math.arizona.edu

[2] Department of Hydrology and Atmospheric Sciences, 1133 E. James E. Rogers Way, Tucson, AZ 85721, USA
winter@email.arizona.edu

We dedicate this work to Chuck Newman on the occasion of his 70th birthday. Friend, mentor, and colleague

Abstract. Groundwater hydrology had reached a crisis by 1980 that Chuck Newman helped resolve. By then a large number of field observations had showed that contaminant transport was anomalously diffusive: the constant coefficient advection-diffusion equations that had been the basis for predictions of contaminant transport consistently underpredicted their spread on long time-scales. This anomalous diffusion appeared to have two sources. First, observations showed transport on small scales was not classically diffusive, as expected. Second, the material properties of natural porous media were not uniform in space or deterministic, but were in fact heterogeneous and uncertain. Both were fundamental assumptions of the classic approach. Chuck proposed and guided the development of an alternative model that views the large-scale behavior of transport in porous media as the convergent limit of a mesoscopic stochastic advection-diffusion process. This addressed both aspects of the crisis since the limit explained the transition from non-Fickian transport on mesoscopic scales to anomalous macroscale transport and the stochastic equation provided a means for quantifying uncertainty. We describe the development of the model, its setting in hydrology, and further developments in hydrology that sprung from it.

Keywords: Stochastic hydrogeology · Anomalous dispersion · Exponential operators · Binary random media · Anisotropy

Even long-time colleagues of Chuck Newman may forget that he was once a closet groundwater hydrologist. In the 1980's he worked with Shlomo Neuman and me on two papers that addressed a fact hydrogeologists hadn't expected: long-term field observations seemed to reveal that mass transport in groundwater aquifers was anomalously super-diffusive at early times. When nonreactive

V. Sidoravicius (Ed.): *Sojourns in Probability Theory and Statistical Physics - II*, PROMS 299, pp. 90–100, 2019.
https://doi.org/10.1007/978-981-15-0298-9_3

solutes were injected in typical aquifers and monitored downstream, their concentrations initially spread at rates greater than $t^{1/2}$. Hydrologists began calling the phenomenon "anomalous dispersion" because an initial inflationary period was not a feature of the constant coefficient advection-diffusion equations (ADE) commonly used to model mass transport in natural porous media. In addition to its scientific interest, the discovery of anomalous dispersion had practical implications for human health and the environment. Simple ADEs were the basis for engineered systems designed to protect groundwater quality, and it was disturbing to learn that predictions based on them might be in error.

An aquifer is a volume of natural porous medium that is saturated and large enough to be used by humans. Although aquifers are composed of linked networks of small pores embedded in solid matrices, they are generally observed, modeled, and managed on continuum scales of 1–100 km using simple linear partial differential equations. Since system parameters are defined at every point of an aquifer in principle, the modeling task would be easy if parameter values were known to a reasonable level of accuracy, but instead they are uncertain. The uncertainty arises because aquifers are spatially heterogeneous and hard to observe. As hydrogeologists sometimes say, "It's really dark down there."

Our work on anomalous dispersion coincided with a growing recognition that predictions of aquifer states needed to account for uncertainty. Hydrogeologists had long recognized that knowledge about the subsurface was incomplete. That had not seemed to seriously affect predictions of groundwater flow, but phenomena like anomalous dispersion seemed to demand more. What could be estimated from data, although with difficulty, were spatial statistics of parameters, in particular second moments like spatial correlations. Uncertainty was not just a nuisance that needed to be quantified to provide confidence intervals for predictions, it also seemed that stochastic models might provide a useful framework, and perhaps a necessary one, for scaling up phenomena that led to physical effects like anomalous dispersion.

At the risk of oversimplification, applications of groundwater hydrology fall into two linked categories: predictions of (1) flow and (2) mass transport. Both are usually based on Darcy's Law,

$$q = -K\nabla h$$

which states that the flux of water at a point in a porous medium, $q(x)$, which has units $[LT^{-1}]$, is proportional to the gradient of hydraulic head, h $[L]$ that is equivalent to a pressure. Henry Darcy, a Frenchman with an English name, called the constant of proportionality hydraulic conductivity, $K > 0$, in conformity with Fourier's recent theory of heat conduction [2]. Darcy based his theory on observations of flow through columns of sand (1–10 m) that he made around the middle of the 19th century in Dijon, France. In his experiments K $[LT^{-1}]$ was essentially constant, although in field applications it is actually $K(x)$. When combined with conservation of mass, Darcy's Law yields a diffusion equation for $h(x)$ that can be inverted to obtain K.

Once hydrologists discovered means of estimating K on larger scales, groundwater hydrology (or hydrogeology) was born. By now K is routinely estimated in the field by pumping a well at a constant rate and observing the drawdown of water in the pumped well and neighboring wells. Such in situ tests are conducted over support volumes ranging over lengths of 10–1000 m. Alternatively, smaller scale field estimates or even laboratory results are sometimes used to estimate K in the field. Typical values reported for conductivities of natural porous media range from 100–1000 ft/day in well-sorted gravel aquifers to 1 ft/day or less in relatively impermeable clay formations [1]. Darcy's Law with constant K proved adequate to predict flow through relatively homogeneous groundwater aquifers at field scales. These included most problems of groundwater water supply, among them the design of municipal and domestic well fields.

Although it is evident from road cuts and other visible cross-sections of natural porous media that conductivity must vary locally on small spatial scales, water supply applications did not seem to require any more complications than occasionally representing K as a tensor, still with constant elements, when material anisotropy had to be accounted for. If any further explanation of the apparent robustness of Darcy's Law seemed needed, the robustness was attributed to averaging flows over large volumes in aquifers that were locally heterogeneous, but not too much. Many groundwater supply systems tap aquifers that are composed of a reasonably uniform geological material, one that was laid down by more or less the same geological processes at more or less the same time.

Mass transport, the other major focus of groundwater hydrology, became a concern by the middle of the 20th century when it was realized industrial and agricultural contaminants were leaking from sources like waste storage sites into groundwater systems in concentrations, C, that were potentially dangerous to health and the environment. The units of concentration are density, mass per unit volume $[ML^{-3}]$. Aquifers were large enough to assume zero boundary conditions applied at infinity and in many cases the initial spill occurred over a short enough period of time in a small enough area to be modeled as an impulse injection at a point. Transport was initially modeled with constant coefficient ADEs,

$$\frac{\partial C}{\partial t} = \nabla \cdot D\nabla C - \nabla \cdot vC.$$

Computational resources were such that the closed-form Gaussian solutions of such problems were extremely convenient. The diffusive coefficient of groundwater flow, D $[L^2T^{-1}]$ generally depends on mechanical dispersion through networks of pores whose length scales are too small to observe in the field, but are generally much larger than the scale of molecular diffusion. The advective component came from solving Darcy's Law with a given K. When divided by π $(0 < \pi < 1)$ the (dimensionless) porosity of a piece-wise continuous porous medium, the steady-state Darcy flux, $q(x)$, yielded the pore, or seepage, velocity, $v(x) = q(x)/\pi$, which is the velocity at which mass is transported. The relative importance of the diffusive and advective components can be compared using the dimensionless Peclet number, $Pe = ||v||\lambda/D$ with λ a fixed characteristic length, e.g., the correlation length of v. When Pe is large, transport is dominated by advection and to a good approximation $D = \delta||v||$. The dispersivity parameter δ $[L]$, is proportional to the reciprocal of Pe.

The need to monitor accidental releases of hazardous chemical species, and the approaching need to store radioactive materials, soon required observations of transport on space-time scales that had previously been unknown in groundwater studies, and it was not long before early-time anomalous diffusion was observed [8]. It was also observed that contaminant plumes seemed to settle down at later times and behave asymptotically like Gaussian plumes. This pair of observations was attributed to a kind of limit behavior: incomplete sampling of a heterogeneous conductivity field at early times evolving to less biased sampling as the plume progressed. A complementary physical explanation was that some parts of a plume might initially follow fast paths that eventually petered out. Thus it seemed two separable scales of space-time might be relevant to transport: a small scale where transport initially responds to local heterogeneities of conductivity and is not Gaussian, converging to a larger scale where transport is Gaussian.

In response a number of groups began in the 1970's and 1980's to develop what eventually became known as stochastic hydrogeology. These included Gedeon Dagan and his students and colleagues at Tel Aviv University, Lynn Gelhar and his colleagues and students at MIT and NM Institute of Technology, Shlomo Neuman's group at University of Arizona, and probably earliest, Georges Matheron and his students and colleagues in France, particularly Ghislain de Marsily. Although these groups took different routes to stochastic hydrogeology, they shared an orientation that was based on treating the basic equations, specifically Darcy's Law and the ADE, as stochastic PDEs with coefficients that were random fields on fine scales of resolution. The problem was to use such techniques to bound uncertainty and, in the case of anomalous diffusion, to understand the physical process by taking an average and scaling it up. The scale-up problem had practical importance because most estimates of hydrogeological system states occurred at aquifer scales much larger than the characteristic lengths of the statistics of system parameters.

As noted, anomalous diffusion seemed especially suitable for an asymptotic approach because the observed physical system seemed to have two distinct scales: a small initial scale where dispersion was anomalous and a large scale that appeared to be classically Gaussian. Since the proposed dynamics for each scale were formally similar, both being advective-dispersive, it seemed reasonable that the larger scale system might be obtained by taking a space-time limit of the smaller scale equation and equating coefficients. It would also be necessary to convert the uncertain small-scale system, which depended on stochastic seepage velocities, to one that was deterministic prior to taking the limit.

An adequate theoretical foundation required solving two problems: logically the first was to demonstrate that the limit of such a small-scale uncertain system was actually Gaussian, and the second was to derive the coefficients of the asymptotic system by scaling the small-scale system $x \times t$ up to large-scale space and time, $\chi \times \tau$. The problems were solved in reverse order with two groups deriving expressions for the coefficients a few years before Albert Fannjiang and George

Papanicolaou determined conditions under which the limit could be taken after averaging [3].

In 1983 Lynn Gelhar and Carl Axness derived large-scale coefficients for transport by an incompressible fluid using spatial (ergodic) averaging and Fourier techniques [4]. Their treatment of time was ad hoc. A year later Chuck, Shlomo, and I published asymptotic coefficients for a general compressible fluid by rescaling space and time so that $t = \tau/\lambda$ and $x = \chi - \bar{v}\tau/\sqrt{\lambda}$ where λ is large [9]. In addition to scaling, this eliminates the advective part from the large-scale equation and replaces the initial impulse with $C_0(\chi) = C_0(\sqrt{\lambda}x)$ but does not affect the large-scale dispersion tensor, $\bar{D}$. We could then write the asymptotics as a limit of exponential operators

$$\lim_{\lambda \to \infty} E[e^{t(A+\varepsilon B)-\sqrt{\lambda}t(\bar{v}\cdot\nabla)}] = e^{t(\nabla\cdot\bar{D}\nabla)}$$

applied to the rescaled initial condition on the left and a delta function on the right. E is the statistical expectation operator. The small-scale velocity vector $V(x) = \mu + \varepsilon U(x)$ is decomposed into a constant mean, μ, and a zero-mean stationary field $U(x)$. It is convenient to think of $\mathrm{var}[U] = 1$ with $\varepsilon > 0$ a small dimensionless parameter measuring fluctuations in velocity. Incidentally, and parenthetically, there was also usually uncertainty about initial and boundary conditions, but their effect on perturbations and Monte Carlo simulations—the two preferred approaches to quantifying uncertainty in groundwater hydrology—is additive while the effects of parameters are multiplicative and introduce a closure problem, so most analyses focused on the more difficult problem of parametric uncertainty.

The deterministic part of the small-scale differential operator is $A = (\nabla \cdot D\nabla)/2 + \sqrt{\lambda}\mu \cdot \nabla$ and the stochastic part is $B = \sqrt{\lambda}U(\sqrt{\lambda}x) \cdot \nabla$. Expanding both sides of the operator equation in terms of ε, taking expectations and limits as indicated above, and equating coefficients yields zeroth order terms for the asymptotic parameters that equal the small-scale means μ and D, the first-order terms are zero because they are proportional to $E[U] = 0$, and the second order terms depend on spatial correlations of U, specifically the cross-power spectrum of U. The analysis was formal in the sense that we assumed perturbation expansions for the asymptotic parameters $\bar{D}$ and $\bar{v}$ existed. When restricted to incompressible fluids, our results were identical to second order with those of Gelhar and Axness, which was a bit surprising since the methods were so different (asymptotic limits of exponential semi-groups of random differential operators vs. spatial averaging of an ADE).

It was Chuck's idea to write the equations in terms of exponential operators and to take the scaling limit. I turned the crank, which was large and moved a lot of parts. I was astounded by the method Chuck had suggested, and working on it was, as it should have been, the seminal experience in my graduate education as an applied mathematician. It had never occurred to me that an exponential operator could be formally treated like an ordinary exponential, and there was not much literature available for guidance at that time. Shlomo provided the hydrology that was the core of our second paper [7]. Since we had second order approximations for the large scale advective and dispersive parts, and respectively, we could estimate hydrologic dispersivities, δ, using $Pe = \lambda\bar{v}/\bar{D}$.

After getting me off on the right foot, Chuck left the country for the better part of a year and Shlomo did too. Chuck had set me to work on generalized functions and Feller, and I found Stakgold. At that stage I had worked out the expansions of the exponential operators and was starting to think about averaging and the limits. This was 1980–1981 so communication was a bit of a challenge. Not only was there no internet, word processors didn't exist and pdf files were unknown. Every few weeks I summarized my progress in about 10 pages of handwritten notes, mimeographed them, and sent the copies to Chuck and Shlomo. I received comments some time later. If I was in doubt about what to do next, I said so and usually Chuck replied pretty quickly. Otherwise it could take a while. In retrospect, it was a great way to interact with a couple of marvelous mentors. Since they weren't right around the corner, I couldn't go to them for a solution to every detail or even confirmation of what I'd done. In most cases I worked things out and saw what happened. If the result was progress, I went ahead from there and if it wasn't, I wrote Chuck. As time went by I had to write less, but Chuck's hand is all over that dissertation. I saved my notes and letters, for many years, and I still have some of the notes I sent Chuck and Shlomo, but I regret to say I've lost their replies. I would never have had the courage or imagination to tackle that problem without Chuck, or later, the hydrology without Shlomo.

The upshot was that we contributed a useful element to a second-order theory of anomalous dispersion in stationary velocity fields, and I got a PhD from Chuck in Applied Math with a minor in Hydrology supervised by Shlomo. Shlomo and I extended these and related ideas to more complicated settings later, and Chuck dabbled in hydrology for a while longer by working on our second paper. Eventually Shlomo developed his method of moment equations for approximating the moments of solutions of stochastic versions of the basic equations of groundwater hydrology, and he made many important contributions to the hydrologic community's understanding of system parameters that extend across continuous ranges of scales.

Research in groundwater hydrology was dominated by uncertainty analyses based on stationary parameter fields (primarily conductivity) for the next decade or two, but after awhile problems with the models based on stationarity became obvious. Most relevant to this paper, some important hydrologic systems are not stationary on scales of interest to hydrologists. In particular, parameters of some systems vary on more than one separable scale of interest. Standard theories like ours and Gelhar and Axness can be adjusted for continuous spatial trends that exist in some porous media, but not for separable scales. One thing hasn't changed: practitioners of groundwater hydrology still want to upscale flow and transport models so that they only depend on constant effective coefficients, in particular an effective conductivity, K_e.

Among the most obvious examples of systems that vary over more than one separable spatial scale are aquifers composed of multiple disjoint volumes of different materials, for instance clay inclusions in a gravel or sand aquifer. A couple of simple examples illustrate the problem. Many porous media exhibit parallel bedding where beds alternate between relatively permeable and impermeable materials that are more or less uniform within beds, for instance the upper levels of Grand Canyon where horizontal bedding is obvious. Hydrogeology is full of such examples, although few are as dramatic. When beds are aligned parallel to the overall direction of a uniform flow, the effective conductivity of the aquifer is the arithmetic mean of the conductivities. If the medium is composed of two materials that (1) occupy volume fractions Q and $(1 - Q)$ and (2) have constant conductivities κ_1 and κ_2, the average conductivity is

$$\kappa = Q\kappa_1 + (1 - Q)\kappa_2$$

regardless of the number of spatial dimensions. However, that does not take account of geometry. If for instance, parallel layers are perpendicular to the direction of flow, the effective mean is dominated by the less permeable material(s), and effective conductivity is harmonic. Matheron [6] provided important bounds on K_e when he showed it is always bounded above by the arithmetic mean and below by the harmonic mean no matter the number of spatial dimensions or the specific properties of the spatial distribution of the materials. More than 50 years later, however, the effect that less regular spatial distributions of materials have on average conductivity is still largely an open question, so I want to close by discussing some results that one of my students, Colin Clark, has recently obtained for effective conductivities in composite porous media. In Colin's media the spatial distribution of sub-volumes composed of two materials is fairly general, certainly far more general than the regular bedding of parallel planes that provide the bounds for Matheron's theorem. Extensions to more than two materials are straightforward.

Many multi-material hydrologic systems can be characterized by two scales of uncertainty: (1) a large-scale random process, $M(x)$, defines the spatial distribution of disjoint material sub-volumes, or units, and (2) another process $K_m(x)$ defines the distribution of conductivity within volumes composed of specific materials m [10]. As noted, we are assuming there are just two kinds of materials, μ and μ'. The spatial distribution of the two units can be defined by joint probabilities

$$P[M(x_1) = m_1, \ldots, M(x_n) = m_n],$$

that points x_i are in units of given material $m_i = \mu$ or μ'. In general $K_m(x)$ is a stationary random process conditioned on material type with mean and variance κ_i and σ_i^2. The joint distribution of M and K is $p(k, m) = p(k|m)P[M = m]$. In this discussion, where the focus is on the effects of the spatial distribution of materials, it's also convenient to suppose $K_m(x) = \kappa_m$, a constant, so $P[Km(x) = \kappa_m] = P[M(x) = m]$. In that case, the unconditioned variance of conductivity

$$\sigma^2 = P[M(x) = \mu]P[M(x) = \mu'](\kappa_1 - \kappa_2)^2$$

can still be quite large because expected conductivities in different materials often differ by orders of magnitude, for instance the afore-mentioned ranges of conductivities for clay and sand.

Colin fills in many details of Matheron's theory for a large class of realistic aquifers with a computational science experiments. He generates a large number of realizations (148,500 to be precise) of two-material porous media and simulates flow through each of them. Each flow simulation is based on numerically solving Darcy's Law coupled with conservation of mass. Effective conductivities, κ_E, are determined by solving Darcy's Law for steady-state flow in realizations of composite media defined over rectilinear domains. Fixed head boundary conditions, $h_{\text{Top}} > h_{\text{Bottom}}$ are set at the top and bottom of the domain and no-flow conditions are maintained on lateral boundaries, so flow is predominately downward. Colin can then solve for κ_E by inverting Darcy's Law.

The sample porous media come from thresholded Gaussian topographies defined over a finite domain that corresponds to the volume of an aquifer [5]. A realization of a topography is created by convolving an initial field of i.i.d. random variables with a kernel that determines the field's correlation structure. A fixed threshold is applied everywhere on the topography, and points of the domain that correspond to elevations above the threshold are assigned to the less conductive material, μ_1, while points below the threshold are in the more conductive material, μ_2 (Fig. 1). Conductivities within the materials are constant κ_1 and κ_2 respectively. As the threshold is lowered the volume occupied by the less conductive material expands and eventually percolates across the domain. Percolation occurs at a critical volume fraction of $Q_{\text{crit}} = 0.5$ in 2D and about $Q_{\text{crit}} = 0.11$ in 3D. The following discussion is based on Colin's 3D results.

Fig. 1. Realizations of porous media created by Gaussian thresholding. White indicates permeable material, μ_2. Thresholds range low to high from left-right.

Colin generates 100 realizations of a random topography defined by a symmetric kernel whose support is small compared to the aquifer volume. He applies 99 different thresholds corresponding to volume fractions of the less conductive material of $0.01, \ldots, 0.99$ to each realization. Each threshold defines a random geometry of the two materials with relative volumes given by the value of the threshold. For each of the random geometries, he then creates 15 different conductivity fields by assigning the conductivity of the more conductive material to unity ($\kappa_2 = 1$), and assigning the conductivity of the less conductive material to a fixed value in the set $\kappa_1 \in \{5 \cdot 10^{-1}, 2 \cdot 10^{-1}, 1 \cdot 10^{-1}, \ldots, 1 \cdot 10^{-5}\}$. It can be seen that $\kappa_E \leq \kappa_2$ with equality occurring only if $\kappa_1 = \kappa_2$, and also $\kappa_E = 0$ if $\kappa_1 = 0$ and the volume fraction $Q_1 < Q_{\text{crit}}$. Otherwise $\kappa_2 > \kappa_E > \kappa_1$.

Behavior of a typical example is shown by a semi-log plot of κ_E against Q_2, the volume fraction of the more conductive material (Fig. 2). The dashed line indicates the location of Q_{crit} for this realization. In the region around Q_{crit} simulations transition from flow dominated by the low permeability material on the left to the high permeability material on the right. Flow in the transition region is subject to small-scale variations in direction where the fluid minimizes energy depending on whether it is cheaper to cross obstructions posed by the less permeable material or to go around them. A large part of this part of Colin's thesis is dedicated to analyzing that choice and to modeling κ_E within the transition region. It is known that the fluid follows power laws on either side of the transition region, and Colin's simulations illustrate that.

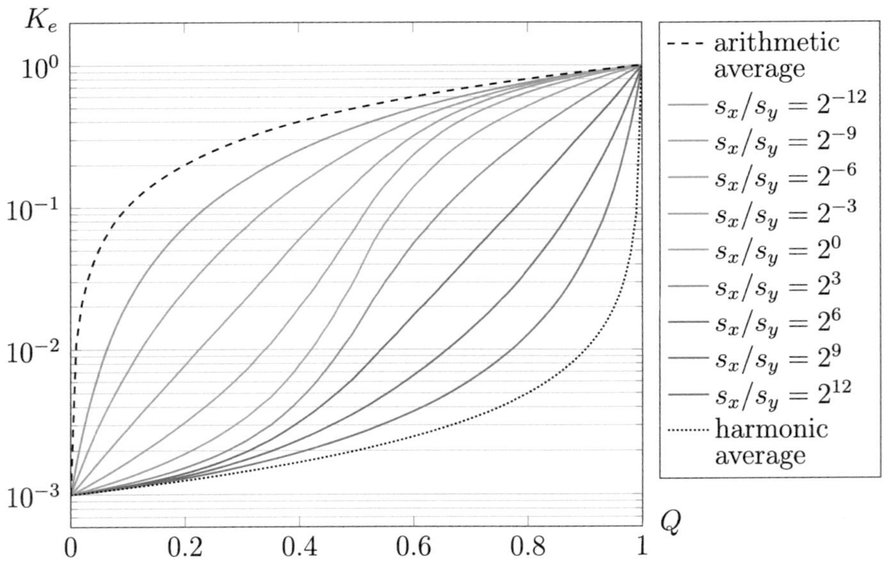

Fig. 2. Influence of Volume Fraction on Effective Conductivity. Curves depend on conductivity of the less conductive material, κ_1

Colin's results also provide insight into Matheron's theory of conductivity (Fig. 3). The results are for simulations in 2D with $\kappa_1 = 10^{-3}$. Now $Q_{\text{crit}} = 0.5$. In this case the generating kernels of the topographies are anisotropic with axes s_x and s_y in the vertical (x) and horizontal (y) directions respectively, so material distributions are anisotropic as well. The anisotropies induce anisotropic spatial correlations. The dashed lines above and below the simulations correspond to Matheron's envelope of conductivity fields with horizontal parallel bedding where

κ_E is arithmetic (above) and κ_E is harmonic (below). The figure indicates that flow to the left of Q_{crit} is quasi-harmonic and flow to the right is quasi-arithmetic, which could be expected from the earlier discussion of the extreme cases of perfect vertical bedding vs. perfect horizontal bedding. When conductivity fields are approximately isotropic, there appears to be a gradual transition from harmonic to arithmetic behavior. Near $Q_2 \sim 0$ there is not much high permeability material, so κ_E is generally small. More simulations are close to the harmonic envelope when anisotropies are aligned across the general direction of flow, but are relatively far from the arithmetic envelope when anisotropies are aligned with the general flow. This is in keeping with the fact that conductivity is dominated by low permeability layers when flow is perpendicular to the layering. Similar, but opposite comments apply near $Q_2 \sim 1$ due to the apparent symmetry. Effective conductivity of roughly isotropic fields jumps near Q_{crit} most likely due to percolation/non-percolation of the less conductive material. More anisotropic materials appear less affected by percolation. In addition to filling in previously unsuspected features of the behavior of κ_E these kinds of details may also be useful in the design of engineered filters.

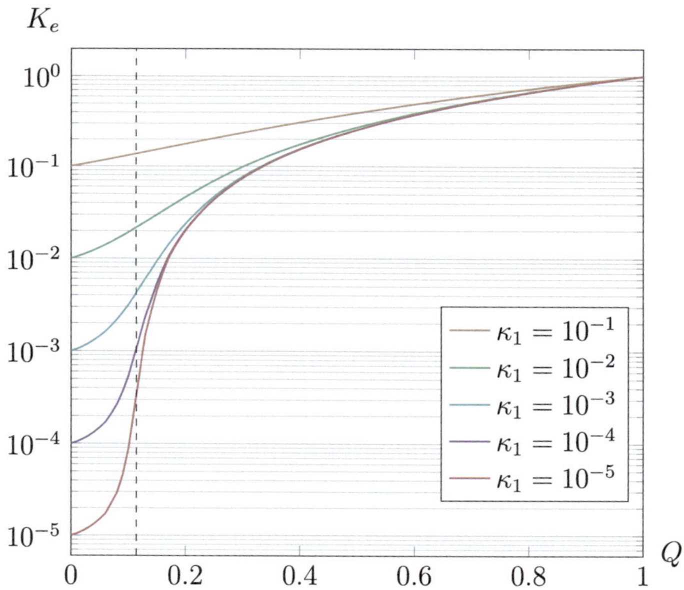

Fig. 3. Effective conductivity for anisotropic fields

I'm grateful to have had the opportunity to present the work of an outstanding student as I write this note. It sort of closes a circle. I'm equally grateful to have had the chance to reflect on the path Chuck set me on. And to express my gratitude. Chuck, and Shlomo too, didn't just provide a technical foundation for my career—although they surely did that—they exemplified the commitment that science requires of its practitioners and the opportunities it gives. I'm not a Chuck, but in a lifetime of knowing him I've learned to take pleasure in my work for its challenges and to take it seriously. In particular, I've had joy in my students and the pure fun of solving puzzles that matter. I can't imagine what I'd be doing if I hadn't met my mentors or students, but I'm pretty sure it wouldn't be this.[1]

References

1. Bear, J.: Dynamics of Fluids in Porous Media. Elsevier, New York (1962)
2. Darcy, H.: Recherches expérimentales relatives au mouvement de l'eau dans les tuyaux. Mallet-Bachelier, Paris (1857)
3. Fannjiang, A., Papanicolaou, G.: Convection enhanced diffusion for random flows. J. Stat. Phys. **88** (1998)
4. Gelhar, L.W., Axness, C.L.: Three-dimensional stochastic analysis of macrodispersion in aquifers. Water Resour. Res. **19**(1), 161–180 (1983)
5. Hyman, J.D., Winter, C.L.: Stochastic generation of explicit pore structures by thresholding Gaussian random fields. J. Comput. Phys. **277**, 16–31 (2014)
6. Matheron, G.: Éléments pour une théorie des milieux poreux. Masson et Cie, Paris (1967)
7. Neuman, S.P., Winter, C.L., Newman, C.M.: Stochastic theory of field-scale Fickian dispersion in anisotropic porous media. Water Resour. Res. (1987)
8. Sudicky, E., Illman, W.: Lessons learned from a suite of CFB Borden experiments. Groundwater **49**(5), 630–648 (2011)
9. Winter, C., Newman, C.M., Neuman, S.: Perturbation expansion for diffusion in a random velocity field. SIAM J. Appl. Math. **44**, 411–424 (1984)
10. Winter, C.L., Tartakovsky, D.M.: Mean flow in composite porous media. Geophys. Res. Lett. **27**(12), 1759–1762 (2000)

[1] Personal remarks in this article are by Larry Winter.

Superlinearity of Geodesic Length in 2D Critical First-Passage Percolation

Michael Damron[1(✉)] and Pengfei Tang[2]

[1] School of Mathematics, Georgia Institute of Technology, 686 Cherry Street, Atlanta, GA 30332, USA
mdamron6@math.gatech.edu

[2] Department of Mathematics, Indiana University, Bloomington, 831 East 3rd Street, Bloomington, IN 47405, USA
tangp@indiana.edu

Abstract. First-passage percolation is the study of the metric space $(\mathbb{Z}^d, T)$, where T is a random metric defined as the weighted graph metric using random edge-weights $(t_e)_{e\in\mathcal{E}^d}$ assigned to the nearest-neighbor edges $\mathcal{E}^d$ of the d-dimensional cubic lattice. We study the so-called critical case in two dimensions, in which $\mathbb{P}(t_e = 0) = p_c$, where p_c is the threshold for two-dimensional bond percolation. In contrast to the standard case ($<p_c$), the distance $T(0, x)$ in the critical case grows sub linearly in x and geodesics are expected to have Euclidean length which is superlinear. We show a strong version of this super linearity, namely that there is $s > 1$ such that with probability at least $1 - \mathrm{e}^{-\|x\|_1^c}$, the minimal length geodesic from 0 to x has at least $\|x\|_1^s$ number of edges. Our proofs combine recent ideas to bound T for general critical distributions, and modifications of techniques of Aizenman–Burchard to estimate the Hausdorff dimension of random curves.

Keywords: Percolation · Critical FPP · Fractal dimension · Geodesic

1 Introduction

1.1 Main Result

We study critical first-passage percolation (FPP) in two dimensions. This is a special case of general FPP, which is a stochastic growth model introduced in the '60s by Hammersley and Welsh [7]. On this occasion of Chuck Newman's 70-th birthday, we honor his deep and extensive contributions to FPP. Through both studying standard FPP and introducing rotationally-invariant models, he brought many new concepts into the subject including curvature bounds, geodesic forests, and directional Busemann functions. His work is the foundation for many of the major developments in the last twenty years.

Research is supported by NSF grant DMS-0901534 and an NSF CAREER award.

V. Sidoravicius (Ed.): *Sojourns in Probability Theory and Statistical Physics - II*, PROMS 299, pp. 101–122, 2019.
https://doi.org/10.1007/978-981-15-0298-9_4

The setup is as follows: on the square lattice $\mathbb{Z}^2$ with nearest-neighbor edges $\mathcal{E}^2$, we assign i.i.d. nonnegative passage times $(t_e)_{e \in \mathcal{E}^2}$ to the edges and define the induced random metric

$$T(x, y) = \inf_{\pi : x \to y} T(\pi), \text{ for } x, y \in \mathbb{Z}^2,$$

where the infimum is over all lattice paths π from x to y, and $T(\pi) = \sum_{e \in \pi} t_e$. If none of the t_e's are zero, then T is a metric (generally a pseudometric), and FPP is the study of geometric and probabilistic properties of the metric space $(\mathbb{Z}^2, T)$.

Instead of assuming that no t_e's are zero, one typically assumes that the common distribution function F of the weights does not give too much mass to zero: $F(0) < 1/2$, as $1/2$ is the critical threshold for Bernoulli percolation in two dimensions, and this ensures that a.s. there is no infinite component of zero-weight edges (edges which we can traverse in zero time). Under this assumption and some mild integrability constraint, there is a type of law of large numbers for T, called the shape theorem, which states that $T(0, x)$ grows linearly as $\|x\|_1 \to \infty$: there is a deterministic norm g on $\mathbb{R}^2$ such that a.s.,

$$\limsup_{\|x\|_1 \to \infty} \frac{|T(0, x) - g(x)|}{\|x\|_1} = 0. \tag{1}$$

So on large scales, the metric T is comparable to the Euclidean one.

The focus of this paper is geodesics, specifically their (Euclidean) lengths. A geodesic from x to y is a minimizer for T: a lattice path π from x to y with $T(\pi) = T(x, y)$. It has been shown [18] that for any F, a.s. there is a geodesic between any x and y, although uniqueness of geodesics is equivalent to continuity of F. In the general case stated above, $F(0) < 1/2$, it is known that the comparability of T to the Euclidean norm extends in a sense to geodesics, which have a linear number of edges. Specifically, [2, Theorem 4.6] if $F(0) < 1/2$, then there are $c_1, c_2 > 0$ such that for any x,

$$\mathbb{P}(m(x) \geq c_1 \|x\|_1) \leq c_1 \mathrm{e}^{-c_2 \sqrt{\|x\|_1}}, \tag{2}$$

where $m(x)$ is the maximal number of edges in any geodesic from 0 to x.

If $F(0) > 1/2$, there is a.s. an infinite component of zero-weight edges, and $T(0, x)$ is stochastically bounded in x, so the function g in (1) is identically zero. In this case, one can also show that the minimal length geodesic between two points has a linear number of edges [19, Theorem 4]. The so-called critical case, when $F(0) = 1/2$, is considerably more complicated, and has only recently been significantly explored. Although there is no infinite cluster of zero-weight edges, the clusters are large enough to force $g \equiv 0$. The precise behavior of $T(0, x)$ as $\|x\|_1 \to \infty$ was quantified in [4], with necessary and sufficient conditions on F for stochastic boundedness of $T(0, x)$ in x (and whether boundedness indeed holds depends on F in the critical case, as discovered by Zhang [20]). Because geodesics can take paths in large critical zero-weight clusters, and these clusters have irregular structure, Kesten [9, p. 259] was led to ask a version of the following (see also [17, p. 1029]):

Question 1. In the critical case, is there $s > 1$ such that a.s., $N_{0,x} \le \|x\|_1^s$ holds for only finitely many $x \in \mathbb{Z}^2$?

In this question, we are using the following notation. For $x, y \in \mathbb{Z}^2$, $N_{x,y}$ is the minimal number of edges in any geodesic between x and y.

In this paper, we give a positive answer to this question, with a stretched exponential estimate similar to (2). From this point forward, we assume that F is critical:

$$F(0^-) = 0 \text{ and } F(0) = 1/2. \tag{3}$$

Theorem 1. *Assuming* (3), *there exist* $c > 0, s > 1$ *such that for all nonzero* $x \in \mathbb{Z}^2$,

$$\mathbb{P}(N_{0,x} \le \|x\|_1^s) \le (1/c) \exp\left(-\|x\|_1^c\right). \tag{4}$$

We end this section with various remarks about the main result. First, our proof works on lattices where near-critical percolation estimates hold, and this includes most regular planar lattices, including edge or site FPP on the hexagonal lattice, square lattice, triangular lattice, etc. The main difference is that one must assume that $F(0) = p_c$, where p_c is the critical threshold for percolation on that lattice. Next, it is important to point out that using the work of Aizenman–Burchard, we can give a simple proof for Question 1 (see Sect. 1.3), but this argument only gives a small polynomial decay of the probability in (4). Our main inequality is sufficient (but polynomial decay is not) to, for example, bound the length of all geodesics simultaneously between points sufficiently far apart in a box.

Last we briefly remark on the proof; a full outline of it appears in Sect. 1.3. The strategy is to combine a block argument from Pisztora [15] with the Aizenman–Burchard technique. The block argument is quite similar to [15]. The main difficulty is in the Aizenman–Burchard technique: the hypothesis of [1] does not hold for our model. Their method relies on a strong independence assumption: to apply their theorem one would need to know that there is $\rho < 1$ such that for any number of thin cylinders $C_1, \ldots, C_k$ which are sufficiently separated, the probability that a geodesic crosses all of these cylinders in the long direction (has a "straight run" in each cylinder) is at most ρ^k. Under this assumption, we could copy their arguments to conclude that straight runs are sufficiently sparse globally to deduce a superlinear lower bound for geodesic length. Although FPP is built on i.i.d. weights, segments of geodesics are highly correlated, and such an independence assumption is not obviously true (and is actually false in the non-critical case). So we need to show differently that hierarchies of nested cylinders cannot contain too many straight runs by geodesics. The approach is to show that if geodesics do cross too many such cylinders, there is a high probability that many of these cylinders are "slow" (in a sense described by near-critical percolation paths) and force the passage time of geodesics to be large. We combine this with new upper bounds on passage times of geodesic segments using ideas from [4] to conclude that straight runs are sparse.

1.2 Notation and Tools from Percolation

We will couple the FPP model to various percolation models. To do this, we let (ω_e) be a collection of i.i.d. uniform $(0,1)$ random variables and $t_e = F^{-1}(\omega_e)$, where F^{-1} is the generalized inverse

$$F^{-1}(t) = \sup\{s : F(s) < t\},\ t \in (0,1).$$

Then the variables (t_e) are i.i.d. with distribution F. For $p \in [0,1]$, an edge e is called p-open if $\omega_e \le p$ and p-closed otherwise. A path is a sequence of edges (or their endpoints, or both) such that each consecutive pair of edges shares an endpoint, and a circuit is a path which starts and ends at the same point. If Γ is a path, we write $\#\Gamma$ for the number of edges in it. For $n \ge 1$, the box $B(n)$ is defined as $[-n,n]^2$.

Next we define the dual lattice, which is used in Sect. 2. It is $(\mathbb{Z}^2)^* = \mathbb{Z}^2 + (1/2,1/2)$ with its set of nearest-neighbor edges $(\mathcal{E}^2)^*$. An edge e has exactly one dual edge e^* that bisects it. We define variables (ω_{e^*}) by the rule $\omega_{e^*} = \omega_e$ and correspondingly use the terms p-open and p-closed. Thus a p_c-closed dual path is a path of edges e^* on the dual lattice each with $\omega_{e^*} > p_c$.

Last we give properties of correlation length, which will be vital for our work. For $\epsilon > 0$ and $p > p_c$, we define

$$L(p,\epsilon) = \min\{m \ge 1 : \mathbb{P}(\sigma(p,m,m)) > 1-\epsilon\},$$

where $\sigma(p,m,m)$ is the event that the box $B(m)$ has a p-open left-right crossing. This is a path, all of whose edges are p-open and in $B(m)$, which touches the left and right sides of the box. It is shown in [10, Eq. (1.24)] that for some ϵ_0 and any $\epsilon_1, \epsilon_2 \in (0,\epsilon_0]$, one has

$$L(p,\epsilon_1) \asymp L(p,\epsilon_2) \text{ as } p \downarrow p_c,$$

so we just set $L(p) = L(p,\epsilon_0)$. (The notation $\asymp$ means that $L(p,\epsilon_1)/L(p,\epsilon_2)$ is bounded away from 0 and ∞ as $p \downarrow p_c$.) We will use the following properties of correlation length. Setting

$$p_m = \min\{p : L(p) \le m\},$$

- (see [8, Eq. (2.10)]) there exists $c_1 \in (0,1)$ such that for all $m \ge 1$,

$$c_1 m \le L(p_m) \le m, \tag{5}$$

- (see [5, Sec. 2.1]) for positive integers k and l, there exists $\delta_{k,l} > 0$ such that for any positive integer n and for all $p \in [p_c, p_n]$,

$$\mathbb{P}(\text{there is a } p\text{-open horizontal crossing of } [0,kn] \times [0,ln]) > \delta_{k,l}$$

and

$$\mathbb{P}(\text{there is a } p\text{-closed horizontal dual crossing of } [0,kn] \times [0,ln]) > \delta_{k,l}. \tag{6}$$

Using these inequalities and a gluing construction involving the FKG inequality [6, Theorem 2.4], one can construct open or closed circuits in annuli around 0. For example, uniformly in n and $p \in [p_c, p_n]$,

$$\mathbb{P}(\text{there is a } p\text{-open circuit around 0 in } B(2n) \setminus B(n)) > 0.$$

We will use similar statements throughout the paper. See [6] for the relevant techniques and background.

– in the proof of Lemma 2 we will use "quasi-multiplicativity" of certain events. Generally this means that probabilities of arm events factor up to a constant. We will use this for 3-, 4-, and 5-arm events in both the full- and half-planes, for near critical values of p. One example of this property is the following: for $n_1 \leq n_2$, and p, q with $L(p), L(q) \geq n_2$, let $A(n_1, n_2, p, q)$ be the event that there are two p-open disjoint paths from $\partial B(n_1)$ to $\partial B(n_2)$ and two q-closed dual paths from $\partial B(n_1)$ to $\partial B(n_2)$ so that the open and closed paths alternate. Then there is a constant c such that for all $0 \leq n_1 \leq n_2 \leq n_3$ and $q, p > p_c$ with $L(p), L(q) \geq n_3$,

$$\mathbb{P}(A(n_1, n_3, p, q)) \geq c\mathbb{P}(A(n_1, n_2, p, q))\mathbb{P}(A(n_2, n_3, p, q)). \tag{7}$$

(See [14] for more background on arm events and their properties, like quasi-multiplicativity.)

1.3 Sketch of Proof and Outline of Paper

Argument for Question 1. We begin with a simple proof of the following statement: there exists $s > 1$ such that a.s.,

$$N_{0,x} \leq \|x\|_1^s \text{ for only finitely many } x \in \mathbb{Z}^2. \tag{8}$$

This essentially follows from the FKG inequality, the Russo–Seymour–Welsh theorem [6, Ch. 11] and the work of Aizenman–Burchard. The latter implies (see a discussion in [3, p. 3597]) that there is $s > 1$ such that

$$\lim_n \mathbb{P}\left(\exists \text{ a } p_c\text{-open path } \Gamma \text{ crossing } B(3^4 n) \setminus B(n) \text{ with } \#\Gamma \leq n^s\right) = 0. \tag{9}$$

Next, for $k \geq 0$, let E_k be the event that the follow conditions hold: putting $\mathrm{Ann}_m = B(3^{m+1}) \setminus B(3^m)$,

1. there is an p_c-open circuit around 0 in Ann_{3k},
2. there is an p_c-open circuit around 0 in Ann_{3k+2},
3. there is an p_c-open path crossing the rectangle $[3^{3k}, 3^{3k+3}] \times [-3^{3k}, 3^{3k}]$ from the left side to the right, and
4. any p_c-open path Γ crossing the annulus Ann_{3k+1} satisfies $\#\Gamma \geq 3^{3ks}$.

(See Fig. 1 for an illustration.) By the RSW theorem, the FKG inequality, and (9), there exists a constant $c_1 > 0$ such that for all k

$$\mathbb{P}(E_k) \geq c_1.$$

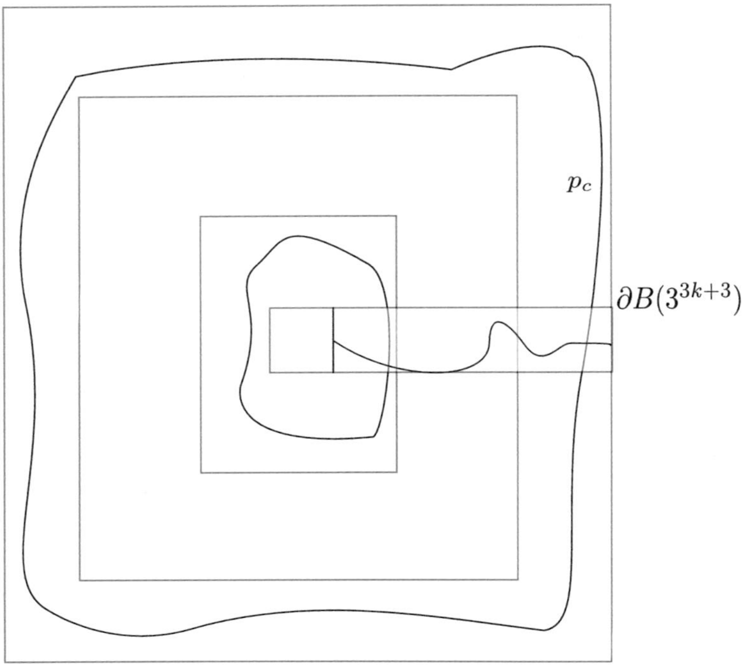

Fig. 1. Illustration of the event E_k. The boxes $B(3^{3k}), \ldots, B(3^{3k+3})$ are in increasing order. The curves represent the p_c-open circuits from items 1, 2, and the p_c-open path from item 3. By planarity, each of these curves overlap, and any geodesic from $\partial B(3^{3k})$ to $\partial B(3^{3k+3})$ must therefore cross Ann_{k+1} through a p_c-open circuit.

For k with $3^{3k+3} \le n$, if E_k occurs, then any geodesic from the origin to $x \notin B(n)$ must use the p_c-open path from item 3 to cross the annulus Ann_{3k+1}. Thus if Γ is any such geodesic,

$$\#\Gamma \ge \sum_{k=1}^{\lfloor \frac{1}{3} \log_3 n \rfloor - 1} 3^{3ks} \mathbf{1}_{E_k}. \tag{10}$$

Note that the E_k's are independent for different k's. Fixing $1 < s' < s$ then, there exists $c > 0$ such that

$$\mathbb{P}\left(\min_{x \notin B(n)} N_{0,x} \le n^{s'} \right) \le \mathbb{P}\left(\bigcap_{k=\lceil \frac{s'}{3s} \log_3 n \rceil + 1}^{\lfloor \frac{1}{3} \log_3 n \rfloor - 1} E_k^c \right) \le (1-c_1)^{\frac{s-s'}{3s} \log_3 n - 2}$$
$$\le n^{-c}.$$

The bounds on k in the above intersection are to ensure that $3^{3k+3} \le n$ and $3^{3ks} > n^{s'}$. Applying this to n which are powers of 3 and using Borel–Cantelli, we find that a.s. for all large m,

$$\min_{x \notin B(3^m)} N_{0,x} > 3^{ms'}.$$

For $x \in \mathbb{Z}^2$ with $\|x\|_\infty$ large, picking $k = \lfloor \log_3 \|x\|_\infty \rfloor - 1$, we have (noting $2 \geq s > s'$)

$$N_{0,x} \geq \min_{y \notin B(3^k)} N_{0,y} > 3^{ks'} \geq c_2 \|x\|_\infty^{s'}$$

for some $c_2 > 0$. Decreasing s' and relabeling it as s gives (8).

Outline of Proof of Theorem 1. Here we give a brief outline of the proof of Theorem 1. There are two parts: a block estimate, stated in Theorem 2, and a block argument which upgrades this estimate to the stretched-exponential convergence in Theorem 1. The block argument is an adaptation of the idea of Pisztora from his extension (and improvement) of ideas of Aizenman–Burchard in the near-critical percolation setting [15]. Since this part is more standard, we focus here on the block estimate, Theorem 2.

Two important things to notice about the statement in Theorem 2 are that (a) the geodesics referred to there are $T^{(n)}$-geodesics; that is, they are minimal-weight paths restricted to a box $B(3^{n+3})$ and crossing the annulus $B(3^{n+3}) \setminus B(3^n)$ (this restriction is needed in the block argument), and (b) the estimate holds almost surely for all large n. This second point is an upgrade from the argument given in the last subsection for (8): the event E_k there only holds with positive probability. This means that, for example, we need to control geodesic lengths even on rare events in which no p_c-open paths (or even near-critical paths) cross the annulus. This makes a considerable difficulty, since the arguments of Aizenman–Burchard do not apply for highly supercritical paths.

The proof of Theorem 2 is itself split into two parts. The first is an almost sure bound on the passage time of paths that cross annuli, stated in Proposition 2. The main estimate there is that a type of maximal passage time $T_{max}(n)$ of paths crossing the annulus $\mathrm{Ann}_{n+1} = B(3^{n+2}) \setminus B(3^{n+1})$ satisfies

$$T_{max}(n) \leq CF^{-1}(q_n) \log^3 n \text{ for large } n, \tag{11}$$

where $q_n > p_c$ is a certain near-critical percolation parameter. Similar bounds were shown in [4] for $T(0, \partial B(n))$, the minimal passage time from 0 to points on the boundary of $B(n)$, but we cannot use them for an annulus estimate. The reason is that if F is critical and such that $T(0,x)$ is stochastically bounded in x, then the passage time of segments far from 0 are dominated heavily by those of segments near the origin. So we need to adapt their arguments to the annulus setting.

The proof of (11) involves showing that with high probability, there are q_n-open circuits around 0 in Ann_n and Ann_{n+2}, with a q_n-open path connecting them. Any geodesic crossing $B(3^{n+3}) \setminus B(3^n)$ can be modified, replacing a portion of it with the union of these q_n-open paths, and we obtain an upper bound for $T_{max}(n)$ by the passage time of these paths. To bound the passage time of these paths, we show in Lemma 1 (as in [4]) that the only edges contributing to the passage time of these paths are ones which are q_n-open but p_c-closed, and which are associated to certain 4-arm events. Then we adapt moment bounds from [12] for arm events in Lemma 2 to bound the number of such edges.

Once we have (11), we dive into the machinery of Aizenman–Burchard, analyzing the number of "straight runs" for geodesics. Roughly speaking, for the length of a geodesic to be linear, it must pass clear (have a straight run) through many long thin cylinders at successively decreasing scales (starting from scale 3^n). The main inequality from Aizenman–Burchard, stated as Proposition 4, gives a lower bound for the *$s;\ell$-capacity* of a path given that it does not have too many straight runs. This capacity is related to the length of the curve in Lemma 3, and thus what we must show is that with high probability, no geodesic crossing the annulus Ann_{n+1} has straight runs through cylinders at more than half of the scales of the form $L_k = \gamma^k 3^n$ from 3^n down to 1, where $\gamma > 1$ is some large number. The difficulty here is that we do not have any precise description of geodesics in this model (including their geometry), and the only information we have is the bound on $T_{max}(n)$ above.

To show that geodesics have "sparse" straight runs, in Sect. 2.3, we prove that if a geodesic passes through a long thin cylinder, this cylinder has a high probability to be "slow." In other words, it is likely that the cylinder is crossed in the short direction by at least 4 dual paths which are well-separated and have weight at least p_m, where m is related to the scale of the rectangle. By choosing this p_m properly, we show in Proposition 5 that such a path would pass through enough slow cylinders at successive scales to have total passage time at least $(1/8)(\log^4 n)F^{-1}(q_n)$ (see (19)), where q_n is the near-critical parameter above in (11), giving a contradiction. In short, if a geodesic does not have sparse straight runs, its passage time violates (11). We combine these tools in Sect. 2.4 to conclude the block estimate.

2 Block Estimate

For $n \geq 1$ and $x, y \in B(3^{n+3}) = [-3^{n+3}, 3^{n+3}]^2$, let $T^{(n)}(x, y)$ be the minimal passage time of all paths from x to y that stay in $B(3^{n+3})$. Let $N_{x,y}^{(n)}$ be the minimal number of edges in any $T^{(n)}$-geodesic from x to y. We aim to show here:

Theorem 2. *There exists $s > 1$ such that almost surely, the following holds for all large n:*

$$N_{x,y}^{(n)} \geq 3^{ns} \text{ for all } x \in B(3^n),\ y \in \partial B(3^{n+3}).$$

The proof is split into several sections. In Sect. 2.1, we use tools from [4] to estimate the minimal passage time across cylinders and then in Sect. 2.2, paste these together to get bounds for the passage time across annuli. In Sect. 2.3, we use the machinery of Aizenman–Burchard to get lower bounds on the dimension of geodesics by estimating the number of "straight runs" they have: if they have too few, then they are forced to go through too many edges of nonzero passage time and they violate the passage time estimates from Sect. 2.2. We bring this all together in Sect. 2.4 to prove Theorem 2.

2.1 A Bound for Cylinder Times

In this section we would like to estimate the minimal passage time of any path crossing the rectangle $R(n) = [-2n, 2n] \times [-n, n]$ in the first coordinate direction. So let $T(n)$ be the minimal passage time among all paths which remain in $R(n)$ and connect the left side $\{-2n\} \times [-n, n]$ to the right side $\{2n\} \times [-n, n]$. The main result of this section is a bound on $T(n)$:

Proposition 1. *There exists $C > 0$ such that for all $n \geq 1$ and $p > p_c$ with $L(p) \leq n$,*

$$\mathbb{P}\left(T(n) \geq \lambda F^{-1}(p)\left(\frac{n}{L(p)}\right)^2\right) \leq \mathrm{e}^{-C\lambda} + \exp\left(-C\frac{n}{L(p)}\right) \quad \text{for } \lambda \geq 0.$$

For any p, let $E_p(n)$ be the event that there is a p-open path in $S(n) := [-4n, 4n] \times [-n, n]$ connecting the left side $\{-4n\} \times [-n, n]$ to the right side $\{4n\} \times [-n, n]$. By [8, Eq. (2.8)] and the RSW theorem [6, Sec. 12.7], one has the bound

$$\mathbb{P}(E_p^c) \leq \exp\left(-C\frac{n}{L(p)}\right) \tag{12}$$

for some $C > 0$ and all $p > p_c$, $n \geq 1$. The above proposition follows immediately from this bound, (13), and the two lemmas below.

On $E_p(n)$, we define $T_p(n)$ as the minimal passage time among all paths which remain in $S(n)$, are p-open, and connect the left side to the right side. Then we put

$$\hat{T}_p(n) = \max_{\Gamma} \sum_{e \in \Gamma \cap R(n)} t_e,$$

where the maximum is over paths Γ in $S(n)$ connecting the left side to the right side, which are p-open and have $T(\Gamma) = T_p(n)$. Note that on $E_p(n)$, one has

$$T(n) \leq \hat{T}_p(n). \tag{13}$$

The next result characterizes the nonzero-weight edges that contribute to $\hat{T}_p(n)$. For e with both endpoints in $R(n)$ and $p > p_c$, let $A_n(p, e)$ be the event that all of the following occur:

1. $\omega_e \in (p_c, p]$,
2. there are two (vertex) disjoint p-open paths in $S(n)$ from e to $\partial B(e, n/2)$, the translate of the box $B(n/2)$ centered at the midpoint of e, and
3. there are two (vertex) disjoint p_c-closed dual paths from e^* to $\partial S(n)$, one touching the top side and one touching the bottom.

Define

$$N_n(p) = \sum_{e \subset R(n)} \mathbf{1}_{A_n(p,e)}.$$

Lemma 1. *For all $p > p_c$ and $n \geq 1$,*

$$\hat{T}_p(n)\mathbf{1}_{E_p(n)} \leq F^{-1}(p)N_n(p)\mathbf{1}_{E_p(n)}.$$

Proof. The proof is quite similar to [4, Lemma 5.2]. If the event $E_p(n)$ does not happen, then both sides are zero, the inequality holds. Thus we suppose the $E_p(n)$ happens.

Suppose Γ is a path which remains in $S(n)$, is p-open, connects the left side to the right side of $S(n)$ and is such that $\hat{T}_p(n) = \sum_{e\in\Gamma\cap R(n)} t_e$. Note if $\omega_e \leq p_c$, then $t_e = F^{-1}(\omega_e) = 0$. Hence

$$\hat{T}_p(n) = \sum_{e\in\Gamma\cap R(n),\omega_e>p_c} t_e$$

Since Γ is p-open, we have $t_e \leq F^{-1}(p)$ for all $e \in \Gamma \cap R(n)$. Then we have

$$\hat{T}_p(n) \leq F^{-1}(p)\#\{e \in \Gamma \cap R(n) : \omega_e \in (p_c, p]\}$$

Thus it suffices to show that for each $e \in \Gamma \cap R(n)$ with $p_c < \omega_e \leq p$, the event $A_n(p, e)$ occurs.

The first condition is obvious, and the second follows from that $e \in \Gamma$ and Γ connects the left side to the right side of $S(n)$. For the third condition, if it does not hold, then by duality we can find a p_c-open path that connects Γ to itself, and the path Γ' formed by replacing the portion of Γ with this p_c-open path will avoid e. Since these p_c-open edges have zero passage time, this contradicts extremality of Γ. □

The next lemma gives a tail bound on the distribution of the number $N_n(p)$.

Lemma 2. *There exists $C > 0$ such that for all $n \geq 1$ and $p > p_c$ with $L(p) \leq n$,*

$$\mathbb{P}\left(N_n(p) \geq \lambda\left(\frac{n}{L(p)}\right)^2\right) \leq \mathrm{e}^{-C\lambda}.$$

Proof. We follow the argument of Kiss [12, Sections 2, 3] and much of what follows is copied from there. Since the proof is similar to that of [12, Eq. (3.11)], we just outline the main modifications necessary. For $n \geq 1$ and $p > p_c$ with $L(p) \leq n$ given, let V_n be the set of edges e in $R(n)$ such that $A_n(p, e)$ occurs. Let $k \in \mathbb{N}$ and

$$X = \{e_1, \ldots, e_k\} \subset R(n).$$

We give a bound on the probability of the event $\{V_n \supseteq X\}$, but first some definitions. Let T_0 denote the empty graph on X. Let us start blowing a box at each edge $e \in X$ at unit speed (starting at the midpoint of the edge). That is, at time $t \geq 0$, we have the boxes $B_t(e) = B(e, t)$, $e \in X$. We will stop at time $t = 2n$, and at this time, all boxes touch.

For small values of t, these boxes are pairwise disjoint. As t increases, more and more of these boxes intersect each other. Let r_1 denote the smallest t when the first pair of boxes touch. We pick one such pair of boxes in some deterministic way, with central edges $e_1, f_1 \in X$. We draw an edge $\hat{e}_1$ between e_1 and f_1 and label it with $l(\hat{e}_1) := r_1$, and get the graph T_1. Note that $\mathrm{dist}(e_1, f_1) = 2r_1$. (Here, dist refers to the ℓ_∞ distance between the midpoints of the edges.) Then we continue with the growth process, and stop at time r_2 if we find a pair of edges $e_2, f_2 \in X$ such that e_2, f_2 are in different connected components of T_1 and $B_{r_2}(e_2)$ and $B_{r_2}(f_2)$ touch. Then we draw an edge $\hat{e}_2$ between one such deterministically chosen pair with the label $l(\hat{e}_2) := r_2$ and get T_2. Note that it can happen that $r_1 = r_2$. We continue with this procedure until we arrive to the tree T_{k-1}. Let $\mathcal{R}(X)$ denote the multiset (a set where elements can appear multiple times) containing $r_i \leq 2n$ for $i = 1, \ldots, k-1$.

The induction argument of [13, Prop. 14] implies the following product statement about our 4-arm events. To state it, we need to define a slightly modified 4-arm event. For $r \leq 2n$ and an edge $e \subset R(n)$, let $\hat{\pi}_4(p; e, r)$ be the probability that the following conditions hold:

1. e is connected inside $S(n)$ to $\partial B(e, s_1)$ by two disjoint p-open paths, where
$$s_1 = \min\{L(p), r\},$$
2. e^* is connected inside $S(n)$ to $\partial B(e, s_2)$ by a p_c-closed dual path, where
$$s_2 = \min\{\mathrm{dist}(e, \partial S(n)), r\},$$
3. and e^* is connected inside $S(n)$ to $\partial B(e, s_3)$ by another disjoint p_c-closed dual path, where
$$s_3 = \min\{r, n\}.$$

Furthermore the paths in items 2 and 3 are alternating (open, closed, open). Set $\hat{\pi}_4(p; r) = \max_{e \subset R(n)} \hat{\pi}_4(p; e, r)$. Then for some C_3 independent of n, k, p, and the e_i's,
$$\mathbb{P}(V_n \supseteq X) \leq C_3(p - p_c)^k \hat{\pi}_4(p; n) \prod_{r \in \mathcal{R}(X)} \left(C_3 \hat{\pi}_4(p; r)\right). \tag{14}$$

The proof of this statement is similar to that of [12, Prop. 2.2], and the main ingredient is that our connection probabilities $\hat{\pi}_4$ have a quasi-multiplicative property that holds for general arm events. (See the discussion around (7) above.)

Continuing from (14), one can show that there is $C_6 > 0$ independent of n, p, such that for $r \leq 2n$,
$$\hat{\pi}_4(p; r) \leq C_6 \pi_4(s_1), \tag{15}$$
where $\pi_4(s)$ is critical four-arm probability; that is, the probability of the event that the edge $f = \{0, \vec{e_1}\}$ has two disjoint p_c-open paths to distance s (to $\partial B(f, s)$) and f^* has two disjoint p_c-closed dual paths to distance s. To prove (15), note that since $L(p) \leq n$, the event in $\hat{\pi}_4(p; e, r)$ for $e \subset R(n)$ implies that e is connected to distance s_1 by two disjoint p-open paths, e^* is connected to

distance s_1 by one disjoint p_c-closed dual path, and e^* is connected to distance $\min\{\text{dist}(e, \partial S(n)), s_1\}$ by another disjoint p_c-closed dual path (alternating). By independence, this probability is bounded by

$$\pi_4'(p, \min\{\text{dist}(e, \partial S(n)), s_1\})\pi_3^H(p, \min\{\text{dist}(e, \partial S(n)), s_1\}, s_1),$$

where $\pi_4'(p, m)$ is the probability that f is connected by two disjoint p-open paths to distance m and f^* is connected by two disjoint p_c-closed dual paths to distance m (alternating), and $\pi_3^H(p, m_1, m_2)$ is the probability that $\partial B(m_1)$ is connected to $\partial B(m_2)$ in the upper half-plane by two disjoint p-open paths and a disjoint p_c-closed dual path (alternating). By [5, Lemma 6.3], there is $D_1 > 0$ such that

$$\pi_4'(p, \min\{\text{dist}(e, \partial S(n)), s_1\}) \leq D_1\pi_4(\min\{\text{dist}(e, \partial S(n)), s_1\}).$$

A similar argument as in [5, Lemma 6.3] also holds for half-plane 3-arm (annulus) events, and we find

$$\pi_3^H(p, \min\{\text{dist}(e, \partial S(n)), s_1\}, s_1) \leq D_2\pi_3^H(\min\{\text{dist}(e, \partial S(n)), s_1\}, s_1),$$

where $\pi_3^H(m_1, m_2)$ is the probability that $\partial B(m_1)$ is connected by two disjoint p_c-open paths and a p_c-closed dual path to $\partial B(m_2)$ (alternating). Using [14, Theorem 24 (2)] and quasi-multiplicativity, one has for some D_3, $\pi_3^H(m_1, m_2) \leq D_3(m_1/m_2)^2$, and by quasi-multiplicativity and [14, Theorem 24 (3)], one has $\pi_4(m_1, m_2) \geq D_4(m_1/m_2)^2$. In total, we can bound $\pi_3^H(p, m_1, m_2)$ above by a multiple of $\pi_4(m_1, m_2)$, giving uniformly in e, a constant D_5 such that

$$\pi_4(p; e, r) \leq D_5\pi_4(\min\{\text{dist}(e, \partial S(n)), s_1\})\pi_4(\min\{\text{dist}(e, \partial S(n)), s_1\}, s_1),$$

which by quasi-multiplicativity is bounded by $D_6\pi_4(s_1)$, showing (15).

Given (15) and (14), we obtain

$$\mathbb{P}(V_n \supseteq X) \leq C_7(p - p_c)^k\pi_4(L(p)) \prod_{r \in \mathcal{R}(X)} (C_7\pi_4(\min\{L(p), r\}))\,.$$

To give a bound on moments of $N_n(p) = \#V_n$, we need to bound the number of sets X such that $\mathcal{R}(X) = R$ for a given R. By arguments analogous to the proof of [13, Prop. 15] we get the following. There is a universal constant C_8 such that for all multisets R with $k - 1$ elements, we have

$$\#\{X \subset R(n) : |X| = k,\ \mathcal{R}(X) = R\} \leq C_8\mathcal{O}(R)n^2 \prod_{r \in R}(C_8 r),$$

where $\mathcal{O}(R)$ denotes the number of different ways the elements of R can be ordered.

Now compute

$$\begin{aligned}
\mathbb{E}\binom{|V_n|}{k} &= \sum_{X \subseteq R(n)} \mathbb{P}(V_n \supseteq X) \\
&= \sum_{X \subseteq R(n)} \sum_{R} \mathbb{P}(V_n \supseteq X) \mathbf{1}_{\mathcal{R}(X)=R} \\
&\le \sum_{R} \left[C_8 \mathcal{O}(R) n^2 C_7 (p-p_c)^k \pi_4(L(p)) \prod_{r \in R} (C_8 C_7 r \pi_4(\min\{L(p), r\})) \right] \\
&\le C_9^k n^2 (p-p_c)^k \pi_4(L(p)) \sum_{R} \left[\mathcal{O}(R) \prod_{r \in R} r\pi_4(\min\{L(p), r\}) \right] \\
&= C_9^k n^2 (p-p_c)^k \pi_4(L(p)) \sum_{\tilde{R}} \prod_{\tilde{r} \in \tilde{R}} \tilde{r} \pi_4(\min\{L(p), \tilde{r}\} \\
&= C_9^k n^2 (p-p_c)^k \pi_4(L(p)) \left(\sum_{r=1}^{n} r\pi_4(\min\{L(p), r\}) \right)^{k-1},
\end{aligned}$$

where R is a multiset of with $k-1$ elements from the set $\{1/2, 1, \ldots, 2n\}$ and $\tilde{R}$ is a sequence of length $k-1$ from the set $\{1/2, 1, \ldots, 2n\}$. Last, we estimate

$$\sum_{r=1}^{n} r\pi_4(\min\{L(p), r\}) \le \sum_{r=1}^{L(p)} r\pi_4(r) + n^2 \pi_4(L(p)).$$

For any $r \le k$, one has $\frac{\pi_4(r)}{\pi_4(k)} \le C_{12}(k/r)^\alpha$ for some $\alpha < 2$ (this follows from Reimer's inequality [16] and the known value of the 5-arm exponent (from [14, Theorem 24(3)], which references [11, Lemma 5])), so

$$\begin{aligned}
\sum_{r=1}^{k} r\pi_4(r) = \pi_4(k) \sum_{r=1}^{k} r \frac{\pi_4(r)}{\pi_4(k)} &\le C_{12} \pi_4(k) \sum_{r=1}^{k} r(k/r)^\alpha \\
&= C_{12} k^\alpha \pi_4(k) \sum_{r=1}^{k} r^{1-\alpha} \\
&\le C_{13} k^2 \pi_4(k).
\end{aligned}$$

We thus obtain

$$\sum_{r=1}^{n} r\pi_4(\min\{L(p), r\}) \le C_{14} n^2 \pi_4(L(p)),$$

and therefore

$$\mathbb{E}\binom{|V_n|}{k} \le \left(C_{15} n^2 (p-p_c) \pi_4(L(p)) \right)^k.$$

Since the product $L(p)^2(p-p_c)\pi_4(L(p))$ is bounded uniformly in $p > p_c$ [14, Prop 34], we finish with

$$\mathbb{E}\binom{|V_n|}{k} \le \left(C_{16}\frac{n}{L(p)}\right)^{2k}.$$

Now we turn the above into a tail bound. For $t = 1 + a > 1$,

$$\mathbb{E}t^{|V_n|} = \sum_{k=1}^{\infty}(t-1)^k \mathbb{E}\binom{|V_n|}{k} \le \sum_{k=1}^{\infty}\left(aC_{16}\left(\frac{n}{L(p)}\right)^2\right)^k \le 1,$$

if a is chosen to be $\left(2C_{16}\left(\frac{n}{L(p)}\right)^2\right)^{-1}$. This implies that for some $C_{17} > 0$,

$$\mathbb{E}\exp\left(C_{17}\frac{|V_n|}{\left(\frac{n}{L(p)}\right)^2}\right) \le 1,$$

and so by Markov,

$$\mathbb{P}\left(N_n(p) \ge \lambda\left(\frac{n}{L(p)}\right)^2\right) \le \mathrm{e}^{-C_{17}\lambda}.$$

□

Corollary 1. *Given an integer $K \ge 2$, there exists $C > 0$ such that for all n and p with $L(p) \le n$,*

$$\mathbb{P}\left(T_K(n) \ge \lambda F^{-1}(p)\left(\frac{n}{L(p)}\right)^2\right) \le \mathrm{e}^{-C\lambda} + \exp\left(-C\frac{n}{L(p)}\right) \text{ for } \lambda \ge 0,$$

where $T_K(n)$ is the corresponding minimal passage time between the left and right sides of $[-Kn, Kn] \times [-n, n]$ among all paths that remain in this rectangle.

Proof. Let $\Gamma_1, \ldots, \Gamma_{K-1}$ be paths such that Γ_i is in $[-Kn + 2(i-1)n, -Kn + 2(i+1)n] \times [-n, n]$, connects the left side of the rectangle to the right side, and has minimal passage time among all such paths. Let $\hat{\Gamma}_1, \ldots, \hat{\Gamma}_{K-2}$ be paths such that $\hat{\Gamma}_i$ is in $[-Kn + 2in, -Kn + 2(i+1)n] \times [-n, 3n]$, connects the top side of the rectangle to the bottom side, and has minimal passage time among all such paths. By planarity, there is a path remaining in $[-Kn, Kn] \times [-n, n]$ which starts on the left side of this rectangle, ends on the right, and is contained in the union $\left(\cup_{i=1}^{K-1}\Gamma_i\right) \cup \left(\cup_{i=1}^{K-2}\hat{\Gamma}_i\right)$. Applying Proposition 1 and a union bound completes the proof. □

2.2 Bounds for Annulus Crossing Times

Using the results of the last section, we will give our main bound on types of maximal annulus crossing times.

For any $n \geq 1$, $x, y \in B(3^{n+3})$, and S a subset of the edges of $B(3^{n+3})$, define

$$T^{(n)}(x,y,S) = \max\left\{\sum_{e \in \Gamma \cap S} t_e : \Gamma \subset B(3^{n+3}),\ T(\Gamma) = T^{(n)}(x,y),\ \Gamma : x \to y\right\}.$$

We will be concerned with a type of annulus-crossing time. For $n \geq 2$, define

$$T_{max}(n) = \max_{x \in \partial B(3^n), y \in \partial B(3^{n+3})} T^{(n)}(x,y,\mathrm{Ann}_{n+1}).$$

(Here, we think of $\mathrm{Ann}_{n+1} := B(3^{n+2}) \setminus B(3^{n+1}))$ as an edge set by considering all edges with both endpoints in Ann_{n+1}.)

The main result is:

Proposition 2. *There exist $C, \mathbf{C}$ such that almost surely*

$$T_{max}(n) \leq CF^{-1}(q_n)\log^3 n \text{ for large } n,$$

where $q_n = p_{\lfloor \frac{3^n}{\mathbf{C}\log n} \rfloor}$.

Proof. We will build two circuits around the origin – one in Ann_n and one in Ann_{n+2}, and a path connecting them, using only crossings in the "long direction" of minimal passage time of translates and rotates of rectangles of the form $[-3^{n+3}, 3^{n+3}] \times [-3^n, 3^n]$ (they start on the left side, end on the right, and remain in the rectangle). One can do this using 9 such crossings. (Refer back to Fig. 1 for a similar construction. One uses 8 such crossings to build the circuits, and one to build the path connecting them.) So letting S_n be the union of all the edges in these crossings, a union bound along with Corollary 1 shows for some $C, \mathbf{C}$ large enough and all n,

$$\mathbb{P}\left(\sum_{e \in S_n} t_e \geq CF^{-1}(q_n)\log^3(n)\right) \leq n^{-2}.$$

In deriving this, one needs to use (5) applied to $L(q_n)$. Borel–Cantelli implies that almost surely,

$$\sum_{e \in S_n} t_e < CF^{-1}(q_n)\log^3 n \text{ for all large } n.$$

If $x \in \partial B(3^n)$ and $y \in \partial B(3^{n+3})$, then let Γ be a $T^{(n)}$-geodesic from x to y. The path Γ has a first intersection z with S_n (which must be in Ann_n) and a last intersection w with S_n (which must be in Ann_{n+2}). One then has

$$T^{(n)}(x,y,\mathrm{Ann}_{n+1}) \leq T^{(n)}(z,w) \leq \sum_{e \in S_n} t_e.$$

This is true for all x, y, so almost surely, for all large n

$$T_{max}(n) \leq \sum_{e \in S_n} t_e \leq CF^{-1}(q_n) \log^3 n.$$

□

2.3 Dimension Bounds

Tools from Aizenman–Burchard. In this section, we recall and use some results from Aizenman–Burchard. To do this, we will think of the box $B(3^{n+3})$ scaled down to unit size; that is, to the box $B(1)$. The lattice spacing in $B(1)$ will be $1/3^{n+3}$, so that our geodesics are polygonal paths of step-size $1/3^{n+3}$. The lower bounds on dimension of random curves of [1] are derived using a truncated form of capacity.

Definition 1. *For $s > 0$ and $\ell \geq 0$, the capacity* $\mathrm{Cap}_{s;\ell} A$ *of a subset A of $\mathbb{R}^d$ is*

$$\frac{1}{\mathrm{Cap}_{s;\ell} A} = \inf_{\mu \geq 0 : \int_A \mathrm{d}\mu = 1} \int \int_{A \times A} \frac{\mu(\mathrm{d}x)\mu(\mathrm{d}y)}{\max\{|x-y|, \ell\}^s}.$$

(The standard definition of capacity does not include the term ℓ in the denominator, but this term helps to deal with the fact that our paths have stepsize > 0.) We will take A to be a geodesic Γ in $B(1)$ from $x \in B(1/27)$ to $y \in \partial B(1)$. The relationship between the length of Γ and its capacity is given by the following lemma.

Lemma 3. *For every collection of sets $\{B_j\}$ covering A with* $\min_j \mathrm{diam}(B_j) \geq \ell$,

$$\sum_j (\mathrm{diam}\ B_j)^s \geq \mathrm{Cap}_{s;\ell} A.$$

Taking $\{B_j\}$ to be a collection of boxes of size $C/3^n$ centered on the edges of Γ, we then obtain

$$\frac{\#\Gamma C^s}{3^{ns}} = \sum_j (\mathrm{diam}\ B_j)^s \geq \mathrm{Cap}_{s;\ell} \Gamma. \tag{16}$$

Therefore Theorem 2 follows directly from this inequality and the following proposition, which we will show in Sect. 2.4:

Proposition 3. *There exist $C_1, C_2, C_3 > 0$ and $s > 1$ such that the following holds for all large n. For all $T^{(n)}$-geodesics Γ in $B(1)$ connecting a point in $B(1/27)$ to a point in $\partial B(1)$,*

$$\mathrm{Cap}_{s;\ell} \Gamma' \geq C_1 \exp(-C_2 \log^4 n),$$

where $\ell = \ell(n)$ satisfies $1/3^{n+3} \leq \ell \leq C_3/3^{n+3}$ and Γ' is the portion of Γ from its last intersection with $B(1/9)$ to its first intersection with $\partial B(1/3)$.

Geodesics Have Sparse Straight Runs. Aizenman–Burchard gave a general theorem to lower bound the capacity of curves using the idea of "straight runs." Roughly speaking, if a path does not cross too many thin cylinders (on successively decreasing scales) in the long direction, then its capacity, and therefore length, is large. We begin by recalling the definition of sparse straight runs.

Given $\gamma > 1$ (which will be taken to be large), define a sequence of successive scales L_k by

$$L_k = \gamma^{-k}, \text{ for } k = 0, \ldots, k_{max},$$

where $k_{max} = k_{max}(\gamma, n)$ is chosen so that $L_{k_{max}}$ is of order $1/3^n$. That is, we set k_{max} as

$$k_{max} = \max\{k \geq 0 : L_k \geq 1/3^{n+3}\}.$$

Definition 2. *A path Γ in $B(1)$ is said to exhibit a straight run at scale L ($= L_k$ for some k) if it traverses some cylinder of length L and cross sectional diameter $(9/\sqrt{\gamma})L$ in the "length" direction, joining the centers of the corresponding sides. Two straight runs are* nested *if one of the defining cylinders contains the other.*

For a given integer k_0 and $\gamma > 1$, we say that straight runs for Γ are (γ, k_0)-sparse, down to the scale ℓ, if Γ does not exhibit any nested collection of straight runs on a sequence of scales $L_{k_1} > \cdots > L_{k_N}$ with $L_{k_N} \geq \ell$ and

$$N \geq \frac{1}{2} \max\{k_N, k_0\}.$$

The next result says that to show our needed capacity bound for $T^{(n)}$-geodesics Γ, it suffices to prove that straight runs for Γ are sparse. Note that in the next proposition, ℓ satisfies

$$1/3^{n+3} \leq \ell = L_{k_{max}} \leq C/3^{n+3} \tag{17}$$

for some $C = C(\gamma)$ independently of n, as required in Proposition 3.

Proposition 4. *Let Γ be a path in $B(1)$, let $\gamma > 1$ and set $m \in [\gamma/2, \gamma)$ with $\epsilon = \gamma/m - 1$. If straight runs for Γ are (γ, k_0)-sparse down to the scale $\ell = L_{k_{max}}$, then for $s > 0$ such that $\gamma^s < \beta := \sqrt{m(m+1)}$, one has*

$$\mathrm{Cap}_{s;\ell}\Gamma \geq \epsilon^s \left[\gamma^{sk_0} + \frac{\beta}{1 - \beta^{-1}\gamma^s}\right]^{-1}.$$

Proof. This is the bound [1, Eq. (5.14)] applied in our context. (See also a similar bound and explanation above [1, Eq. (5.22)] with the same choice of β.) □

We now address sparsity of straight runs for geodesics. In the next section we use the above proposition to show Proposition 3 and conclude Theorem 2. To show that geodesics must leave cylinders, we will show that many cylinders are slow in the following sense:

Definition 3. *For $a \in (0,1)$ and $L \le 1$, an $L \times aL$ cylinder in $B(1)$ is said to be* slow *if it is traversed in the aL-direction by two $p_{aL3^{n+3}}$-closed dual paths P_1, P_2 such that*

$$\min\{|x-y| : x \in P_1, y \in P_2\} \ge aL.$$

A cylinder that is not slow is fast.

Lemma 4. *For any large enough $\gamma > 1$, the following occurs almost surely for all large n. One cannot find a nested collection of fast cylinders $R_1, \ldots, R_N$ at scales $L_{k_1} > \cdots > L_{k_N}$ with $k_N \le k_{max}$ and*

$$N \ge \frac{1}{4}\max\{k_N, k_0\},$$

where $k_0 = \lceil \log^4 n \rceil$.

Proof. We follow the proof in Aizenman–Burchard. We first give a bound on the probability that for any fixed sequence $k_1 < \cdots < k_N \le k_{max}$ there is a sequence $R_1, \ldots, R_N$ of nested cylinders at scales $L_{k_1} > \cdots > L_{k_N}$ all of which are fast. Specifically, we will first show:

$$\mathbb{P}(\text{there is a nested sequence of fast cylinders at scales } L_{k_1}, \ldots, L_{k_N}) \le C_1 \gamma^{4k_N} \exp\left(-cN\sqrt{\gamma}\right). \quad (18)$$

If an $L \times (9/\sqrt{\gamma})L$ cylinder is fast, then if γ is large, (independent of L), we can find a cylinder of width $(10/\sqrt{\gamma})L$ and length $L/2$ centered at a line segment joining discretized points in $L'\mathbb{Z}^d$ (with $L' \le L/\gamma$) that cannot be traversed in the $(10/\sqrt{\gamma})L$-direction by two disjoint $p_{(9/\sqrt{\gamma})L3^{n+3}}$-closed dual paths P_1, P_2 with $\min\{|x-y| : x \in P_1, y \in P_2\} \ge (9/\sqrt{\gamma})L$. As in [1, Eq. (6.2)], the number of positions of N nested cylinders at scales $L_{k_1}, \ldots, L_{k_N}$ is bounded by

$$C_1 \gamma^{4k_1} \gamma^{4(k_2-k_1)} \cdots \gamma^{4(k_N-k_{N-1})} \le C_1 \gamma^{4k_N}.$$

Fix now such a sequence R_i, $i = 1, \ldots, N$ of nested cylinders of length $L_{k_i}/2$ and width $(10/\sqrt{\gamma})L_{k_i}$. Cut each of the cylinders into $\lfloor \sqrt{\gamma}/18 \rfloor$ shorter cylinders of dimensions $(9/\sqrt{\gamma})L_{k_i} \times (10/\sqrt{\gamma})L_{k_i}$ (plus a possible remaining one of smaller length which we do not consider) and pick a maximal number of disjoint cylinders from this collection. For γ large, each R_{i+1} intersects at most two of the shorter cylinders obtained by subdividing R_i, so the number of cylinders at scale L_{k_i} in a maximal collection is at least $\lfloor \sqrt{\gamma}/18 \rfloor - 2$. By (6), the probability that a $(9/\sqrt{\gamma})L_{k_i} \times (10/\sqrt{\gamma})L_{k_i}$ cylinder is traversed in the $(10/\sqrt{\gamma})L_{k_i}$-direction by a $p_{(9/\sqrt{\gamma})L_{k_i}3^{n+3}}$-closed dual path is bounded below by some constant uniformly in n, γ, and the choice of the k_i's. By standard large deviations for sums of Bernoulli random variables, there is a universal constant $c > 0$ such that probability that at least four cylinders from the maximal collection at scale L_{k_i} are traversed by such closed dual paths is at least $1 - \exp(-c\sqrt{\gamma})$. If four distinct such cylinders have this property at scale L_{k_i} and γ is large, then the original cylinder R_i is slow. These events are independent at distinct scales, so

$$\mathbb{P}(R_1, \ldots, R_N \text{ are fast}) \le \exp\left(-cN\sqrt{\gamma}\right).$$

Summing over positions of the original cylinders gives the bound (18).

Now we sum (18) over choices of cylinders to show the lemma. Namely, for a given n and $k = k_N \le k_{max}$, the probability that there is a nested sequence of fast cylinders $R_1, \ldots, R_N$ at some scales $L_{k_1} > \cdots > L_{k_N}$ with $k_N \ge N \ge k_N/4$ is bounded by

$$\sum_{N=\lceil k/4 \rceil}^{k} \binom{k}{N} C_1 \gamma^{4k} \exp\left(-cN\sqrt{\gamma}\right) \le C_1 k 2^k \gamma^{4k} \exp\left(-ck\sqrt{\gamma}/4\right).$$

Taking γ large, this is bounded by $c_2 \exp(-c_2 k)$. Summing over $k = k_N \ge N \ge \frac{1}{4}\lceil \log^4 n \rceil$ gives a probability which is summable in n and Borel–Cantelli finishes the proof. □

From the existence of many slow cylinders, we can prove that geodesics have sparse straight runs.

Proposition 5. *For any sufficiently large $\gamma > 1$, almost surely, the following occurs for all large n. For any $T^{(n)}$-geodesic Γ from a vertex $x \in B(1/27)$ to a vertex $y \in \partial B(1)$, Γ' has $(\gamma, \lceil \log^4 n \rceil)$-sparse straight runs down to the scale $\ell = L_{k_{max}}$. Here Γ' is the portion of Γ from its last intersection with $B(1/9)$ to its first intersection with $\partial B(1/3)$.*

Proof. Take $\gamma > 1$ large enough so that the event (call it E_n) in Lemma 4 holds almost surely for all large n. Take $\omega \in E_n$, $k_0 = \lceil \log^4 n \rceil$, and suppose for the sake of contradiction that $R_1, \ldots, R_N$ is a nested collection of cylinders at scales $L_{k_1} > \cdots > L_{k_N}$ with $k_N \le k_{max}$ and $N \ge \frac{1}{2}\max\{k_N, k_0\}$ for which a Γ' has straight runs in each of the R_i's. Suppose that some $\lceil N/2 \rceil$ of these cylinders are fast and label them in order of decreasing scales as $R_{j_1}, \ldots, R_{j_{\lceil N/2 \rceil}}$. Then $j_{\lceil N/2 \rceil} \le k_N \le k_{max}$ and

$$N/2 \ge \frac{1}{4}\max\{k_N, k_0\} \ge \frac{1}{4}\max\{j_{\lceil N/2 \rceil}, k_0\},$$

contradicting $\omega \in E_n$. Thus at least $\lceil N/2 \rceil$ of these cylinders are slow. Let $\hat{R}_1, \ldots, \hat{R}_{\lceil N/4 \rceil}$ be the $\lceil N/4 \rceil$ slow cylinders at the smallest scales in the sequence $R_1, \ldots, R_N$. In each slow cylinder, there are two dual paths as in the definition of slow. If γ is large enough, then each cylinder $\hat{R}_{i+1}$ can intersect at most one of the two closed dual paths in $\hat{R}_i$. Therefore, as Γ' crosses each of these cylinders, it must intersect a distinct edge from at least one dual path in each $\hat{R}_i$. This means that if $e_1, \ldots, e_{N/4}$ are such edges, then $T(\Gamma') \ge t_{e_1} + \cdots + t_{e_{\lceil N/4 \rceil}}$. If the cylinder $\hat{R}_i$ is at scale $L_{\hat{k}_i}$, then $\hat{k}_i \ge \lfloor N/4 \rfloor$, so because $\hat{R}_i$ is slow,

$$t_{e_i} \ge F^{-1}\left(p_{(9/\sqrt{\gamma})L_{\hat{k}_i}3^{n+3}}\right) \ge F^{-1}\left(p_{(9/\sqrt{\gamma})L_{\lfloor N/4 \rfloor}3^{n+3}}\right).$$

As $N \ge k_0/2 \ge \frac{1}{2}\lceil \log^4 n \rceil$, one has for large n

$$\frac{9}{\sqrt{\gamma}} L_{\lfloor N/4 \rfloor} 3^{n+3} \le \left\lfloor \frac{3^n}{\mathbf{C}\log n} \right\rfloor,$$

where $\mathbf{C}$ is from Proposition 2. Therefore

$$T(\Gamma') \geq \lceil N/4 \rceil F^{-1}(p_{\lfloor \frac{3^n}{\mathbf{C} \log n} \rfloor}) \geq \frac{1}{8}(\log^4 n) F^{-1}(q_n), \tag{19}$$

where $q_n = p_{\lfloor \frac{3^n}{\mathbf{C} \log n} \rfloor}$, and this contradicts Proposition 2 for large n. □

2.4 Proof of Theorem 2

We prove Theorem 2 by showing Proposition 3. As before, we shrink the lattice so that $B(3^{n+3})$ is shrunk to $B(1)$. Choose $\gamma > 1$ large enough for Proposition 5 and so that if $m = \lfloor \gamma \rfloor$, then $\sqrt{m(m+1)} > \gamma$. This means in particular that if we define s by $\gamma^s = \frac{\sqrt{m(m+1)}+\gamma}{2}$, then $s > 1$.

We now apply Proposition 5 along with the capacity lower bound Proposition 4. The first says that if Γ is a $T^{(n)}$-geodesic from a vertex $x \in B(1/27)$ to a vertex $y \in \partial B(1)$, then Γ' has $(\gamma, \lceil \log^4 n \rceil)$-sparse straight runs down to scale $\ell = L_{k_{max}}$, a number satisfying (17). In this setting, Proposition 4 gives for $\epsilon = \gamma - m$, $L_0 = 1$, and $\beta = \sqrt{m(m+1)}$,

$$\mathrm{Cap}_{s;\ell}\Gamma' \geq \epsilon^s \left[\gamma^{s\lceil \log^4 n \rceil} + \frac{\beta}{1 - \beta^{-1}\gamma^s} \right]^{-1} \geq C_1 \exp(-C_2 \log^4 n).$$

This proves Proposition 3.

Last we combine this bound with Eq. (16), applied to Γ'. Putting $\{B_j\}$ as a collection of boxes of size $C_3/3^{n+3}$ centered on the edges of Γ' (where C_3 is from Proposition 3), one has $\#\Gamma \geq \frac{C_1}{C_3^s} 3^{ns} \exp\left(-C_2 \log^4 n\right)$, so

$$\min_{x \in B(3^n), y \in \partial B(3^{n+3})} N_{x,y}^{(n)} \geq \frac{C_1}{C_3^s} 3^{ns} \exp(-C_2 \log^4 n).$$

Since $\exp(C_2 \log^4 n) = o(3^{\delta n})$ for each $\delta > 0$, we can slightly decrease $s > 1$ to obtain Theorem 2.

3 Block Argument: Proof of Theorem 1

Let m be such that $3^{m-1} \leq \|x\|_\infty < 3^m$ and set $s > 1$ as the constant in Theorem 2. Fix $s' \in (1, s)$, and let $n = \lceil m \frac{s'}{s} \rceil$, so that $3^{ns} \geq 3^{ms'}$.

Definition 4. *For $y \in \mathbb{Z}^2$, define the annulus $A(y,n) := 2 \cdot 3^n y + B(3^{n+3}) \setminus B(3^n)$. For $z, w \in y + B(3^{n+3})$, define $T_y^{(n)}(z,w)$ as the minimal passage time from z to w among paths remaining in $y + B(3^{n+3})$ and $N_y^{(n)}(z,w)$ the minimal number of edges in any $T_y^{(n)}$-geodesic from z to w. Call $A(y,n)$ bad if*

$$\min_{z \in y + \partial B(3^n), w \in y + \partial B(3^{n+3})} N_y^{(n)}(z,w) < 3^{ns}$$

By stationarity, $\hat{p}_n := \mathbb{P}(A(y,n) \text{ is bad})$ depends only on n, and by Theorem 2, it approaches 0 as $n \to \infty$.

For a geodesic Γ from 0 to x, we may follow Γ, marking each box of the form $y+B(3^n)$ that it touches inside the box $B(3^m)$. Note that if $\|x\|_\infty$ is large enough, Γ must cross the annulus $A(y,n)$ surrounding the box. By standard arguments, we can extract a sequence $\gamma = (A_1, \dots, A_r)$ of these "crossed" annuli satisfying the following properties: for universal constants $c_1, c_2 > 0$,

1. A_i and A_j are disjoint for $i \neq j$,
2. if $|i-j| = 1$ and $A_i = A(y_i, n)$, $A_j = A(y_j, n)$, then $\|y_i - y_j\|_\infty \le c_1$,
3. $\|y_1\|_\infty \le c_1$, and
4. $r \ge c_2 3^{m-n}$.

Note that if any one of these annuli $A(y,n)$ is not bad, then defining z to be the first entrance of Γ to $y + B(3^n)$ and w the last entrance of Γ to $y + B(3^{n+3})$ before z, then

$$\#\Gamma \ge N_y^{(n)}(z,w) \ge 3^{ns} \ge 3^{ms'}.$$

Hence, letting $c_3 > 0$ be such that, given y, there are at most c_3 choices of y' with $\|y - y'\|_\infty \le c_1$, one has

$$\begin{aligned}\mathbb{P}(N_{0,x} < \|x\|_\infty^{s'}) \le \mathbb{P}(N_{0,x} < 3^{ms'}) &\le \sum_{r \ge c_2 3^{m-n}} \sum_{\#\gamma = r} \mathbb{P}(\text{all } A_i \in \gamma \text{ are bad}) \\ &\le \sum_{r \ge c_2 3^{m-n}} \sum_{\#\gamma = r} \hat{p}_n^r \\ &\le \sum_{r \ge c_2 3^{m-n}} (c_3 \hat{p}_n)^r\end{aligned}$$

By Theorem 2, there exists constant $N > 0$ such that when $n \ge N$, $c_3 \hat{p}_n \le 1/2$. So choosing $\|x\|_\infty$ large enough so that $n \ge N$, we obtain constants $c_4, c_5 > 0$ such that

$$\mathbb{P}(N_{0,x} < \|x\|_\infty^{s'}) \le \sum_{r \ge c_2 3^{m-n}} 2^{-r} \le c_4 \mathrm{e}^{-c_5 3^{m-n}}. \tag{20}$$

This implies Theorem 1, since $3^{m-1} \le \|x\|_\infty < 3^m$.

References

1. Aizenman, M., Burchard, A.: Hölder regularity and dimension bounds for random curves. Duke Math. J. **99**, 419–453 (1999)
2. Auffinger, A., Damron, M., Hanson, J.: 50 years of first-passage percolation. University Lecture Series, vol. 68. American Mathematical Society, Providence (2017)
3. Damron, M., Hanson, J., Sosoe, P.: Subdiffusivity of random walk on the 2*D* invasion percolation cluster. Stoch. Proc. Appl. **123**, 3588–3621 (2013)
4. Damron, M., Lam, W., Wang, X.: Asymptotics for 2D critical first passage percolation. Ann. Probab. **45**, 2941–2970 (2017)

5. Damron, M., Sapozhnikov, A., Vágvölgyi, B.: Relations between invasion percolation and critical percolation in two dimensions. Ann. Probab. **37**, 2297–2331 (2009)
6. Grimmett, G.R.: Percolation. Grundlehren der matematischen Wissenschaften, vol. 321, 2nd edn. Springer, Berlin (1999)
7. Hammersley, J., Welsh, D.: First-passage percolation, subadditive processes, stochastic networks, and generalized renewal theory. In: Proc. Internat. Res. Semin., Statist. Lab., Univ. California, Berkeley, pp. 61–110. Springer, New York (1965)
8. Járai, A.A.: Invasion percolation and the incipient infinite cluster in $2D$. Commun. Math. Phys. **236**, 311–334 (2003)
9. Kesten, H.: Aspects of first-passage percolation. In: École d'été de probabilités de Saint-Flour, XIV–1984. Lecture Notes in Mathematics, vol. 1180, pp. 125–264. Springer, Berlin (1986)
10. Kesten, H.: Scaling relations for $2D$-percolation. Commun. Math. Phys. **109**, 109–156 (1987)
11. Kesten, H., Sidoravicius, V., Zhang, Y.: Almost all words are seen in critical site percolation on the triangular lattice. Electron. J. Probab. **3**, paper no. 10 (1998)
12. Kiss, D.: Large deviation bounds for the volume of the largest cluster in $2D$ critical percolation. Electron. Commun. Probab. **19**, 1–11 (2014)
13. Kiss, D., Manolescu, I., Sidoravicius, V.: Planar lattices do not recover from forest fires. Ann. Probab. **43**, 3216–3238 (2015)
14. Nolin, P.: Near-critical percolation in two dimensions. Electron. J. Probab. **13**, 1562–1623 (2008)
15. Pisztora, A.: Scaling inequalities for shortest paths in regular and invasion percolation. Carnegie–Mellon CNA preprint. http://www.math.cmu.edu/CNA/Publications/publications2000/001abs/00-CNA-001.pdf
16. Reimer, D.: Proof of the van den Berg-Kesten conjecture. Combin. Probab. Comput. **9**, 27–32 (2000)
17. Steele, J.M., Zhang, Y.: Nondifferentiability of the time constants of first-passage percolation. Ann. Probab. **31**, 1028–1051 (2003)
18. Wierman, J., Reh, W.: On conjecture in first passage percolation theory. Ann. Probab. **6**, 388–397 (1978)
19. Zhang, Y.: Supercritical behaviors in first-passage percolation. Stoch. Proc. Appl. **59**, 251–266 (1995)
20. Zhang, Y.: Double behavior of critical first-passage percolation. In: Perplexing problems in probability. Progress in Probability, vol. 44, pp. 143–158. Birkhäuser, Boston (1999)

A Note on Schramm's Locality Conjecture for Random-Cluster Models

Hugo Duminil-Copin[1,2](✉) and Vincent Tassion[3]

[1] Institut des Hautes Études Scientifiques, Bures-Sur-yvette, France
duminil@ihes.fr
[2] Université de Genève, Genève, Switzerland
hugo.duminil@unige.ch
[3] ETH Zurich, Zürich, Switzerland
vincent.tassion@math.ethz.ch

To Chuck, for his 70th birthday

Abstract. In this note, we discuss a generalization of Schramm's locality conjecture in the case of random-cluster models. We give some partial (modest) answers, and present several related open questions. Our main result is to show that the critical inverse temperature of the Potts model on $\mathbb{Z}^r \times (\mathbb{Z}/2n\mathbb{Z})^{d-r}$ (with $r \geq 3$) converges to the critical inverse temperature of the model on $\mathbb{Z}^d$ as n tends to infinity. Our proof relies on the infrared bound and, contrary to the corresponding statement for Bernoulli percolation, does not involve renormalization arguments.

Keywords: Phase transition · Lattice model · Potts model · Sharpness · Exponential decay

1 Motivation

In [6], Benjamini and Schramm initiated the theory of Bernoulli percolation on general transitive graphs, thus motivating many new questions and problematics in the field. Among them, one conjecture, now known under the name of Schramm's locality conjecture, was asked in [5]. Roughly speaking, it can be stated as follows: the critical parameter of Bernoulli percolation is continuous in the local topology on transitive locally finite graphs with $p_c < 1$ (see below for a formal definition in the context of random-cluster models). Since the formulation of the conjecture, a number of results have been proved [5,19,21,26], yet a final answer is still lacking.

We would like to advertise that Schramm's conjecture can be extended to random-cluster models and the associated Potts models. This paper contains the proofs of two results in that direction. The results are modest but we believe that they raise some interesting questions. Even though we will not discuss it here, let us mention that in recent years, Schramm's conjecture was also stated for self-avoiding walks in [4], and that some partial results were obtained in [16,17].

V. Sidoravicius (Ed.): *Sojourns in Probability Theory and Statistical Physics - II*, PROMS 299, pp. 123–134, 2019.
https://doi.org/10.1007/978-981-15-0298-9_5

2 About Schramm's Conjecture for Random-Cluster Models

Random-Cluster Model on Transitive Graphs. Through this note, all the graphs are assumed to be connected and locally-finite. Consider an infinite transitive graph $\mathbb{G} = (\mathbb{V}, \mathbb{E})$ (recall that a graph is said to be transitive if its group of automorphisms acts transitively on $\mathbb{V}$). We call one of the vertices the *origin* and denote it by 0. A percolation configuration ω on a finite subgraph $G = (V, E)$ of $\mathbb{G}$ is a subset of E, which will be seen as a subgraph of $G = (V, E)$ with vertex-set V and edge-set ω. Let $|\omega|$ be the number of edges in ω and $k_1(\omega)$ the number of connected components in ω, when all connected components intersecting the vertex boundary $\partial G := \{x \in V : \exists y \in \mathbb{V} \setminus V \text{ such that } \{x, y\} \in \mathbb{E}\}$ are counted as one.

For $p \in [0, 1]$ and $q > 0$, the random-cluster measure on G with edge-weight p and cluster-weight q is defined by the formula

$$\phi^1_{G,p,q}[\omega] = \frac{1}{Z^1(G,p,q)}\Big(\frac{p}{1-p}\Big)^{|\omega|} q^{k_1(\omega)},$$

where $Z^1(G, p, q)$ is such that the measure has mass one.

The model may be extended to $\mathbb{G}$ by taking weak limits of measures on finite subgraphs G tending to $\mathbb{G}$; see [15, Thm 4.19]. From now on, set $\phi^1_{\mathbb{G},p,q}$ for the measure on $\mathbb{G}$.

Below, we write $0 \longleftrightarrow \infty$ for the event that the connected component of 0 in ω is infinite. Define the *critical point* of the model on $\mathbb{G}$ by the formula

$$p_c(\mathbb{G}, q) := \inf\{p \in [0, 1] : \phi^1_{\mathbb{G},p,q}[0 \longleftrightarrow \infty] > 0\}.$$

Local Convergence and Schramm's Conjecture. Given an infinite transitive graph $\mathbb{G}$, we consider the ball of radius R (for the graph distance) around the origin 0. Up to isomorphism of rooted graphs, it does not depend on the choice of the origin, and we simply refer to it as the ball of radius R in $\mathbb{G}$. We say that a sequence of infinite transitive graphs $(\mathbb{G}_n)$ converges (locally) to an infinite transitive graph $\mathbb{G}$ if for any $R > 0$, there exists $N = N(R) > 0$ such that for all $n \geq N$, the balls of radius R in $\mathbb{G}_n$ and $\mathbb{G}$ are isomorphic (as rooted graphs). Schramm's conjecture for random-cluster models can be stated as follows.

Conjecture 1 (Schramm's conjecture for random-cluster model). Fix $q > 0$ and consider a sequence of infinite transitive $\mathbb{G}_n$ converging to $\mathbb{G}$. If $\sup_{n\geq 1} p_c(\mathbb{G}_n, q) < 1$, then,

$$\lim_{n\to\infty} p_c(\mathbb{G}_n, q) = p_c(\mathbb{G}, q). \tag{1}$$

For Bernoulli percolation, Pete [23, Section 14] noticed that

$$\liminf_{n\to\infty} p_c(\mathbb{G}_n, q) \geq p_c(\mathbb{G}, q) \tag{2}$$

can be deduced from the mean-field lower bound. This inequality can be obtained in a number of other (elementary) ways for Bernoulli percolation; see e.g. [27, Thm 5.3] or [12, Sec 1.2] for finite criteria approaches. The same inequality was also known in the $q = 2$ case; see [12].

For random-cluster models with $q \geq 1$, the mean-field lower bound was recently established by the authors and Aran Raoufi [11]. The first result of this note is to derive (2) from the mean-field lower bound.

Proposition 1. *Fix $q \geq 1$. Consider a sequence of infinite transitive graphs $\mathbb{G}_n$ converging to $\mathbb{G}$. Then,*

$$\liminf_{n\to\infty} p_c(\mathbb{G}_n, q) \geq p_c(\mathbb{G}, q).$$

Proof. Fix $q \geq 1$ and drop it from the notation. In [11] (see also [10] for a statement in the nearest neighbor case), one proves that for an infinite transitive graph $\mathbb{G}$, and $p \in [0, 1]$,

$$\phi^1_{\mathbb{G},p}[0 \longleftrightarrow \infty] \geq c(p - p_c(\mathbb{G})), \tag{3}$$

where the constant $c > 0$ depends a priori on p and $\mathbb{G}$, but can be bounded uniformly from below as p remains away from 0 and 1, and the degree of $\mathbb{G}$ remains bounded. If the liminf is equal to 1, there is nothing to prove and we now choose p such that

$$1 > p > \liminf_{n\to\infty} p_c(\mathbb{G}_n) =: \tilde{p}.$$

For any fixed $R > r > 0$, pick n large enough such that the ball B_R of size R in $\mathbb{G}_n$ is the same as in $\mathbb{G}$. By comparison between boundary conditions for the random-cluster model [15, Lemma 4.14], we deduce that

$$\phi^1_{B_R,p}[0 \longleftrightarrow \partial B_r] \geq \phi^1_{\mathbb{G}_n,p}[0 \longleftrightarrow \partial B_r] \geq \phi^1_{\mathbb{G}_n,p}[0 \longleftrightarrow \infty] \geq c(p - p_c(\mathbb{G}_n)).$$

(Above, B_r denotes the ball of size r around the origin.) In the last inequality, we used (3) together with the observation that for every n large enough, the ball of size 1 in $\mathbb{G}_n$ is isomorphic to the ball of size 1 in $\mathbb{G}$, and therefore the degrees of $\mathbb{G}_n$ and $\mathbb{G}$ are the same. As a consequence, the constant $c > 0$ can be chosen independent of n.

Taking the liminf implies that

$$\phi^1_{B_R,p}[0 \longleftrightarrow \partial B_r] \geq c(p - \tilde{p}).$$

Letting R tend to infinity, the convergence of $\phi^1_{B_R,p}$ to $\phi^1_{\mathbb{G},p}$ implies that

$$\phi^1_{\mathbb{G},p}[0 \longleftrightarrow \partial B_r] \geq c(p - \tilde{p}).$$

Letting r tend to infinity concludes that

$$\phi^1_{\mathbb{G},p}[0 \longleftrightarrow \infty] \geq c(p - \tilde{p}) > 0,$$

which implies that $p \geq p_c(\mathbb{G})$. The claim follows.

□

Exactly as in the case of Bernoulli percolation, the difficult part is to prove that

$$\limsup_{n\to\infty} p_c(\mathbb{G}_n, q) \le p_c(\mathbb{G}, q).$$

In particular, this raises the following natural question.

Question 1. Extend the locality results known for Bernoulli percolation to the random-cluster model with cluster-weight $q \ge 1$.

Quotient Graphs. Quotient graphs appear naturally when studying local limits of graphs. Let Γ be a normal subgroup of a group of automorphisms of $\mathbb{G} = (\mathbb{V}, \mathbb{E})$ acting transitively on $\mathbb{V}$. The quotient graph $\mathbb{G}/\Gamma$ is the (transitive, locally-finite) graph with vertex-set given by the equivalence classes $\{\Gamma v : v \in \mathbb{V}\}$ and edge-set given by the $\{\Gamma u, \Gamma v\}$ such that there exist $u_0 \in \Gamma u$ and $v_0 \in \Gamma v$ with $\{u_0, v_0\} \in \mathbb{E}$.

In [6], Benjamini and Schramm mentioned several questions regarding inequalities between the critical parameters of a graph and its quotient. We would like to highlight the fact that here the inequalities are not obvious at all in our case.

Question 2. Is it always the case that $p_c(\mathbb{G}, q) \le p_c(\mathbb{G}/\Gamma, q)$? If not, find a counter-example. If yes, when is the inequality strict?

In order to go back to our question on locality, notice that we may produce sequences of graphs converging to $\mathbb{G}$ by considering quotients by smaller and smaller groups of automorphisms. Actually, when restricted to Cayley graphs of finitely *presented* groups, the local convergence is always by quotient (this is explained in detail in [21] for abelian groups). For this reason, understanding the relation between percolation on a graph and its quotient is important toward a better understanding of Schramm's locality conjecture.

3 A Special Case of Schramm's Conjecture

A simple example of sequence $(\mathbb{G}_n)$ converging to a graph $\mathbb{G}$ is provided by the graphs $\mathbb{Z}^r \times (\mathbb{Z}/n\mathbb{Z})^{d-r}$ that converge to $\mathbb{Z}^d$ as n tends to infinity. For this graph, we obtain the following result.

Theorem 1. *Fix an integer $q \ge 2$ and $d > r \ge 3$, then*

$$\lim_{n\to\infty} p_c(\mathbb{Z}^r \times (\mathbb{Z}/2n\mathbb{Z})^{d-r}, q) = p_c(\mathbb{Z}^d, q).$$

On the one hand, there are two noteworthy restrictions to our theorem: we do not treat the natural case of $r = 2$, and more importantly we are bound to integer values of $q \ge 2$. This second restriction is important. Indeed, the random-cluster models with integer values of $q \ge 2$ enjoy some additional properties that are wrong for Bernoulli percolation. We believe these additional properties to be very interesting, and we would therefore like to encourage the reader to pursue potential applications (see the discussion below the proof of the theorem).

On the other hand, the proof of the theorem is very short. This is quite surprising since the similar result for Bernoulli percolation (i.e. $q = 1$) is not that simple to obtain: it relies on Grimmett and Marstrand's celebrated result [18] (see [20] for the explanation of how this result implies the one above for $q = 1$).

This also suggests that Schramm's locality conjecture may be simpler to obtain for integer values of $q \geq 2$ than for $q = 1$. This immediately raises the following question.

Question 3. Is Schramm's locality conjecture for a certain value of $q \geq 1$ implying the conjecture for other values of q?

To motivate this question, let us mention that Benjamini and Schramm [6] mentioned several questions regarding the existence of a phase transition for Bernoulli percolation on general graphs, i.e. whether $p_c(\mathbb{G}, 1) < 1$ or not. It can be easily proved using monotonocity arguments (see [15, Eq. (5.5)]) that for all $q \geq 1$, $p_c(\mathbb{G}, 1) < 1$ if and only if $p_c(\mathbb{G}, q) < 1$. In this case, the original questions of [6] can be translated directly into equivalent questions for random-cluster models.

4 Proof of Theorem 1

We will rely heavily on the connection between the random-cluster model with integer cluster-weight and the Potts model. The Potts model is one of the most fundamental examples of a lattice spin model and studying its properties near its phase transition is an active topic of research; see e.g. [10] for a recent account. Fix an integer $q \geq 2$ and introduce a polyhedron $\Omega \subset \mathbb{R}^{q-1}$ with q elements (often interpreted as colors) satisfying that for any $\mathrm{a}, \mathrm{b} \in \Omega$, $\mathrm{a} \cdot \mathrm{b}$ is equal to 1 if $\mathrm{a} = \mathrm{b}$ and $-1/(q-1)$ otherwise, where $\cdot$ denotes the scalar product on $\mathbb{R}^{q-1}$.

Let $G = (V, E)$ be a finite subgraph of $\mathbb{G}$ and $\beta > 0$. The *q-state Potts model on G at inverse-temperature $\beta > 0$ with monochromatic boundary conditions* $\mathrm{b} \in \Omega$ is defined as follows. The *energy* of a configuration $\sigma = (\sigma_x : x \in G) \in \Omega^V$ is given by the Hamiltonian

$$H_G^{\mathrm{b}}(\sigma) \ := \ - \sum_{\substack{x,y \in V \\ \{x,y\} \in E}} \sigma_x \cdot \sigma_y - \sum_{\substack{x \in V, y \notin V \\ \{x,y\} \in \mathbb{E}}} \sigma_x \cdot \mathrm{b}$$

and the probability measure $\langle \cdot \rangle_{G,\beta,q}^{\mathrm{b}}$ is defined by

$$\langle X \rangle_{G,\beta,q}^{\mathrm{b}} \ := \ \frac{\sum_{\sigma \in \Omega^V} X(\sigma) \exp[-\beta H_G^{\mathrm{b}}(\sigma)]}{\sum_{\sigma \in \Omega^V} \exp[-\beta H_G^{\mathrm{b}}(\sigma)]}$$

for every $X : \Omega^V \longrightarrow \mathbb{R}$.

The Potts model on $\mathbb{G}$ can be defined by taking the weak limit of measures on a nested sequence of finite graphs, exactly as for the random-cluster model. The infinite-volume measure is denoted by $\langle\cdot\rangle^{\mathrm{b}}_{\mathbb{G},\beta,q}$. The model undergoes a phase transition between absence/existence of long-range order at the so-called critical inverse temperature $\beta_c(\mathbb{G},q)$ defined by

$$m(\beta,\mathbb{G},q) := \langle\sigma_0\cdot \mathrm{b}\rangle^{\mathrm{b}}_{\mathbb{G},\beta,q}\begin{cases}=0 & \text{if } \beta<\beta_c(\mathbb{G},q),\\ >0 & \text{if } \beta>\beta_c(\mathbb{G},q).\end{cases}$$

The Potts model is coupled to the random-cluster model, see [15, Sec 1.4] in such a way that

$$m(\beta,\mathbb{G},q) = \phi^1_{\mathbb{G},p,q}[0\longleftrightarrow\infty]$$

when $\beta = -\frac{q-1}{q}\log(1-p)$. We therefore deduce that

$$\beta_c(\mathbb{G},q) := -\tfrac{q-1}{q}\log(1-p_c(\mathbb{G},q)) \tag{4}$$

so that Theorem 1 follows from the following result.

Theorem 2. *For integers $q\ge 2$ and $d>r\ge 3$,*

$$\lim_{n\to\infty}\beta_c(\mathbb{Z}^r\times(\mathbb{Z}/2n\mathbb{Z})^{d-r},q) \;=\; \beta_c(\mathbb{Z}^d,q).$$

The main tool used for the proof of this theorem is the so-called infrared bound. Define $\mathbb{T}_{N,n} := (\mathbb{Z}/2N\mathbb{Z})^r\times(\mathbb{Z}/2n\mathbb{Z})^{d-r}$. Following [7, Eq. (3.18)], introduce the Green function $\mathsf{G}_{N,n}$ on the torus $\mathbb{T}_{N,n}$ by

$$\forall x,y\in\mathbb{T}_{N,n}\qquad \mathsf{G}_{N,n}(x,y) := \frac{1}{|\mathbb{T}_{N,n}|}\sum_{\mathbf{k}\in\mathbb{T}^*_{N,n}\setminus\{0\}}\frac{e^{\mathbf{k}\cdot(\mathbf{x}-\mathbf{y})}}{1-\phi(\mathbf{k})}, \tag{5}$$

where $\phi(\mathbf{k}) = \frac{1}{d}\sum_{j=1}^d\cos(k_j)$ is the characteristic function associated with the simple random walk in dimension d, and

$$\mathbb{T}^*_{N,n} := [\tfrac{2\pi}{N}(\mathbb{Z}/2N\mathbb{Z})]^r\times[\tfrac{2\pi}{n}(\mathbb{Z}/2n\mathbb{Z})]^{d-r}.$$

Lemma 1. *Consider $\beta>0$ and two integers n and N. For any $v\in\mathbb{R}^{\mathbb{T}_{N,n}}$ with $\sum_x v_x = 0$,*

$$\sum_{x,y\in\mathbb{T}_{N,n}} v_xv_y\,\langle\sigma_x\cdot\sigma_y\rangle_{\mathbb{T}_{N,n},\beta}\;\le\;\tfrac{q-1}{2\beta}\sum_{x,y\in\mathbb{T}_{N,n}} v_xv_y\,\mathsf{G}_{N,n}(x,y).$$

This inequality follows from reflection positivity. It originated in the works [13, 14] and has since then been extremely useful in statistical physics. We do not include the proof of this result nor discuss it any further. We refer to [7] for a good review on the subject.

Note that the fact that the widths of the torus in the previous lemma have to be even is the reason why our theorem is restricted to $\mathbb{Z}^r \times (\mathbb{Z}/2n\mathbb{Z})^{d-r}$ and not simply to $\mathbb{Z}^r \times (\mathbb{Z}/n\mathbb{Z})^{d-r}$.

Even though we do not use them here, we mention for completeness that [1] provides bounds on $\mathsf{G}_{N,n}(x, y)$ that are valid for finite N and n.

Proof of Theorem 2. We use the notation $\mathbb{T}_{N,n}$ from above and set $\mathbb{T}_{\infty,n} := \mathbb{Z}^r \times (\mathbb{Z}/2n\mathbb{Z})^{d-r}$. Below, we fix β and q and drop them from the notation.

Fix a finite subset E of $\mathbb{Z}^d$ and see E as a subset of $\mathbb{T}_{N,n}$ provided that N and n are sufficiently large to contain it (injectively). Similarly, we see 0 as a vertex of $\mathbb{T}_{N,n}$.

Note that

$$\sum_{y \in \mathbb{T}_{N,n}} G_{N,n}(\cdot, y) = \sum_{x \in \mathbb{T}_{N,n}} G_{N,n}(x, \cdot) = 0$$

and

$$\sum_{y \in \mathbb{T}_{N,n}} \langle \sigma_0 \cdot \sigma_y \rangle_{\mathbb{T}_{N,n}} = \sum_{y \in \mathbb{T}_{N,n}} \langle \sigma_x \cdot \sigma_y \rangle_{\mathbb{T}_{N,n}}$$

for every $x, y \in \mathbb{T}_{N,n}$. We may therefore apply Lemma 1 in $\mathbb{T}_{N,n}$ to $v \in \mathbb{R}^{\mathbb{T}_{N,n}}$ defined for every $x \in \mathbb{T}_{N,n}$ by

$$v_x = \frac{1}{|\mathbb{T}_{N,n}|} - \frac{1}{|E|} \mathbb{1}[x \in E]$$

to get that

$$\underbrace{\frac{1}{|E|^2} \sum_{x,y \in E} \langle \sigma_x \cdot \sigma_y \rangle_{\mathbb{T}_{N,n}}}_{\mathsf{S}_{N,n}} - \underbrace{\frac{1}{|\mathbb{T}_{N,n}|} \sum_{y \in \mathbb{T}_{N,n}} \langle \sigma_0 \cdot \sigma_y \rangle_{\mathbb{T}_{N,n}}}_{\mathsf{T}_{N,n}}$$
$$\leq \tfrac{q-1}{2\beta} \underbrace{\frac{1}{|E|^2} \sum_{x,y \in E} \mathsf{G}_{N,n}(x, y)}_{\mathsf{U}_{N,n}}. \quad (6)$$

We first wish to take the limit of (6) as N and n tend to infinity. We start by the terms on the left. Since E is fixed, the terms $\mathsf{S}_{N,n}$ and $\mathsf{T}_{N,n}$ can be treated by taking limits term by term. It will be easier to use the random-cluster model to understand the different dominations. Set p such that

$$\beta = -\tfrac{q-1}{q} \log(1 - p).$$

Let $\phi_{\mathbb{T}_{N,n}}$ be the random-cluster model on $\mathbb{T}_{N,n}$ and $\phi^0_{\mathbb{Z}^d}$ be the random-cluster measure on $\mathbb{Z}^d$ with free boundary conditions (both with parameters p and q) [15, Thm 4.19]. Note that by coupling (see again [15, Sec 1.4]) we have

$$\langle \sigma_x \cdot \sigma_y \rangle_{\mathbb{T}_{N,n}} = \phi_{\mathbb{T}_{N,n}}[x \longleftrightarrow y].$$

Any sub-sequential limit (as n and N tend to infinity) of the family $(\phi_{\mathbb{T}_{N,n}})$ is a limit-random-cluster measure on $\mathbb{Z}^d$ [15, Def 4.15] and is therefore stochastically dominating $\phi^0_{N,n}$ [15, Thm 4.19]. As a consequence, we obtain that

$$\begin{aligned}\liminf_{n\to\infty}\liminf_{N\to\infty}\ \phi_{\mathbb{T}_{N,n}}[x\longleftrightarrow y] &\geq \phi^0_{\mathbb{Z}^d}[x\longleftrightarrow y]\\ &\geq \phi^0_{\mathbb{Z}^d}[x,y\longleftrightarrow\infty]\\ &\geq \phi^0_{\mathbb{Z}^d}[0\longleftrightarrow\infty]^2,\end{aligned}$$

where in the third inequality we used the uniqueness of the infinite-connected component [15, Theorem 4.33], and in the last, the FKG inequality [15, Thm 3.8] and the invariance under translation of $\phi^0_{\mathbb{Z}^d}$ [15, Thm 4.19]. Since this is true for every $x,y\in E$, we deduce that

$$\liminf_{n\to\infty}\liminf_{N\to\infty}\ \mathsf{S}_{N,n}\geq \phi^0_{\mathbb{Z}^d}[0\longleftrightarrow\infty]^2, \tag{7}$$

For the limit of $\mathsf{T}_{N,n}$, we use again the random-cluster model. Fix n. For every y in $\mathbb{T}_{N,n}$ at graph distance larger than $2R\leq N$ from 0, the comparison between boundary conditions [15, Lem 4.14] gives

$$\phi_{\mathbb{T}_{N,n}}[x\longleftrightarrow y]\leq \phi^1_{B_R}[0\longleftrightarrow\partial B_R]\times\phi^1_{B'_R}[x\longleftrightarrow\partial B'_R]\leq\phi^1_{B_R}[0\longleftrightarrow\partial B_R]^2,$$

where we recall that B_R is the ball of radius R around 0 in $\mathbb{T}_{\infty,n}$ and B'_R is its translate by y. Bounding $\phi_{\mathbb{T}_{N,n}}[0\longleftrightarrow y]$ using the inequality above when 0 and y are at a distance larger than $2R$, and otherwise by 1 for remaining values of y, gives

$$\frac{1}{|\mathbb{T}_{N,n}|}\sum_{y\in\mathbb{T}_{N,n}}\phi_{\mathbb{T}_{N,n}}[0\longleftrightarrow y]\leq\frac{|B_R|}{|\mathbb{T}_{N,n}|}+\phi^1_{B_R}[0\longleftrightarrow\partial B_R]^2.$$

Taking the limit as N tends to infinity (recall that $\phi^1_{B_R}$ tends to $\phi^1_{\mathbb{Z}^d}$), and then letting R tend to infinity, gives

$$\limsup_{N\to\infty}\frac{1}{|\mathbb{T}_{N,n}|}\sum_{y\in\mathbb{T}_{N,n}}\langle\sigma_0\cdot\sigma_y\rangle_{\mathbb{T}_{N,n}}\leq\phi^1_{\mathbb{T}_{\infty,n}}[0\longleftrightarrow\infty]^2.$$

Taking the limit as n tends to infinity gives that

$$\limsup_{n\to\infty}\limsup_{N\to\infty}\mathsf{T}_{N,n}\leq\limsup_{n\to\infty}\phi^1_{\mathbb{T}_{\infty,n}}[0\longleftrightarrow\infty]^2. \tag{8}$$

Let us now turn to $\mathsf{U}_{N,n}$ on the right-hand side. As N tends to infinity, the torus Green function $\mathsf{G}_{N,n}(x,y)$ converges to the "slab" Green function $\mathsf{G}_{\infty,n}(x,y)$ associated with the random walk in $\mathbb{T}_{\infty,n}$ (as a Riemann sum, the right hand side of (5) converges to the Fourier representation of $\mathsf{G}_{\infty,n}(x,y)$, this uses that $r\geq 3$ since one needs the walk in the slab to be transient). Then, the limit of

$\mathsf{G}_{\infty,n}(x,y)$ converges as n tends to infinity to the Green function G associated to the simple random walk in $\mathbb{Z}^d$. Therefore

$$\lim_{n\to\infty}\lim_{N\to\infty}\mathsf{U}_{N,n}=\frac{1}{|E|^2}\sum_{x,y\in E}\mathsf{G}(x,y). \tag{9}$$

We can now plug (7), (8) and (9) in (6) to get that for every E,

$$\phi^0_{\mathbb{Z}^d}[0\longleftrightarrow\infty]^2-\limsup_{n\to\infty}\phi^1_{\mathbb{T}_{\infty,n}}[0\longleftrightarrow\infty]^2\le\frac{1}{|E|^2}\sum_{x,y\in E}\mathsf{G}(x,y). \tag{10}$$

Now, the random walk in $\mathbb{Z}^d$ is transient, so that $\mathsf{G}(x,y)$ tends to 0 as x and y become far apart. We deduce that, as $|E|$ tends to infinity, the right-hand side tends to 0. Overall, we find

$$\phi^0_{\mathbb{Z}^d}[0\longleftrightarrow\infty]\le\limsup_{n\to\infty}\phi^1_{\mathbb{T}_{\infty,n}}[0\longleftrightarrow\infty]. \tag{11}$$

Since $p_c(\mathbb{Z}^d)$ is also defined as the supremum of the values of p for which $\phi^0_{\mathbb{Z}^d,p}[0\longleftrightarrow\infty]=0$ [15, (5.4)], this immediately implies that

$$p_c(\mathbb{Z}^d)\ge\limsup_{n\to\infty}p_c(\mathbb{T}_{\infty,n})$$

or equivalently by (4),

$$\beta_c(\mathbb{Z}^d)\ge\limsup_{n\to\infty}\beta_c(\mathbb{T}_{\infty,n}).$$

Since the inequality $\beta_c(\mathbb{Z}^d)\le\liminf\beta_c(\mathbb{T}_{\infty,n})$ is given by Proposition 1, this concludes the proof. □

Note that the estimate given by the infrared bound is absolutely crucial here and that we do not know how to go around it for non-integer values of q. This perfectly illustrates that the random-cluster models associated with the Ising and Potts models may be simpler to handle than Bernoulli percolation in some cases. Other instances of such a phenomenon include proofs of conformal invariance in two dimensions for the random-cluster model with cluster-weight $q=2$ on the square lattice [25] (the corresponding result is still open for Bernoulli percolation) or the proof of continuity of the phase transition for the Ising model in dimension 3 and therefore the random-cluster model with $q=2$ in dimension 3 [2] (the corresponding statement is one of the main conjectures in the theory of Bernoulli percolation).

We wish to advertise the following question, which would require to find an argument not relying on the infrared bound:

Question 4. Prove that $p_c(\mathbb{Z}^r\times(\mathbb{Z}/2n\mathbb{Z})^{d-r},q)$ converges to $p_c(\mathbb{Z}^d,q)$ for any (non-necessarily integer valued) $q\ge1$ and any $d>r\ge2$.

5 Comparison with the Slab Percolation Threshold

In this section, we link the previous result to the Grimmett–Marstrand result [18]. Define the random-cluster model on the infinite graph $\mathbb{Z}^r \times [\![0,n]\!]^{d-r}$. Even though this graph is not transitive, a non-ambiguous notion of a critical point can be defined as before (asking for the smallest value of p for which 0 is connected to infinity with positive probability).

Question 5. Show that $p_c(\mathbb{Z}^r \times [\![0,n]\!]^{d-r}, q)$ converges to $p_c(\mathbb{Z}^d, q)$ for $q \geq 1$ and $d > r \geq 2$.

Even for integers $q \geq 2$ and $r \geq 3$, this question, first raised in [22], is open. Note that combined with the following simple proposition, it would imply that finite connected components have exponential tails for $p > p_c(\mathbb{Z}^d)$ (when $d = 2$, this result follows from duality and [3]).

Proposition 2 ([15, Thm. 5.104])**.** *Fix $q \geq 1$ and $d > r \geq 2$. If $p_c(\mathbb{Z}^r \times [\![0,n]\!]^{d-r}, q)$ converges to $p_c(\mathbb{Z}^d, q)$, then for any $p > p_c(\mathbb{Z}^d, q)$, there exists $c = c(p,q) > 0$ such that*

$$\phi^1_{\mathbb{Z}^d,p,q}[0 \longleftrightarrow x, 0 \not\longleftrightarrow \infty] \leq \exp(-c\|x\|).$$

Let us conclude that the convergence of $p_c(\mathbb{Z}^r \times [\![0,n]\!]^{d-r}, q)$ to $p_c(\mathbb{Z}^d, q)$ is equivalent to the convergence of $\beta_c(\mathbb{Z}^r \times [\![0,n]\!]^{d-r}, q)$ to $\beta_c(\mathbb{Z}^d, q)$ as n tends to infinity. This result was proved for $q = 2$ (i.e. for the Ising model) in [8]. In this paper, Bodineau obtained a slightly stronger result related to the *slab percolation threshold* $\hat{\beta}_c(\mathbb{Z}^d)$ of the Ising model defined by

$$\hat{\beta}_c(\mathbb{Z}^d) := \inf\big\{\beta \text{ for which } \exists n \geq 0 \text{ such that } \inf_N \inf_{x,y\in\Lambda_{N,n}} \langle\sigma_x{\cdot}\sigma_y\rangle_{\Lambda(2N,n),\beta,2} > 0\big\},$$

where $\Lambda_{N,n} := [\![-N,N]\!]^{d-1} \times [\![-n,n]\!]$ and $\langle\cdot\rangle_{\Lambda(2N,n),\beta,2}$ is the $q = 2$ Potts measure on $\Lambda_{N,n}$ with free boundary conditions (we omit the definition here).

This notion, introduced by Pisztora in [22], enables one to perform a powerful renormalization scheme to derive a number of properties of the regime $\beta > \hat{\beta}_c(\mathbb{Z}^d)$. Motivated by [22], we propose to enhance Question 5 into the following one. One can easily extend the notion of the slab percolation threshold to Potts model by replacing $q = 2$ with an arbitrary integer $q \geq 2$.

Question 6. Show that $\hat{\beta}_c(\mathbb{Z}^d, q)$ is equal to $\beta_c(\mathbb{Z}^d, q)$ for integer values of $q \geq 3$.

Let us recall for completeness that despite the fact that the regime $\beta > \hat{\beta}_c(\mathbb{Z}^d, q)$ is well understood, several fundamental questions remain open even under this additional assumption on β (and sometimes even in the case of the Ising model): to mention but a few, the continuity of the magnetization parameter, exponential decay of truncated correlations, description of the translational invariant Gibbs states (see [9] and [24] for the case of the Ising model), etc. Investigating these questions further is of prime importance.

Acknowledgments. This research was supported by the IDEX grant from Paris-Saclay, the ERC grant CriBLaM, and the NCCR SwissMAP. We thank Angelo Abächerli for useful discussions, and Sébastien Martineau and Aran Raoufi for their careful reading of our manuscript and their insightful comments and suggestions.

References

1. Abächerli, A.: Local picture and level-set percolation of the Gaussian free field on a large discrete torus. To appear in Stochastic processes and Applications. arXiv:1707.05935 (2017)
2. Aizenman, M., Duminil-Copin, H., Sidoravicius, V.: Random currents and continuity of Ising model's spontaneous magnetization. Commun. Math. Phys. **334**, 719–742 (2015)
3. Beffara, V., Duminil-Copin, H.: The self-dual point of the two-dimensional random-cluster model is critical for $q \geq 1$. Probab. Theory Related Fields **153**(3–4), 511–542 (2012)
4. Benjamini, I.: Euclidean vs Graph metric. In: Erdos Centennial, pp. 35–57. Springer, Heidelberg (2013)
5. Benjamini, I., Nachmias, A., Peres, Y.: Is critical percolation local? Probab. Theory Related Fields **149**, 261–269 (2011)
6. Benjamini, I., Schramm, O.: Percolation beyond $\mathbb{Z}^d$, many questions and a few answers and a few answers. Electron. Commun. Prob. **1**, 71–82 (1996)
7. Biskup, M.: Reflection positivity and phase transitions in lattice spin models. In: Methods of Contemporary Mathematical Statistical Physics, pp. 1–86. Springer (2009)
8. Bodineau, T.: Slab percolation for the Ising model. Prob. Theory Related Fields **132**(1), 83–118 (2005)
9. Bodineau, T.: Translation invariant Gibbs states for the Ising model. Prob. Theory Related Fields **135**(2), 153–168 (2006)
10. Duminil-Copin, H.: Lectures on the lsing and Potts models on the hypercubic lattice. Preprint arXiv:1707.00520 (2017)
11. Duminil-Copin, H., Raoufi, A., Tassion, V.: Sharp phase transition for the random-cluster and Potts models via decision trees. Annal. Math. **189**(1), 75–99 (2019)
12. Duminil-Copin, H., Tassion, V.: A new proof of the sharpness of the phase transition for Bernoulli percolation and the Ising model. Commun. Math. Phys. **343**(2), 725–745 (2016)
13. Fröhlich, J., Simon, B., Spencer, T.: Infrared bounds, phase transitions and continuous symmetry breaking. Comm. Math. Phys. **50**(1), 79–95 (1976)
14. Fröhlich, J., Spencer, T.: The Kosterlitz-Thouless transition in two-dimensional abelian spin systems and the Coulomb gas. Comm. Math. Phys. **81**(4), 527–602 (1981)
15. Grimmett, G.: The random-cluster model. Volume 333 of Grundlehren der Mathematischen Wissenschaften (Fundamental Principles of Mathematical Sciences). Springer, Berlin (2006)
16. Grimmett, G., Li, Z.: Locality of connective constants, I. Transitive graphs. arXiv:1412.0150 (2014)
17. Grimmett, G., Li, Z.: Locality of connective constants, II. Cayley graphs. arXiv:1501.00476 (2015)
18. Grimmett, G.R., Marstrand, J.M.: The supercritical phase of percolation is well behaved. Proc. R. Soc. London Ser. A **430**(1879), 439–457 (1990)

19. Hutchcroft, T.: Locality of the critical probability for transitive graphs of exponential growth. arxiv:188:08940 (2018)
20. De Lima, B.N.B., Sanchis, R., Silva, R.W.C.: Critical point and percolation probability in a long range site percolation model on $\mathbb{Z}^d$. Stoch. Process. Appl. **121**(9), 2043–2048 (2011)
21. Martineau, S., Tassion, V.: Locality of percolation for abelian Cayley graphs. Annal. Prob. **45**(2), 1247–1277 (2017)
22. Pisztora, A.: Surface order large deviations of Ising, Potts and percolation models. Probab. Th. Rel. Fields **104**, 427–466 (1996)
23. Pete, G.: Probability and geometry on groups. Lecture notes for a graduate course, Lecture Notes (2015)
24. Raoufi, A.: Translation Invariant lsing Gibbs States: General Setting (2017, To appear)
25. Smirnov, S.: Conformal invariance in random cluster models. I. Holomorphic fermions in the Ising model. Ann. Math. **172**(2), 1435–1467 (2010)
26. Song, H., Xiang, K.N., Zhu, S.C.H.: Locality of percolation critical probabilities: uniformly nonamenable case. arXiv:1410.2453 (2014)
27. Tassion, V.: Planarity and locality in percolation theory. Theses, Ecole Normale Supérieure de Lyon - ENS LYON, June 2014

Rooted Tree Graphs and the Butcher Group: Combinatorics of Elementary Perturbation Theory

William G. Faris[1,2(✉)]

[1] NYU Shanghai, Shanghai, China
[2] Department of Mathematics, University of Arizona, Tucson, AZ 85721, USA
faris@math.arizona.edu

Abstract. The perturbation expansion of the solution of a fixed point equation or of an ordinary differential equation may be expressed as a power series in the perturbation parameter. The terms in this series are indexed by rooted trees and depend on a parameter in the equation in a way determined by the structure of the tree. Power series of this form may be considered more generally; there are two interesting and useful group structures on these series, corresponding to operations of composition and substitution. The composition operation defines the Butcher group, an infinite dimensional group that was first introduced in the context of numerical analysis. This survey discusses various ways of realizing these rooted trees: as labeled rooted trees, or increasing labeled rooted trees, or unlabeled rooted trees. It is argued that the simplest framework is to use labeled rooted trees.

Keywords: Perturbation expansion · Rooted tree graphs · Butcher group · Combinatorial species · Connes–Kreimer renormalization · Hopf algebras of trees

1 Introduction

This year we celebrate the scientific contributions of Charles Newman. Chuck has worked in almost every aspect of mathematical physics. He can identify a significant problem, locate the appropriate framework, and find an unexpected path to a comprehensible solution. His insights and his generosity in sharing them are extraordinarily valuable to the community. Chuck has been my friend and colleague from years together at the University of Arizona and now at NYU Shanghai. It is a pleasure to dedicate this paper to such a distinguished scientist.

The paper is a largely expository survey of rooted trees and the Butcher group. The Butcher group is an infinite dimensional group associated with rooted trees. See [8] for the current status of this subject. Most expositions are in the framework of unlabeled rooted trees. The main message of the present work is

V. Sidoravicius (Ed.): *Sojourns in Probability Theory and Statistical Physics - II*, PROMS 299, pp. 135–166, 2019.
https://doi.org/10.1007/978-981-15-0298-9_6

that the combinatorics is simpler when formulated in terms of labeled rooted trees. In particular, group structures associated with rooted trees arise naturally from calculations using elementary calculus formulas. This is in the spirit of the theory of combinatorial species [1]. This same point of view is fruitful in the study of graph expansions in statistical mechanics [10] and of diagram expansions in quantum field theory [11].

The subject matter of the present work has two independent origins. One begins in 1963 with work of Butcher in numerical analysis. He discovered that a large class of numerical methods for ordinary differential equations may be expressed as sums indexed by rooted trees. Furthermore, these sums may be combined in a way that defines a group structure. This subject has become part of the lore of numerical analysis [6,13,14]. It remains active; for instance see [4,16] and works cited therein.

The other origin is research on renormalization in quantum field theory, starting with the 1998 contribution of Connes and Kreimer [9]. This work is usually presented in the language of combinatorial Hopf algebras. Many authors, for instance [5,12], have investigated the relation between the Butcher group and problems in quantum field theory. Recently there has been an explosion of mathematics papers treating Hopf algebras associated with rooted trees. See for instance [7] and papers cited there.

The approach here begins by distinguishing basic rooted tree constructions. The starting point is a finite set U known as the *label set* or *vertex set*. A labeled rooted tree on U is a tree graph with vertex set U together with a distinguished point r in U. An increasing rooted tree is one for which the label set U is linearly ordered and the labels increase with distance from the root r. An unlabeled rooted tree is an isomorphism class of labeled rooted trees.

- $\mathcal{A}[U]$ is all rooted trees on label set U.
- $\mathcal{A}^{\uparrow}[U]$ is all increasing rooted trees on linearly ordered label set U.
- $\tilde{\mathcal{A}}[n]$ is all unlabeled rooted trees with n vertices.

The $\mathcal{A}$ notation is from the French "arbre". The relation between these constructions is that if U has n elements, then

$$\mathcal{A}^{\uparrow}[U] \to \mathcal{A}[U] \to \tilde{\mathcal{A}}[n].$$

The first map is an injection, and the second map is a surjection.

The topics discussed include:

- Unlabeled, labeled, and increasing rooted trees
- Fixed point equations and ordinary differential equations
- The composition operation (Butcher group)
- The substitution operation

An appendix reviews calculus formulas as used in combinatorics.

Example 1. Figure 1 illustrates the distinction between labeled rooted trees and unlabeled rooted trees in the case of a vertex set with three elements. There are 9 labeled rooted trees. However there are only 2 unlabeled rooted trees. One of these has a symmetry that exchanges the two non-root vertices.

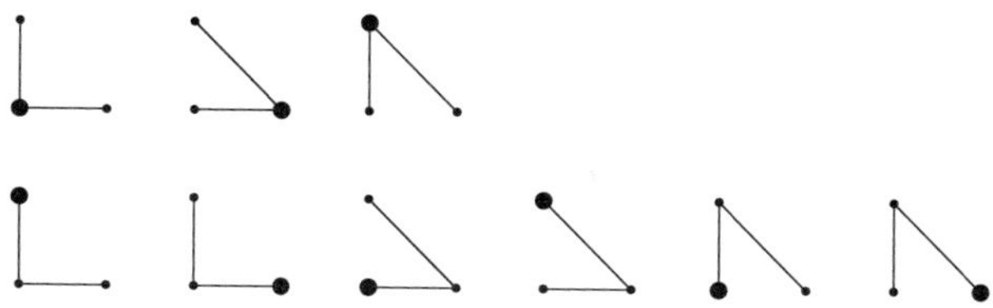

Fig. 1. Labeled rooted trees on three vertices

2 Labeled Rooted Trees

Labeled Rooted Trees as Functions. Let U be a non-empty finite set. Let $f : U \to U$ be a function with a single fixed point r such that $U \setminus \{r\}$ has no non-empty invariant subset. Then f defines a labeled rooted tree. In the following it will be more convenient to consider the function f restricted to $U \setminus \{r\}$. This leads to the official definition of labeled rooted tree used here.

Let U be a non-empty finite set. Let r be a point in U, and let $T : U \setminus \{r\} \to U$ be a function that has no non-empty invariant subset. Then T is called a *labeled rooted tree*. The set U is the label set, and the point r is the *root*. A point in U is called a *label* or a *vertex*. Each ordered pair $(i, T(i))$ with $i \neq r$ is called an *edge*. The point $T(i)$ is called the *predecessor* of i.

The set of labeled rooted trees with label set U is $\mathcal{A}[U]$. If U is empty, then there are no labeled rooted trees on U, so $\mathcal{A}[\emptyset] = \emptyset$. If T is in $\mathcal{A}[U]$, then $[T] = U$ is the label set of T. The number of points in the label set is $|T| = |U|$. The set of *immediate successor* points that map to j in U is $T^{-1}(j)$. The *degree* of j is $|T^{-1}(j)|$, the number of immediate successor points. (The definition of degree used here is special to rooted trees; it is not the usual definition from graph theory.) A *leaf* is a vertex with degree zero. For a one-vertex tree $\bullet$ the root is a leaf.

If $b : W \to U$ is a bijection, then the map $T \mapsto T \circ b$ maps $\mathcal{A}[U]$ to $\mathcal{A}[W]$. Such a map is called a *relabeling*. Most interesting properties of labeled rooted trees are not affected by relabeling. It might seem reasonable to use a standard label set U_n for each n. An obvious candidate is $[1, n]$, that is, the set $\{1, \ldots, n\}$. On the other hand, it is common to consider a labeled rooted tree on a subset of U, and that means that other labels sets are going to arise naturally.

The set $\mathcal{A}$ of all labeled rooted trees may be defined by choosing a label set U with $|U| = n$ for each $n = 1, 2, 3, \ldots$. For some purposes it is useful to adjoin an empty set object associated with the empty label set. The set of all labeled rooted trees with this extra object is written $\mathcal{A}_\emptyset$.

Labeled Rooted Trees as Partially Ordered Sets. A labeled rooted tree T on a finite label set $U \neq \emptyset$ may be viewed as a partial order $\leq_{\text{tree}}$ on U. This is the unique partial order with the property that $T(j) \leq_{\text{tree}} j$ for every vertex $j \neq r$.

For each vertex i there is a rooted tree T_i whose vertex set consists of all vertices that are sent to i by some iterate of T. The tree T_i is the restriction of

T to this set. To say that $i \leq_{\text{tree}} j$ is the same as saying that j is in the vertex set of the tree T_i.

The special feature of this partial order is that for each j in U the set of all $i \leq_{\text{tree}} j$ is linearly ordered with respect to the restriction of $\leq_{\text{tree}}$ to this set. Furthermore, there is a least element in U, the root r. There can be one or more maximal elements; these are the leaves.

Labeled Rooted Trees as Graphs. A labeled tree T on a non-empty finite set U is a simple graph with U as vertex set that is connected and has no cycles. For every pair of vertices $i \neq j$ there is a unique simple path connecting the two points. A labeled rooted tree T on U is equivalent to a labeled tree and a choice of root point in U. For each vertex i other than the root, there is a unique edge from i in the direction of the root and corresponding vertex $T(i)$. The usual way of picturing a labeled rooted tree is as a set together with tree graph and distinguished point.

Forests of Labeled Rooted Trees. Let V be a finite set. Let $f : V \to V$ be a function with a set of fixed points R such that $U \setminus R$ has no non-empty invariant subset. Then consider the function f restricted to $U \setminus R$. This motivates the official definition of forest of labeled rooted trees.

A *forest of labeled rooted trees* on a set V is a subset $R \subseteq V$ and a function $F : V \setminus R \to V$ such that $V \setminus R$ has no non-empty invariant subset. It is possible that $V = \emptyset$, in which case $R = \emptyset$, and F is the empty function.

The most important fact about a forest of labeled rooted trees is that there is a set partition Γ of V with the following property. For each block B of Γ there is a unique r in R such that the restriction F_B of F to $B \setminus \{r\}$ is a labeled rooted tree. When $V = \emptyset$ this is the empty set partition.

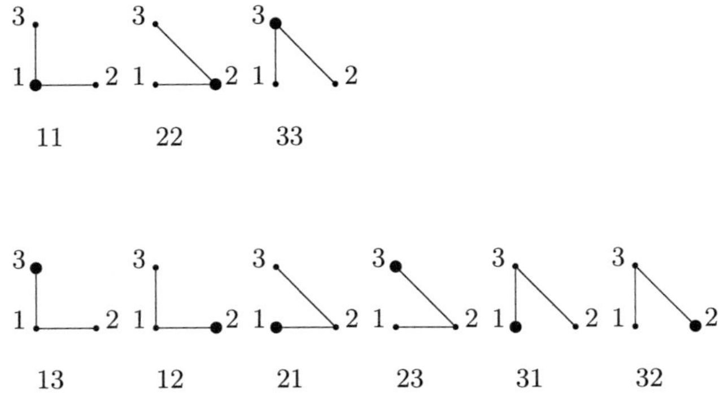

Fig. 2. Labeled rooted trees on three vertices: Prüfer sequences

Labeled Rooted Trees Defined Recursively. A labeled rooted tree T on U has a recursive definition as a point r in U together with a forest of labeled rooted trees on $U \setminus \{r\}$. This forest F is the restriction of T to points in $U \setminus \{r\}$ that are not in $T^{-1}(r)$. The set of roots of the forest is $R = T^{-1}(r)$. The recursive definition ends when the tree consists only of a root; the forest is then empty.

Notation for Labeled Rooted Trees. A labeled rooted tree on a one-point set may be designated by its label j. A labeled rooted tree on a set with two or more points may be denoted by $j[\quad]$, where the forest of immediate successor rooted trees is listed (in some arbitrary order) inside the bracket. As an example, consider the tree with root c and with vertices b, e that are sent to c and vertices a, d that are sent to e. In this notation the tree would be $c[be[ad]]$.

Labeled Rooted Trees as Sequences of Vertices. Consider a non-empty vertex set U with n elements. For the construction to follow it is necessary to impose a linear order on U. The result says that labeled rooted trees on U correspond to sequences of $n-1$ elements.

Example 2. For $n = 3$ the vertices of a labeled rooted tree may be numbered 1,2,3. As shown below, each labeled rooted tree may be coded by a sequence of two numbers. For example, the tree 2[13] is coded by 22, wile the tree 2[1[3]] is coded by 12. There are nine sequences: three of them 11, 22, 33 correspond to one unlableled rooted tree, while six of them 12, 13, 21, 23, 31, 33 correspond to the other unlabeled rooted tree. See Fig. 2 for the picture.

Proposition 1 (Prüfer correspondence). *Given non-empty label set U with a given linear order, there is a bijection between the set of sequences s of length $n-1$ of elements of U and the set of labeled rooted trees T in $\mathcal{A}[U]$. For each j in U the number of times the sequence s assumes the value j is the degree $|T^{-1}(j)|$.*

Proof. It is easiest to see how to go from the labeled rooted tree T to the corresponding sequence s. At each stage remove the smallest leaf and its corresponding edge from the tree. The value of s at this stage is the vertex at the other end of this edge. When this is repeated $k-1$ times the final sequence value is the root. This procedure is illustrated in Fig. 3.

Here is the construction to go from the sequence s to the labeled rooted tree. The edges are restored in the same order as they were removed. At each stage add a new edge as follows. Take the smallest vertex that has not yet been used and that does not occur in the part of the sequence that has not been used. The edge then goes from this vertex to the next element of the sequence. See Fig. 4 for a picture. □

Proposition 2 (Cayley). *The number of labeled rooted trees T with $|T| = n$ vertices is n^{n-1}.*

Proof. This famous result of Cayley results from the Prüfer correspondence. There are n^{n-1} sequences of length $n-1$ in a set U with n elements. □

3 Unlabeled Rooted Trees

The difficulty with unlabeled rooted trees is the general difficulty with unlabeled combinatorial structures (isomorphism classes of structures). This is the presence of symmetry. This is a well-studied topic; there are nice accounts in [1] and in [15].

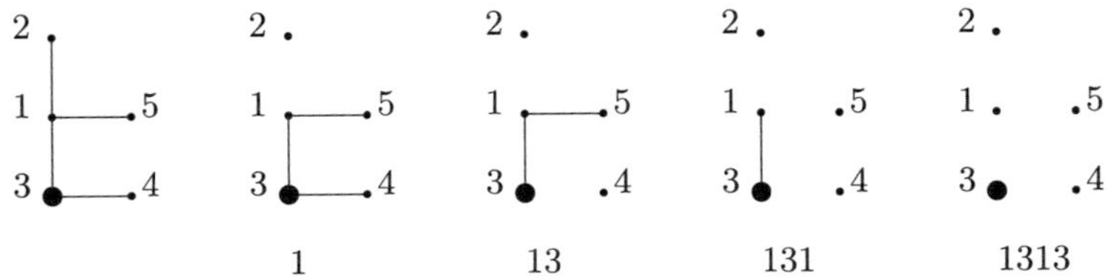

Fig. 3. From labeled rooted tree to Prüfer sequence

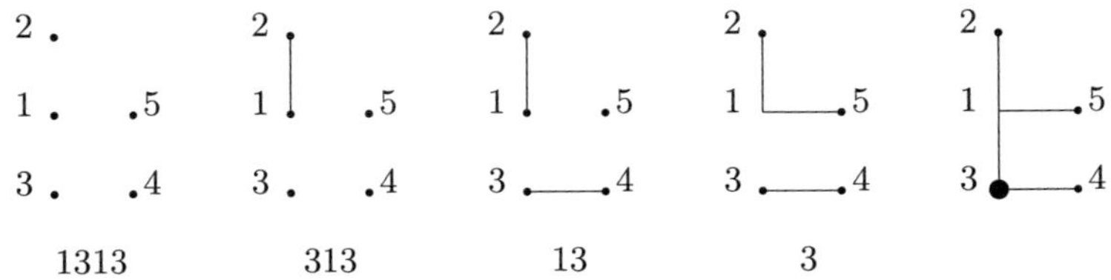

Fig. 4. From Prüfer sequence to labeled rooted tree

Unlabeled Rooted Trees via Orbits of Labeled Rooted Trees. An *unlabeled rooted tree* is an isomorphism invariant of labeled rooted trees. It is convenient to fix the label set U. For each labeled rooted tree T on U there is a corresponding unlabeled rooted tree τ. Here are the details.

Let U be a non-empty set. Let $\mathcal{A}[U]$ be the set of labeled rooted trees on vertex set U. Each such tree is a function $T : U \setminus \{r\}$ to U. For each T in $\mathcal{A}[U]$ and each bijection $b : U \to U$, the composite function $T \circ b$ is another labeled rooted tree in $\mathcal{A}[U]$. It is a function $T' = T \circ b : U' \setminus \{r'\}$ to U, where $br' = r$. Two such labeled rooted trees T, T' are said to be isomorphic.

An *unlabeled rooted tree* τ with n vertices is an object that corresponds to an isomorphism class of labeled rooted trees, where the label set has n elements. The set of unlabeled rooted trees with n vertices is denoted $\tilde{\mathcal{A}}[n]$. The set of all unlabeled rooted trees is denoted $\tilde{\mathcal{A}}$. There is no unlabeled rooted tree with zero vertices, but sometimes it is convenient to introduce an extra empty set object associated with zero vertices. The augmented set is written $\tilde{\mathcal{A}}_{\emptyset}$.

An unlabeled rooted tree τ has no underlying set and thus no vertices and no edges. Nevertheless, it may be pictured by any T in the isomorphism class. There are various invariants of an unlabeled rooted tree τ. Among them are the number of vertices $|\tau|$ and the number of vertices $v_k(\tau)$ of degree k. These are related by

$$\sum_k k v_k(\tau) = |\tau| - 1.$$

There are relatively few unlabeled rooted trees. The simplest (but nevertheless important) ones have 1 root and $n-1$ leaves. In the following such a tree will be denoted $\tau = n-1$. For this tree $|\tau| = n$ and $v_0(\tau) = n-1$, $v_{n-1}(\tau) = 1$. Another simple class of unlabeled rooted trees are the linear trees with $v_0(\tau) = 1$ and $v_1(\tau) = n-1$.

Group theory illuminates the situation. Fix a label set U with n elements. Let $\mathcal{G}$ be the permutation group of this set. This consists of all bijections $b : U \to U$. This group has $|\mathcal{G}| = n!$ elements. The group $\mathcal{G}$ also acts on the set $\mathcal{A}[U]$ of labeled rooted trees with n^{n-1} elements. For each bijection b, the map $T \to T \circ b$ is a map from $\mathcal{A}[U]$ to itself. The set $\tilde{\mathcal{A}}[n]$ of unlabeled rooted trees corresponds to the set of orbits under this action:

$$\tilde{\mathcal{A}}[n] \cong \mathcal{A}[U]/\mathcal{G}.$$

For each labeled rooted tree T, the corresponding unlabeled rooted tree τ is an abstract object corresponding to the orbit $\mathcal{G}T$ of T. The map from labeled rooted trees to unlabeled rooted trees may be summarized by the surjection

$$\mathcal{A}[U] \to \tilde{\mathcal{A}}[n].$$

According to the theory of group actions, the size of the orbit $\mathcal{G}T$ of T is the order $|\mathcal{G}| = n!$ divided by the order $|\mathcal{G}_T|$, where $\mathcal{G}_T$ is the stabilizer subgroup of T. Thus

$$|\mathcal{G}T| = \frac{n!}{|\mathcal{G}_T|}.$$

The order $|\mathcal{G}_T|$ is the same for all T in an orbit and hence may be denoted $\sigma(\tau)$, where τ is the unlabeled rooted tree corresponding to T. The number $\sigma(\tau)$ is the *symmetry factor* of τ. The size of the orbit also depends only on τ; it will be denoted $r(\tau)$. This gives the following basic result:

Proposition 3. *For rooted trees with $|\tau| = n$ vertices the number of labeled rooted trees in $\mathcal{A}[U]$ per unlabeled rooted tree in $\tilde{\mathcal{A}}[n]$ is*

$$r(\tau) = \frac{n!}{\sigma(\tau)}.$$

Example 3. There are labeled rooted trees on $n = 4$ vertices that consist of a root and three leaves. The orbit $\mathcal{G}T$ of such a rooted tree is shown in Fig. 5. For each rooted tree T in this orbit, the corresponding stabilizer subgroup $\mathcal{G}_T$ has 6 elements, corresponding to the 6 permutations of the leaves. The symmetry factor is $\sigma(\tau) = 6$. The number of labeled rooted trees in the orbit is $r(\tau) = 24/6 = 4$.

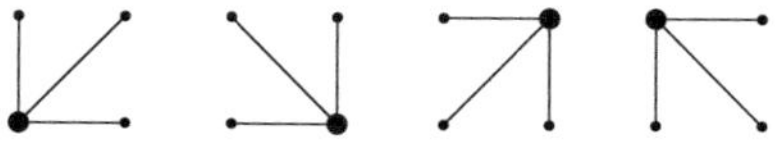

Fig. 5. Orbit of a labeled rooted tree on four vertices: 4 trees, $\sigma(\tau) = 6$

There is an identity that expresses the fact that the sum over unlabeled rooted trees of the corresponding number of labeled rooted trees is the total number of labeled rooted trees.

Proposition 4. *For unlabeled rooted trees τ with $|\tau| = n$ vertices the sum of the corresponding numbers $r(\tau)$ of labeled rooted trees in $\mathcal{A}[U]$ with $|U| = n$ is*

$$\sum_{\tau \in \tilde{\mathcal{A}}[n]} \frac{n!}{\sigma(\tau)} = n^{n-1}.$$

Multisets of Unlabeled Rooted Trees. The analog of a forest of labeled rooted trees is a multiset of unlabeled rooted trees. A multiset of unlabeled rooted trees with m vertices is defined as a function N from unlabeled rooted trees in $\tilde{\mathcal{A}}$ to natural numbers ≥ 0 such that

$$\sum_{\tau} |\tau| N(\tau) = m.$$

Thus $N(\tau)$ represents the number of times the unlabeled rooted tree τ occurs in the multiset. Every forest of labeled rooted trees gives rise to a multiset of unlabeled rooted trees, where $N(\tau)$ is the number of blocks in the forest that correspond to unlabeled rooted tree τ.

Such a multiset has a *symmetry factor* derived from the symmetry factor associated with unlabeled rooted trees. Let F be a forest with corresponding multiset N. The symmetry factor $\sigma(N)$ is the order of the stabilizer subgroup $\mathcal{G}_F$ of the forest. It is

$$\sigma(N) = \prod_{\tau'} N(\tau')! \sigma(\tau')^{N(\tau')}.$$

This expression may be derived using group theory. Let $\mathcal{B}$ be the subgroup of $\mathcal{G}_F$ generated by permutations that leave each block invariant. For each block there is a corresponding symmetry factor $\sigma(\tau')$, so $\mathcal{B}$ is a product group with order $\prod_{\tau'} \sigma(\tau')^{N(\tau')}$. Let $\mathcal{H}$ be the subgroup that permutes blocks with identical unlabeled rooted trees. If unlabeled rooted tree τ' occurs in $N(\tau')$ blocks, then there are $N(\tau')!$ permutations involving that rooted tree. The order of $\mathcal{H}$ is thus $\prod_{\tau'} N(\tau')!$. The group $\mathcal{H}$ is a normal subgroup of $\mathcal{G}_F$, so in particular for every b' in $\mathcal{B}$ and for every h in $\mathcal{H}$ the element $hb'h^{-1}$ is also in $\mathcal{B}$. Every element of $\mathcal{G}_F$ may be uniquely expressed as product bh of an element of $\mathcal{B}$ with an element of $\mathcal{H}$. (This decomposition respects multiplication: $bhb'h' = (bhb'h^{-1})(hh')$. In fact the group $\mathcal{G}_F$ is the semidirect product of the group $\mathcal{B}$ with the group $\mathcal{H}$ [3].) The conclusion is that the order of $\mathcal{G}_F$ is the product of the orders of $\mathcal{B}$ and $\mathcal{H}$.

Here is how to construct a forest corresponding to a given multiset N with m vertices. Find a set V with m elements and a set partition Γ of V. Require that there is a function $\chi : \Gamma \to \mathcal{A}$ such that for each τ the inverse image $\chi^{-1}(\tau)$ consists of $N(\tau)$ blocks of size $|\tau|$. Finally, for each block B in Γ find a labeled rooted tree T that determines unlabeled rooted tree $\chi(B)$.

The number of pairs Γ, χ satisfying these conditions is the coefficient

$$C(N) = \frac{m!}{\prod_{\tau'}(|\tau'|!)^{N(\tau')}} \frac{1}{\prod_{\tau'} N(\tau')!}.$$

The first factor is the multinomial coefficient that determines how many ways of producing blocks of the appropriate sizes in a given order. The second factor has a denominator that describes how many ways there are of permuting the blocks to preserve χ. The number of forests is

$$f(N) = C(N) \prod_{\tau'} r(\tau')^{N(\tau')} = \frac{m!}{\prod_{\tau'} N(\tau')!\sigma(\tau')^{N(\tau')}} = \frac{m!}{\sigma(N)}.$$

The first equality comes from counting the number of ways of putting labeled rooted trees in the appropriate blocks. The second equality results from inserting $r(\tau') = |\tau'|!/\sigma(\tau')$.

Unlabeled Rooted Trees Defined Recursively. There is a recursive definition of unlabeled rooted trees. An unlabeled rooted tree τ with $n \geq 1$ vertices is equivalent to a multiset N of unlabeled rooted trees with $n-1$ vertices. This counts the unlabeled rooted subtrees that result when the root is removed. The recursion terminates with unlabeled rooted trees with one vertex; the corresponding multiset is zero.

The number $r(\tau)$ that counts labeled rooted trees satisfies the recursion

$$r(\tau) = |\tau| f(N) = |\tau| C(N) \prod_{\tau'} r(\tau')^{N(\tau')}.$$

Here τ is an unlabeled rooted tree on n vertices, and N is the corresponding multiset on $n-1$ vertices. This is because a labeled rooted tree is determined by a root point and a forest over the remaining points.

There is also a recursion relation for the symmetry factors. If τ has n vertices and its subtrees define a multiset N with $n-1$ vertices.

$$\sigma(\tau) = \sigma(N) = \prod_{\tau' \in \tilde{\mathcal{A}}} N(\tau')!\sigma(\tau')^{N(\tau')}.$$

This recursion relation has an explicit solution. Let τ be an unlabeled rooted tree. Consider some labeling, so that there is a set U of vertices and a labeled rooted tree on U. For each vertex j, consider the subtree above j, and let N_j count the unlabeled rooted trees above this subtree. Then

$$\sigma(\tau) = \prod_j \prod_{\tau'} N_j(\tau')!$$

Notation for Unlabeled Rooted Trees. The multiset notation gives a convenient way of describing unlabeled rooted trees. A tree with a single vertex is denoted 0. Otherwise, the tree is denoted $N_1[\tau_1]N_2[\tau_2]\ldots N_k[\tau_k]$, where each $N_j \neq 0$ and the τ_j are descriptions of different unlabeled rooted trees. It is convenient to abbreviate $N[0]$ by N. Thus, for example, the labeled rooted tree $c[b[e[ad]]$ would determine the unlabeled rooted tree 1[0]1[2[0]]. In the abbreviated form this would be 11[2]. This says that the root has 1 immediate successor with a single vertex and 1 immediate successor that is a tree with 2 immediate successor vertices.

Example 4. For $n = 1$ the only rooted tree is 0, consisting of a single root point. For $n = 2$ and a given vertex set there are two labeled rooted trees, depending on which point is chosen for the root. There is only one unlabeled rooted tree, denoted 1. For $n = 3$ and a given vertex set there are $3^2 = 9$ labeled rooted trees. These decompose into two orbits, as shown in Fig. 1. These correspond to unlabeled rooted trees τ that may be denoted 2 and 1[1]. The rooted tree 2 has a root with two leaves. The symmetry factor is 2. The 1[1] linear rooted tree has a root and a successor rooted tree 1. The symmetry factor is 1. This gives the correct number of labeled rooted trees as the sum $6/2 + 6/1 = 9 = 3^2$. The two unlabeled rooted trees are shown in Fig. 6.

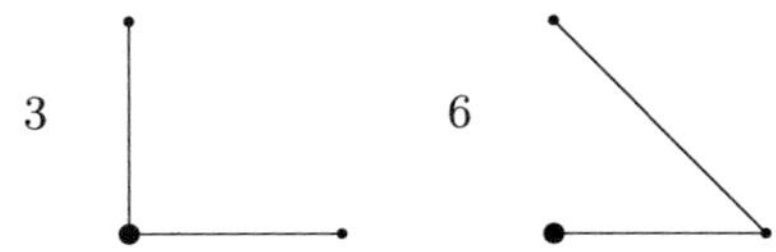

Fig. 6. Unlabeled rooted trees on three vertices: $\sigma(\tau) = 2, 1$

Example 5. The case $n = 4$ is more interesting. There are four unlabeled rooted trees, which may be denoted in multiset notation by 3, 11[1], 1[2], and 1[1[1]]. These rooted trees have symmetry factors $6, 1, 2, 1$. The number of labeled rooted trees is the sum $24/6 + 24/1 + 24/2 + 24/1 = 64 = 4^3$. See Fig. 7 for a picture of the unlabeled rooted trees.

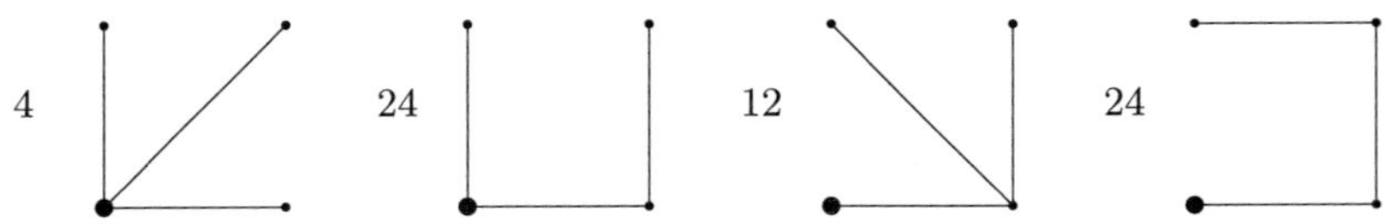

Fig. 7. Unlabeled rooted trees on four vertices: $\sigma(\tau) = 6, 1, 2, 1$

Remark 1. The formula for the number a_n of labeled rooted trees on a vertex set with n vertices is $a_n = n^{n-1}$. The number $\tilde{a}_n$ of unlabeled rooted trees with n vertices is not so easy to compute. This may be seen by contrasting the generating functions.

The exponential generating function for the number a_n of labeled rooted trees with n vertices is $a(t) = \sum_{n=1}^{\infty} \frac{1}{n!} a_n t^n$. The recursive definition of labeled rooted tree from root point and forest gives

$$a(t) = t \exp(a(t)).$$

Labeled enumeration give a simple result: for fixed t the value $x = a(t)$ satisfies the fixed point equation $x = t \exp(x)$.

Contrast this with the generating function for the number $\tilde{a}_n$ of unlabeled rooted trees with n vertices. This is $\tilde{a}(t) = \sum_{n=1}^{\infty} \tilde{a}_n t^n$. The recursive definition of unlabeled rooted tree from multiset gives the identity

$$\tilde{a}(t) = t \exp\left(\sum_{k=1}^{\infty} \frac{1}{k} \tilde{a}(t^k)\right).$$

Unlabeled enumeration produces a much more complicated equation. See [2] or [1] for the full story.

4 Fixed Point Equations and Labeled Rooted Trees

Let $\beta(x)$ be a formal power series in x. Let t and g be parameters. Consider the *fixed point equation*

$$x = g + t\beta(x).$$

Proposition 5. *The fixed point equation $x = g + t\beta(x)$ has the formal solution*

$$x = f(t, g) = \sum_{n=0}^{\infty} \frac{t^n}{n!} f_n(g),$$

where $f_0(g) = g$, and where for $n \geq 1$ the coefficient $f_n(g)$ has the explicit representation

$$f_n(g) = \left(\frac{\partial}{\partial g}\right)^{n-1} \beta(g)^n.$$

Proof. Fix g. The problem is to find the expansion of x as a function of t. We know the inverse function

$$t = (x - g)/\beta(x)$$

giving t as a function of x. Notice that $t = 0$ corresponds to $x = g$. The Lagrange inversion formula applies. The formula is based on the the fact that the residue of a differential form expressed by a formal Laurent series is invariant under change of variable. Start with the identity

$$\frac{1}{n} d\left(\frac{x}{t^n}\right) = -\frac{x}{t^{n+1}}\, dt + \frac{1}{n} \frac{1}{t^n}\, dx.$$

Since the left hand is a perfect differential, it has residue zero. This gives an identity for the residues

$$\operatorname{res}\left(\frac{x}{t^{n+1}}\, dt\right) = \frac{1}{n} \operatorname{res}\left(\frac{1}{t^n}\, dx\right).$$

This is the Lagrange inversion formula.

In the case at hand

$$\frac{1}{n!} f_n(g) = \operatorname{res}\left(\frac{x}{t^{n+1}}\, dt\right) = \frac{1}{n} \operatorname{res}\left(\frac{1}{t^n}\, dx\right) = \frac{1}{n} \operatorname{res}\left(\frac{\beta(x)^n}{(x-g)^n}\, dx\right),$$

where the last residue is computed at the singularity $x = g$. The residue is $1/(n-1)!$ times the $n-1$st derivative of $\beta(g)^n$ with respect to g. This gives the result. □

A combinatorial solution gives more detailed information. This is given by an expansion indexed by rooted trees. For each $n \geq 1$ fix a label set U_n with n elements, and consider rooted trees T with the label set as vertex set. For $n = 0$ there is an empty set object associated with the label set $U_0 = \emptyset$.

Proposition 6. *The fixed point equation $x = g + t\beta(x)$ has the solution given by a formal power series as above, where for $n \geq 1$ the coefficient $f_n(g)$ has the explicit representation*

$$f_n(g) = \sum_{T \in \mathcal{A}[U_n]} f_T(g)$$

For $n \geq 1$ the coefficient

$$f_T(g) = \prod_{j \in U_n} \left(\frac{\partial}{\partial g}\right)^{|T^{-1}(j)|} \beta(g),$$

where $|T^{-1}(j)|$ is the degree of vertex j of rooted tree T. For $n = 0$ the contribution from the empty rooted tree is $f_\emptyset(g) = g$.

Proof. This proof uses calculus formula in the form explained in the appendix. Identify the n factors in $\beta(g)^n$ with the n points in the vertex set U_n. Each derivative corresponds to a point in $[1, n-1] = \{1, \ldots, n-1\}$. Use the product rule to expand

$$\left(\frac{\partial}{\partial g}\right)^{n-1} \prod_{\ell \in U_n} \beta(g) = \sum_{\phi:[n-1] \to U_n} \prod_{\ell \in U_n} \left(\frac{\partial}{\partial g}\right)^{\phi^{-1}(\ell)} \beta(g).$$

The sum is over functions ϕ that pick out for each of the $n-1$ derivatives the factor to which it applies.

Use the Prüfer correspondence. Given a linear order on U_n, there is a corresponding bijection between functions ϕ from $[n-1]$ to U_n and rooted trees with vertex set U_n. Furthermore, given a rooted tree T, the number of times the function assumes value j in U_n is the degree $|T^{-1}(j)|$. □

The above may be expressed in an elegant way as

$$f(t, g) = \sum_{T \in \mathcal{A}_\emptyset} \frac{1}{|T|!} t^{|T|} f_T(g).$$

Here $|T|$ denotes the number of vertices of the rooted tree, and there is one label vertex set for each value of this number.

Remark 2. The problem of counting labeled rooted trees is the special case when $\beta(g) = \exp(g)$. In that case $f_n(g) = n^{n-1} \exp(ng)$ and each $f_T(g) = \exp(ng)$. In this special case the result is Cayley's formula $n^{n-1} = |\mathcal{A}[U_n]|$.

The solution of the fixed point equation may be expressed more economically in terms of unlabeled rooted trees by

$$f(t, g) = \sum_{\tau \in \tilde{\mathcal{A}}_\emptyset} \frac{r(\tau)}{|\tau|!} t^{|\tau|} f_\tau(g) = \sum_{\tau \in \tilde{\mathcal{A}}_\emptyset} \frac{1}{\sigma(\tau)} t^{|\tau|} f_\tau(g).$$

Here $r(\tau)$ is the corresponding number of labeled rooted trees, $|\tau|$ is the number of vertices, and $\sigma(\tau)$ is the symmetry factor.

Example 6. Use the notation $f_\tau(g)$ to denote the factor associated with unlabeled rooted tree τ. Then

$$f(x, t) = g + f_0 t + f_1(g) t^2 + \left(\frac{1}{2} f_2(g) + f_{1[2]}(g) \right) t^2$$
$$+ \left(\frac{1}{6} f_3(g) + f_{11[1]}(g) + \frac{1}{2} f_{1[2]}(g) + f_{1[1[1])}(g) \right) t^4 + \cdots .$$

Explicitly, this is

$$f(x, t) = g + \beta(g) t + \beta'(g)\beta(g) t^2 + \left(\frac{1}{2} \beta''(g)\beta(g)^2 + \beta'(g)^2\beta(g) \right) t^3$$
$$+ \left(\frac{1}{6} \beta'''(g)\beta(g)^3 + \beta''(g)\beta'(g)\beta(g)^2 + \frac{1}{2} \beta''(g)\beta'(g)\beta(g)^2 + \beta'(g)^3\beta(g) \right) t^4$$
$$+ \cdots .$$

To determine the contribution of a rooted tree, all that is needed is to know the number of vertices with given degree.

5 Increasing Rooted Trees

Increasing Rooted Trees as Partially Ordered Sets. The convention used here is that the partial order of a labeled rooted tree increases as one moves away from the root. Consider a non-empty label set U together with a given linear order $\leq_{\text{lin}}$. An *increasing rooted tree* is a labeled rooted tree with the property that the map from U with its rooted tree partial order $\leq_{\text{tree}}$ to U with the given linear order $\leq_{\text{lin}}$ is order-preserving. In other words, if $i \leq_{\text{tree}} j$ in the partial order of the labeled rooted tree, then the $i \leq_{\text{lin}} j$ in the linear order of the labels. For an increasing rooted tree the root r is the least element in both orders. The greatest element in the linear order is maximal in the partial order, so it is a leaf.

Increasing Rooted Trees as Functions. Consider a non-empty label set U together with a given linear order. An increasing rooted tree is a labeled rooted tree $T : U \setminus \{r\} \to U$ that is decreasing with respect to the linear order. In other words, it is required that $T(j) \leq_{\text{lin}} j$ in the linear order for all $j \neq r$. The collection of all increasing rooted trees on U is denoted $\mathcal{A}^\uparrow[U]$.

The relation between the three kinds of rooted trees for a given linearly ordered label set U with $n \geq 1$ elements is

$$\mathcal{A}^{\uparrow}[U] \to \mathcal{A}[U] \to \tilde{\mathcal{A}}[n].$$

The first map is an injection, and the second map is a surjection. The composite map is also a surjection.

Increasing Rooted Trees Defined Recursively from Below. For each k in U the set of all ℓ with $k \leq_{\text{tree}} \ell$ is also an increasing rooted tree T_k, with root k. If r is the root, then it must be the least element of U. So for each immediate successor j of r the rooted tree T_j is an increasing rooted tree. This gives a recursive characterization of an increasing rooted tree on U as a forest of increasing rooted trees on $U \setminus \{r\}$.

Increasing Rooted Trees Defined Recursively from Above. This gives another recursive description. Let m be the greatest element of U in the linear order. An increasing rooted tree on U is a increasing rooted tree on $U \setminus \{m\}$ together with a point in $U \setminus \{m\}$. Thus there is an edge in the tree from m to the chosen point. By taking $m = n, n-1, n-2, \ldots, 2$ this defines a map ϕ from $\{2, 3, \ldots, n\}$ to the set of non-leaf vertices. (This is a variation on the Prüfer correspondence.)

Increasing Rooted Trees as Permutations. Each increasing rooted tree on $\{1, \ldots, n\}$ may be coded as a permutation of $\{2, \ldots, n\}$. Such a permutation may be represented as a list of the elements of $\{2, \ldots, n\}$ in some order. For $n = 2$ the only entry in the list is 2. Suppose $n \geq 3$ and an increasing rooted trees on $\{1, \ldots, n-1\}$ is coded as a list taken from $\{2, \ldots, n-1\}$. Consider a new increasing rooted tree on $\{1, \ldots, n\}$. If the tree sends n to k with $1 \leq k \leq n-1$, then create a new list such that for $j < k$ the jth place entry is the same, the kth place entry is n, and for $j > k$ the entry in the jth place is the original $j-1$ place entry. This represents the new increasing rooted tree as a list taken from $\{2, \ldots, n\}$.

Example 7. The tree 1[2] is encoded by 2. The trees 1[23] and 1[2[3]] are encoded by 32 and 23. The trees 1[234] and 1[32[4]] and 1[23[4]] are encoded by 432 and 342 and 324, while the trees 1[42[3]] and 1[2[34]] and 1[2[3[4]]] are encoded by 423 and 243 and 234.

The permutation representation immediately gives the following result.

Proposition 7. *The number of increasing rooted trees on a label set with n vertices is $(n-1)!$.*

The *rooted tree factorial* $T!$ of a labeled rooted tree T with root r is defined inductively as the number of vertices of T times the product over i with $T(i) = r$ of $T_i!$. (An empty product gives 1.) This is an invariant under isomorphism, so for each unlabeled rooted tree τ there is a rooted tree factorial $\tau!$. The rooted tree factorial satisfies the recursive relation

$$\tau! = |\tau| \prod_{\tau'} (\tau'!)^{N(\tau')},$$

where N counts the subtrees obtained by removing the root.

There is another formula for the rooted tree factorial that is often convenient. For an unlabeled rooted tree τ consider a corresponding labeled rooted tree T. Let T_j be the subtree over vertex j. Then

$$\tau! = \prod_j |T_j|,$$

where the product is over all vertices of T. The quantity $|T_j|$ is the number of vertices of T_j.

The number of increasing rooted trees per unlabeled rooted tree satisfies a recursion relation

$$i(\tau) = C(N) \prod_{\tau'} i(\tau')^{N(\tau')},$$

where $N(\tau')$ counts immediate successor rooted trees τ'. This is similar to the formula $r(\tau)$ for the number of rooted trees per unlabeled rooted tree; the distinction is that there is only one choice of root point. It follows that the ratio $r(\tau)/i(\tau)$ satisfies

$$\frac{r(\tau)}{i(\tau)} = |\tau| \prod_{\tau'} \left(\frac{r(\tau')}{i(\tau')}\right)^{N(\tau')},$$

where N counts the successor rooted trees obtained by removing the root. This leads to the following relation.

Proposition 8. *Fix n and an unlabeled rooted tree τ with n vertices. The ratio of the number of labeled rooted trees to the number of increasing rooted trees is the rooted tree factorial*

$$\frac{r(\tau)}{i(\tau)} = \tau!.$$

As a consequence, the number of increasing rooted trees for given unlabeled rooted tree τ is

$$i(\tau) = \frac{r(\tau)}{\tau!} = \frac{|\tau|!}{\tau!\sigma(\tau)}$$

For τ with n vertices there is an identity that expresses the fact that the sum over unlabeled rooted trees of the corresponding number of increasing labeled rooted trees is the total number of increasing labeled rooted trees.

Proposition 9. *The sum over unlabeled rooted trees with n vertices of the corresponding number of increasing rooted trees gives*

$$\sum_{\tau \in \tilde{A}[n]} \frac{n!}{\sigma(\tau)\tau!} = (n-1)!.$$

Example 8. When $n = 3$ there are only two increasing rooted trees. The symmetry factors are 2 and 1, while the rooted tree factorials are 3 and 6. This is illustrated in Fig. 8.

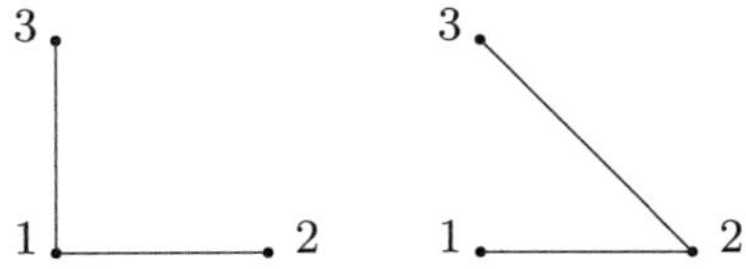

Fig. 8. Increasing rooted trees on three vertices: $\tau! = 3, 6$

Example 9. The case $n = 4$ is more interesting. There are $3! = 6$ increasing rooted trees. These map to the four unlabeled rooted trees, which are

$$3, 11[1], 1[2], 1[1[1]].$$

These four unlabeled rooted trees have symmetry factors $6, 1, 2, 1$ and rooted tree factorials $4, 8, 12, 24$. Three of the increasing rooted trees correspond to the unlabeled rooted tree $11[1]$ with symmetry factor 1 and rooted tree factorial 8. The number of increasing labeled rooted trees is the sum $24/(6 \cdot 4) + 24/(1 \cdot 8) + 24/(2 \cdot 12) + 24/(1 \cdot 24) = 1 + 3 + 1 + 1 = 6$. The picture is in Fig. 9.

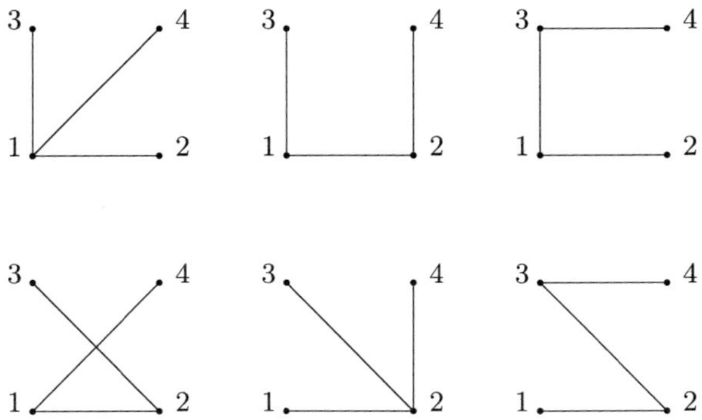

Fig. 9. Increasing rooted trees on four vertices: $\tau! = 4, 8, 8, 12, 24$

Example 10. For $n = 5$ vertices there are 9 unlabeled rooted trees. These are indicated in Table 1. The symmetry group of all permutations has order $n! = 120$. The table lists the symmetry factors $\sigma(\tau)$, the number of labeled rooted trees $r(\tau)$, the tree factorial $\tau!$, and the number of unlabeled rooted trees. The symmetry factor $\sigma(\tau)$ may be read off from the multiset description of τ. The other quantities are related by $r(\tau) = n!/\sigma(\tau)$ and $i(\tau) = r(\tau)/\tau!$. If all is well, the total number of labeled rooted trees should be $n^{n-1} = 625$, while the total number of increasing rooted trees should be $(n-1)! = 24$.

Remark 3. It is instructive to look at the exponential generating function for the number $a_n^{\uparrow}$ of increasing rooted trees with n vertices. The recursive definition gives

$$\frac{d}{dt}a^{\uparrow}(t) = \exp(a^{\uparrow}(t)).$$

This has the easy solution $a^{\dagger}(t) = \log(\frac{1}{1-t})$. The details are in [1].

Table 1. Unlabeled rooted trees on 5 vertices

τ	$\sigma(\tau)$	$r(\tau)$	$\tau!$	$i(\tau)$
4	24	5	5	1
21[1]	2	60	10	6
2[1]	2	60	20	3
11[2]	2	60	15	4
11[1[1]]	1	120	30	4
1[3]	6	20	20	1
1[11[1]]	1	120	40	3
1[1[2]]	2	60	60	1
1[1[1[1]]]	1	120	120	1

6 Ordinary Differential Equations and Increasing Rooted Trees

Let $\beta(x)$ be a formal power series in x. Let t and g be parameters. Consider the *ordinary differential equation*

$$\frac{dx}{dt} = \beta(x)$$

with initial condition $x = g$ at $t = 0$.

Proposition 10. *This ordinary differential equation has the formal solution*

$$x = \bar{f}(t, g) = \sum_{n=0}^{\infty} \frac{t^n}{n!} \bar{f}_n(g),$$

where the coefficient $\bar{f}_n(g)$ has the explicit representation

$$\bar{f}_n(g) = \left(\beta(g)\frac{\partial}{\partial g}\right)^n g.$$

Proof. Let the solution of the initial value problem be $\bar{f}(t, g)$. It is easy to see by induction that

$$\frac{\partial^n \bar{f}(t,g)}{\partial t^n} = h_n(\bar{f}(t,g))$$

for a suitable function $h_n(x)$. Taking one more derivative gives

$$h_{n+1}(\bar{f}(t,g)) = h'_n(\bar{f}(t,g))\beta(\bar{f}(t,g)).$$

Setting $t = 0$ gives $\bar{f}(0, g) = g$, so the result is

$$h_{n+1}(g) = h'_n(g)\beta(g).$$

The conclusion follows immediately. □

The combinatorial solution is given by an expansion indexed by increasing rooted trees. For each $n \geq 1$ fix a linearly ordered label set U_n with k elements. For instance, take $U_n = \{1, \ldots, n\}$ with the usual linear order. Furthermore, consider increasing rooted trees $\bar{T}$ with the label set as vertex set. For $n = 0$ introduce an empty set object.

Proposition 11. *The ordinary differential equation has the solution given by a formal power series as above, where $\bar{f}_n(g)$ has the explicit representation*

$$\bar{f}_n(g) = \sum_{\bar{T} \in \mathcal{A}^{\uparrow}[U_n]} f_{\bar{T}}(g)$$

For $n \geq 1$ the coefficient is

$$f_{\bar{T}}(g) = \prod_{j \in U_n} \left(\frac{\partial}{\partial g}\right)^{|\bar{T}^{-1}(j)|} \beta(g),$$

where $|\bar{T}^{-1}(j)|$ is the degree of vertex j of rooted tree $\bar{T}$. For $n = 0$ the contribution from the empty rooted tree is $f_{\emptyset}(g) = g$.

Proof. Write

$$\bar{f}_n(g) = \left(\beta(g)\frac{\partial}{\partial g}\right)^{n-1} \beta(g).$$

Index the partial derivatives from n down to 2. Index the $\beta(g)$ factors from n down to 1. Then every partial derivative acts only on the $\beta(g)$ factors with strictly smaller index. So

$$\bar{f}_n(g) = \sum_{\phi} \prod_{j=1}^{n} \left(\frac{\partial}{\partial g}\right)^{|\phi^{-1}(j)|} \beta(g),$$

where the sum is over functions ϕ from to $[2, n]$ to $[1, n]$ with the property that $\phi(i) < i$ for all i.

Every such function ϕ is an increasing rooted tree on $[1, n]$, where $\phi(i)$ is the immediate predecessor of i, and $\phi^{-1}(j)$ is the set of immediate successors of j. □

The above may be expressed in an elegant way as

$$\bar{f}(t,g) = \sum_{\bar{T} \in \mathcal{A}^{\uparrow}_{\emptyset}} \frac{1}{|\bar{T}|!} t^{|\bar{T}|} f_{\bar{T}}(g).$$

Here $|\bar{T}|$ denotes the number of vertices of the increasing rooted tree, and there is one label vertex set for each value of this number.

Remark 4. The problem of counting increasing rooted trees is the special case when $\beta(g) = \exp(g)$. The solution of the differential equation is given explicitly by $x = g - \log(1 - \exp(g)t)$. The coefficient in nth order is $\bar{f}_n(g) = (n-1)! \exp(ng)$, and each $f_T(g) = \exp(ng)$. In this special case the result is equivalent to the formula $(n-1)! = |\mathcal{A}^{\uparrow}[U_n]|$.

The solution of the ordinary differential equation may be expressed in terms of unlabeled rooted trees by

$$\bar{f}(t,g) = \sum_{\tau \in \tilde{\mathcal{A}}_{\emptyset}} \frac{i(\tau)}{|\tau|!} t^{|\tau|} f_{\tau}(g) = \sum_{\tau \in \tilde{\mathcal{A}}_{\emptyset}} \frac{1}{\sigma(\tau)\tau!} t^{|\tau|} f_{\tau}(g).$$

Here $r(\tau)$ is the corresponding number of increasing rooted trees, $|\tau|$ denotes the number of vertices of the rooted tree, $\sigma(\tau)$ is the symmetry factor, and $\tau!$ is the rooted tree factorial.

Example 11. Use the notation $f_{\tau}(g)$ to denote the factor associated with unlabeled rooted tree τ. Then

$$\bar{f}(x,t) = g + f_0 t + \frac{1}{2} f_1(g) t^2 + \frac{1}{6}\left(f_2(g) + f_{1[2]}(g)\right) t^2$$
$$+ \frac{1}{24}\left(f_3(g) + 3 f_{11[1]}(g) + f_{1[2]}(g) + f_{1[1[1]]}(g)\right) t^4 + \cdots .$$

Explicitly, this is

$$\bar{f}(x,t) = g + \beta(g)t + \frac{1}{2}\beta'(g)\beta(g)t^2 + \frac{1}{6}\left(\beta''(g)\beta(g)^2 + \beta'(g)^2\beta(g)\right) t^3$$
$$+ \frac{1}{24}\left(\beta'''(g)\beta(g)^3 + 3\beta''(g)\beta'(g)\beta(g)^2 + \beta''(g)\beta'(g)\beta(g)^2 + \beta'(g)^3\beta(g)\right) t^4$$
$$+ \cdots .$$

The factor 3 comes from the 3 increasing rooted trees associated with $\tau = 11(1)$. These ordinary differential equation coefficients are the fixed point equation coefficients divided by tree factorials.

7 The Butcher Group (Composition) for Labeled Rooted Trees

Let T be a labeled rooted tree with vertex set U and root r in U. A *rooted subtree* T_0 is a rooted tree on some non-empty subset $U_0 \subseteq U$ with the same root r in

U_0 that is a restriction of T to this subset. The condition that T_0 is a rooted subtree of T is denoted $T_0 \to T$. The empty subset $U_0 = \emptyset$ corresponds to an empty set object T_0; that case is also abbreviated $T_0 \to T$.

For each T_0 with $T_0 \to T$ there is a corresponding *difference forest* $T \setminus T_0$ of rooted trees on $U \setminus U_0$. The trees in the forest are all subtrees T_j with $j \in U \setminus U_0$ and $T(j) \in U_0$. If $U_0 = \emptyset$ and T_0 is the empty set object, then the difference forest consists of the tree T on U.

Proposition 12. *For each non-empty label set U there is a one-to-one correspondence between rooted tree pairs T_0, T with $T_0 \to T$ and triples T_0, F_1, ϕ, where T_0 is a rooted tree on $U_0 \subseteq U$, F_1 is a forest on $U_1 = U \setminus U_0$, and ϕ is a function from the set partition Γ_1 of the forest to U_0.*

Let $\tilde{\mathcal{A}}_\emptyset$ be the set of unlabeled rooted trees together with the empty set object. Consider the space $\mathbf{R}^{\tilde{\mathcal{A}}_\emptyset}$ of all functions c from $\tilde{\mathcal{A}}_\emptyset$ to the real numbers. These are coefficients $c(\tau)$ that depend on unlabeled rooted trees τ. Since each labeled rooted tree T determines a corresponding unlabeled rooted tree τ, the coefficients $c(T)$ are also defined for labeled rooted trees. For a forest of rooted trees the coefficient $a^\times(F)$ is the product of the $a(T')$ for T' in the forest. In particular, for the empty forest $a^\times(\emptyset) = 1$.

The operation of *subtree convolution* $a*b$ is defined when $a(\emptyset) = 1$ by $c = a*b$, where

$$c(T) = \sum_{T_0 : T_0 \to T} b(T_0) a^\times(T \setminus T_0).$$

When T_0 is the empty set object, the corresponding term in the sum is $b(\emptyset)a(T)$. When $T_0 = T$, the corresponding term in the sum is $b(T)a^\times(\emptyset) = b(T)$. When T is the empty set object, then $c(\emptyset) = b(\emptyset)$.

In the multiplication $a * b$ the forest factor is on the left and the tree factor is on the right. This convention is common in this context [8,14]. The Butcher group multiplication is the special case when both $a(\emptyset) = 1$ and $b(\emptyset) = 1$. This group is denoted G_C, where C stands for composition. The identity $\delta_\emptyset$ in the group has coefficient 1 for the empty set object and 0 for the rooted trees. An inductive argument shows that every element has an inverse element. The group G_C is the character group of the rooted tree Hopf algebra of Connes and Kreimer.

Example 12. It is easy to compute the Butcher group multiplication for reasonably small labeled rooted trees. Here are the results for up to 3 vertices. For notational simplicity label the tree as an increasing tree. Take the root with label 1. For one-vertex rooted trees $c(1) = b(1) + a(1)$. For two-vertex rooted trees the result is $c(1[2]) = b(1[2]) + b(1)a(2) + a(1[2])$. The first interesting case is $n = 3$. Let $1[23]$ be the rooted tree with root at 1 and with leaves at 2 and 3, and let $1[2[3]]$ be the rooted tree with root at 1 and with successor rooted tree $2[3]$. Then

$$c(1[23]) = b(1[23]) + b(1[2])a(3) + b(1[3])a(2) + b(1)a(2)a(3) + a(1[23]).$$

Similarly

$$c(1[2[3]]) = b(1[2[3]]) + b(1[2])a(3) + b(1)a(2[3]) + a(1[2[3]]).$$

The expressions for the two cases are quite different.

Example 13. Here is one calculation for $n = 4$. The rooted tree is $T = 1[3[2[4]]$ with root at 1 and successor rooted trees 3 and 2[4]. This happens to be an increasing rooted tree, but that is just for notational convenience. The result is

$$\begin{aligned} c(T) &= b(T) + b(1[2[4]])a(3) + b(1[23])a(4) \\ &+b(1[3])a(2[4]) + b(1[2])a(3)a(4)) + b(1)a(2[4])a(3) + a(T). \end{aligned}$$

This is illustrated in Fig. 10. The number of vertices in the subtrees are 4, 3, 3, 2, 2, 1, 0. The difference forests consist of 0, 1, 1, 1, 2, 2, 1 rooted trees.

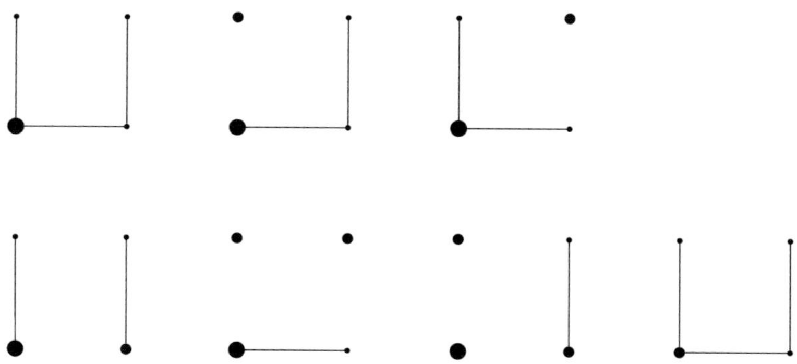

Fig. 10. Subtrees and difference forests

Coefficients depending on rooted trees also define certain functions (defined as formal power series). Each such function is a weighted tree sum in the form of an exponential generating function. For coefficients c the function is

$$f^c(t, g) = \sum_{T \in \mathcal{A}_\emptyset} \frac{1}{|T|!} t^{|T|} c(T) f_T(g).$$

Again for rooted tree T the coefficient is

$$f_T(g) = \prod_{\ell \in U} \left(\frac{\partial}{\partial g}\right)^{|T^{-1}(\ell)|} \beta(g).$$

The zero order term is $c(\emptyset)g$. The remainder of the series is a power series in powers of t and powers of derivatives of $\beta(g)$. Such as series may be further expanded in powers of g, but that is not done here.

Theorem 1 (Composition). *Suppose $a(\emptyset) = 1$, and let $c = a * b$ be the subtree convolution. The corresponding weighted sums are related by*

$$f^c(t, g) = f^b(t, f^a(t, g)).$$

Proof. The proof is based on the calculus formulas in the appendix. The task is to show that $f^c(t, g)$ and $f^b(t, f^a(t, g))$ have the same terms of each order. The nth term is the nth partial derivative with respect to t, evaluated at zero. (The case $n = 0$ is trivial, so assume $n \geq 1$.) Take a set U with $|U| = n$ elements. For $f^c(t, g)$

$$D_t^{|U|} f^c(t, g)|_{t=0} = \sum_{T \in \mathcal{A}[U]} c(T) f_T(g).$$

Product Rule. The composition

$$f^b(t, f^a(t, g)) = \sum_{T \in \mathcal{A}_\emptyset} \frac{1}{|T|!} t^{|T|} b(T) f_T(f^a(t, g)).$$

is a combination of products of two factors $t^{|T|}$ and $f_T(f^a(t, g))$ By the product rule, the nth derivative is a sum over disjoint unions $U = U_0 + U_1$ of the form

$$D_t^{|U|} f^b(t, f^a(t, g))|_{t=0} = \sum_{U = U_0 + U_1} \sum_{T_0 \in \mathcal{A}[U_0]} b(T_0) D_t^{|U_1|} f_{T_0}(f^a(t, g))|_{t=0}.$$

Chain Rule. The next task is to evaluate $D^{|U_1|} f_{T_0}(f^a(t, g))$ by the chain rule. This is a sum over set partitions Γ_1 of U_1

$$D_t^{|U_1|} f_{T_0}(f^a(t, g))|_{t=0} = \sum_{\Gamma_1 \in \mathrm{Part}[U_1]} D_g^{|\Gamma_1|} f_{T_0}(f^a(t, g))|_{t=0} \prod_{B \in \Gamma_1} D^{|B|} f^a(t, g)|_{t=0}.$$

Since $f^a(0, g) = g$, this is

$$D_t^{|U_1|} f_{T_0}(f^a(t, g))|_{t=0} = \sum_{\Gamma_1 \in \mathrm{Part}[U_1]} D_g^{|\Gamma_1|} f_{T_0}(g) \prod_{B \in \Gamma_1} \left(\sum_{H \in \mathcal{A}[B]} a(H) f_H(g) \right).$$

Distributive Law. The distributive law is used to expand the product

$$\prod_{B \in \Gamma_1} \left(\sum_{H \in \mathcal{A}[B]} a(H) f_H(g) \right) = \sum_F \prod_{B \in \Gamma_1} a(F(B)) f_{F(B)}(g),$$

where F is a function defined on Γ_1 such that the value of F on block B in Γ_1 is a rooted tree on B.

Product Rule. When T_0 is non-empty the coefficient $f_{T_0}(g)$ is a product over vertices ℓ in U_0 of factors $\beta^{(d(\ell))}(g)$. By the product rule the derivative of order $|\Gamma_1|$ is a sum over functions $\phi : \Gamma_1 \to U_0$ in the form

$$D_g^{|\Gamma_1|} f_{T_0}(g) = \sum_{\phi : \Gamma_1 \to U_0} f_{T_0, \phi}(g).$$

Here

$$f_{T_0,\phi}(g) = \prod_{\ell \in U_0} \beta^{(|T_0^{-1}(\ell)|+|\phi^{-1}(\ell)|)}(g).$$

For the empty rooted tree $f_\emptyset(g) = g$ and so the only derivative that is non-zero is the first derivative, corresponding to a set partition of $U = U_1$ into one block. It is convenient to define $f_{\emptyset,\phi}(g) = 1$, where ϕ is some unique object whose nature is not important.

Construction of Rooted Tree and Subtree. The end result is

$$D_t^{|U|} f^b(t, f^a(t,g))|_{t=0} =$$
$$\sum_{U=U_0+U_1} \sum_{T_0 \in \mathcal{A}[U_0]} \sum_{\Gamma_1 \in \mathrm{Part}[U_1]} \sum_{\phi:\Gamma_1 \to U_0} \sum_{F} \prod_{B \in \Gamma_1} b(T_0)a(F(B)) f_{T_0,\phi}(g) f_{F(B)}(g).$$

Fix $U = U_0 + U_1$ and rooted tree T_0 on U_0. The remaining data Γ_1, ϕ, F determine a rooted tree T on U that extends T_0. The value $\phi(B)$ may be thought of as the vertex to which the root of the tree $F(B)$ maps. So

$$D_t^{|U|} f^b(t, f^a(t,g))|_{t=0} = \sum_{U=U_0+U_1} \sum_{T_0 \in \mathcal{A}[U_0]} \sum_{T:T_0 \to T} b(T_0) a^\times (T \setminus T_0) f_T(g).$$

The sum may be done in the other order, first the rooted tree T and then the subtree T_0. This gives

$$D_t^{|U|} f^b(t, f^a(t,g))|_{t=0} = \sum_{T \in \mathcal{A}[U]} \sum_{T_0:T_0 \to T} b(T_0) a^\times (T \setminus T_0) f_T(g).$$

In other words, the derivative is $\sum_{T \in \mathcal{A}[U]} (a * b)(T) f_T(g)$. □

One special case of the multiplication law is when the sequence b is zero except for $b(\bullet) = 1$. This corresponds to the function $f^b(t,g) = t\beta(g)$. In this case $c(T) = a^\times(T \setminus \bullet)$, the product of $a(T')$ for all rooted trees T' in the successor forest.

Example 14. Consider $f^b(t,g) = t\beta(g)$ and $f^a(t,g) = g + a(\bullet)t\beta(g)$. The composition is

$$f^c(t,g) = t\beta(g + a(\bullet)t\beta(g)).$$

This type of composition occurs in numerical methods for the solution of ordinary differential equations. In this case $c(T)$ is zero unless T is a rooted tree on a set with $n \geq 1$ points, one root and $n-1$ leaves. There are n such rooted trees. The corresponding successor forests each consist of $n-1$ one point rooted trees. This gives $c(T) = a(\bullet)^{n-1}$. The function $f^c(t,g)$ has the rooted tree expansion

$$f^c(t,g) = \sum_{n=1}^{\infty} n \frac{t^n}{n!} a(\bullet)^{n-1} \beta^{(n-1)}(g) \beta(g)^{n-1}.$$

This is the Taylor expansion of the composite function.

This example underpins the Runge–Kutta methods for the numerical solution of ordinary differential equations. The first order Euler method is to use $g + \beta(g)t$ to approximate the solution. Various second order methods depend on a parameter a; they are of the form $g + (1 - \frac{1}{2a})\beta(g)t + \frac{1}{2a}\beta(g + a\beta(g)t)$. In particular, $a = 1$ is the analog of the trapezoidal rule, and $a = \frac{1}{2}$ is the analog of the midpoint rule. Since $\beta(g + a\beta(g)t)$ agrees with $\beta(g) + a\beta'(g)\beta(t)t$ to second order, the second order Runge–Kutta method agrees with the Taylor method $g + \beta(t)t + \frac{1}{2}\beta'(g)\beta(g)t$ to second order. The Runge–Kutta method has the advantage that it does not require computing the derivative $\beta'(g)$.

The Butcher group is related to composition of power series. Take the case when $\beta(g) = \exp(g)$. In that case the contribution of a labeled rooted tree on $n \geq 1$ vertices is $\exp(ng)$. Suppose $c = a * b$ and $f^c(t, g) = f^b(t, f^a(t, g))$. Each of the individual functions is a power series in powers of $\exp(g)$. Define $h^a(w) = \exp(f^a(1, \log(w)))$ and similarly for the others. The resulting functions are power series in w, and they are related by $h^c(w) = h^b(h^a(w))$. Subtree convolution is mapped to composition of power series.

8 The Butcher Group for Increasing Rooted Trees

The composition law for the Butcher group may also be written in terms of increasing rooted trees. The function corresponding to sequence a is

$$f^a(t, g) = \sum_{\bar{T} \in \mathcal{A}_\emptyset^\uparrow} \frac{t^{|\bar{T}|}}{|\bar{T}|!} a(\bar{T})\bar{T}! f_{\bar{T}}(g).$$

The rooted subtree convolution is defined in the same way, since if T is an increasing rooted tree and $T_0 \to T$, then the subtree T_0 is also increasing.

If $T_0 \to T$, the *rooted tree binomial coefficient* is

$$\binom{T}{T_0} = \frac{T!}{T_0! \prod_{T' \in T \setminus T_0} T'!}.$$

For a linear rooted tree this is the usual binomial coefficient.

The change of variable $\bar{c}_T = c(T)T!$ gives another representation of the Butcher multiplication as

$$\bar{c}(T) = \sum_{T_0 : T_0 \to T} \binom{T}{T_0} \bar{b}(T_0)\bar{a}^\times(T \setminus T_0).$$

The solution $\bar{f}(t, g)$ of the differential equation $dx/dt = \beta(x)$ with initial condition g is the case corresponding to $a(T) = 1/T!$, or $\bar{a}(T) = 1$. This fact leads to an identity for a binomial coefficient associated with rooted trees.

Proposition 13. *For a labeled rooted tree with n vertices*

$$\sum_{T_0 \to T} \binom{T}{T_0} = 2^n.$$

Proof. The solution of the differential equation has the group property $\bar{f}(2t, g) = \bar{f}(t, \bar{f}(t, g))$. Take $\bar{a}(T) = 1$ and the corresponding $\bar{f}^a(t, g)$. Then $\bar{f}^a(2t, g) = \bar{f}^a(t, \bar{f}^a(t, g))$. Now take $\bar{a}(T) = \bar{b}(T) = 1$, so $\bar{c}(T) = 2^{|T|}$. This then translates to $\bar{f}^c(T) = \bar{f}^b(t, \bar{f}^a(t, g))$. This gives the rooted tree binomial coefficient identity for increasing rooted trees. Since rooted tree factorials do not depend on the order on the label set, the identity holds for all labeled rooted trees. □

Example 15. For the labeled rooted tree with four vertices shown in Fig. 10 the binomial identity says

$$16 = 1 + \frac{4}{3} + \frac{8}{3} + 2 + 4 + 4 + 1.$$

The rooted tree binomial coefficients need not be whole numbers, but the sum is always a power of 2.

9 The Butcher Group for Unlabeled Rooted Trees

The Butcher group may also be presented using unlabeled rooted trees, but there is a complication. If τ, τ_0 are unlabeled rooted trees, choose T to be a labeled rooted tree that determines τ. Consider the set of labeled rooted trees T_0 with $T_0 \to T$ and with T_0 determining τ_0. This set depends on the chosen T, but the number of elements in the set is independent of T. Denote this number by $[\tau, \tau_0]$.

The multiplication operation for coefficients may then be written $c = a * b$, where

$$c(\tau) = \sum_{\tau_0} [\tau, \tau_0] b(\tau_0) a^{\times}(\tau \setminus \tau_0).$$

The extra complication is the presence of the multiplicity coefficients $[\tau, \tau_0]$.

For an arbitrary coefficient function c there is a corresponding function

$$f^c(t, g) = \sum_{\tau \in \tilde{A}_\emptyset} \frac{1}{\sigma(\tau)} c(\tau) t^{|\tau|} f_\tau(g).$$

The zero order term is $c(\emptyset)g$, and for the other terms

$$f_\tau(g) = \prod_k (D_g^k \beta(g))^{v_k(\tau)}.$$

This is precisely the same sum as before. The following is a restatement of the previous theorem.

Corollary 1. *Suppose $a(\emptyset) = 1$. Define $c = a * b$ with the multiplicity factor. The corresponding functions satisfy*

$$f^c(t, g) = f^b(t, f^a(t, g)).$$

Example 16. Take $k = 4$ vertices, and let 3 be the unlabeled rooted tree with a root and three leaves. Let 2 be the unlabeled rooted tree with a root and two leaves. Then the multiplicity $[3, 2] = 3$. Similarly, $[3, 1] = 3$. On the other hand, $[3, 0] = 1$, since there is only one way of inserting the root. The conclusion is that

$$c(3) = b(3) + 3b(2)a(0) + 3b(1)a(0)^2 + b(0)a(0)^3 + a(3).$$

In the world of unlabeled rooted trees, multiplicity factors are inescapable.

10 Substitution for Labeled Rooted Trees

The Butcher group is about subtree convolution and composition of functions. There is another algebraic structure based on quotient tree convolution and substitution. The reference [8] has an account of the subject and its history, along with some applications. See also [7] for a Hopf algebra approach and for more background. The following is a brief account in the labeled rooted tree framework.

Given a rooted tree T on U with root r and a set R with $r \in R$, there is a corresponding forest F obtained by restricting T to $U \setminus R$. This is called a *subforest* of the rooted tree. The set partition Γ defined by this forest is in one-to-one correspondence with the set of roots R of the rooted trees in the forest. When F is a subforest of T we can write $F \sqsubseteq T$.

Given a rooted tree T and a subforest F, there is also a *quotient rooted tree* T/F. This is a labeled rooted tree with label set Γ. Let U_0 be the block in Γ such that r is in $F(U_0)$. Then T/F is defined on $\Gamma_1 = \Gamma \setminus \{U_0\}$ as follows. The value of T/F on block B is obtained by finding the root j of rooted tree $F[B]$ and taking the value to be the block B' containing $T(j)$.

There is another representation of the quotient rooted tree T/F that may be easier to picture. This is as a labeled rooted tree with label set R, where R is the set of roots in the forest F. The value of T/F on vertex j in $R \setminus \{r\}$ is obtained by finding $T(j)$ and the block B' that contains it, and taking the value to be the root of rooted tree $F[B']$.

Remark 5. The subtree T_0 and difference forest $T \setminus T_0$ construction used for the Butcher group is a special case. The subforest F is T_0 together with $F_1 = T \setminus T_0$, and in this case the quotient rooted tree consists only of a root and immediate successors.

Example 17. Figure 11 gives an example of the forests associated with a given rooted tree. The number of rooted trees in the forest (and the number of roots of these rooted trees) are 1, 2, 2, 2, 3, 3, 3, 4. Figure 12 gives an example of quotient rooted trees of a given rooted tree on 4 vertices. These correspond to the subforests in Fig. 11. The roots of the rooted trees in the forest give the vertices of the quotient rooted tree. The number of vertices in the quotient rooted trees are 1, 2, 2, 2, 3, 3, 3, 4.

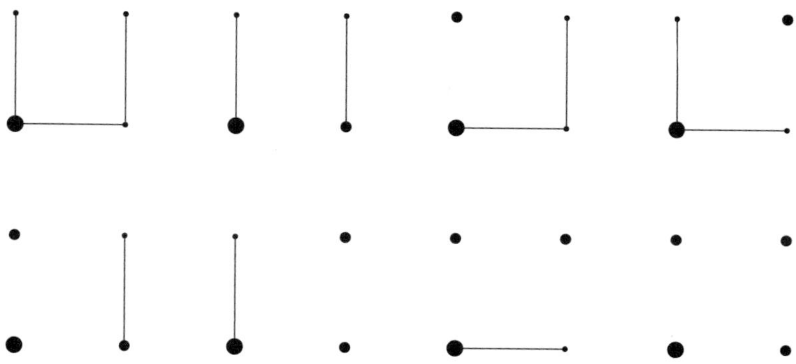

Fig. 11. Subforests

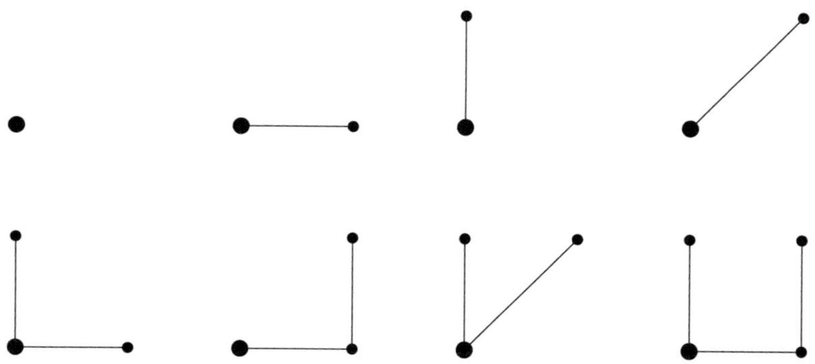

Fig. 12. Quotient rooted trees

The pair consisting of the subforest F and the quotient rooted tree T/F do not completely determine the original rooted tree T. The quotient rooted tree assigns to each block in Γ_1 another block in Γ. It is also necessary to specify a point in that block.

Proposition 14. *For each label set U there is a one-to-one correspondence between rooted tree, subforest pairs T, F and triples F, $\hat{T}$, ϕ, where F is a forest of rooted trees with set partition Γ, $\hat{T}$ is a rooted tree on vertex set Γ with root U_0, and ϕ is a function defined on $\Gamma_1 = \Gamma \setminus \{U_0\}$ such that for every B in Γ_1 the value $\phi(B)$ is in $\hat{T}(B)$.*

In the proposition the function ϕ sends a block to a vertex in another target block. There is a parametrization of such functions by target blocks. If B' is a block in Γ, then define $\Phi(B')$ to be the restriction of ϕ to $T^{-1}(B')$. Thus if $\hat{T}(B) = B'$, then $\Phi(B')$ applied to B is a vertex in B'. Conversely, suppose that there is a function Φ that maps each B' in Γ to a function $\Phi(B')$ from $\hat{T}^{-1}(B')$ to B'. Then there is a corresponding ϕ given on B with $\hat{T}(B) = B'$ by $\phi(B) = \Phi(B')(B)$.

Let $\tilde{A}_\emptyset$ be the set of unlabeled rooted trees augmented with the empty object. There is a multiplication defined on certain elements of $\mathbf{R}^{\tilde{A}_\emptyset}$. This is the *quotient rooted tree convolution* $c = a \star b$, defined whenever $a(\emptyset) = 0$, such that

$$c(T) = \sum_{F \sqsubseteq T} b(T/F) a^{\times}(F).$$

The sum is over subforests F of T. The contribution of a forest is $a^{\times}(F) = \prod_{T' \in F} a(T')$. For the empty forest this product is 1. For the special case of the empty set object $c(\emptyset) = b(\emptyset) a^{\times}(\emptyset) = b(\emptyset)$.

If $R = \{r\}$, then the forest F has only one rooted tree T, and T/F is a rooted tree on a one point vertex set. So the contribution to the sum is $a(T)b(\bullet)$. When R is the entire vertex set, then F is the discrete forest, and $T/F = T$. The contribution to the sum is $a(\bullet)^{|T|} b(T)$. As a special case, $c(\bullet) = a(\bullet) b(\bullet)$.

If the multiplication is restricted to $a(\emptyset) = b(\emptyset) = 0$, then $G_S = \mathbf{R}^{\tilde{A}}$ becomes an algebraic system closed under multiplication. The S stands for substitution. The identity for quotient rooted tree convolution is $\delta_\bullet$, which has coefficient 1 for a one-vertex rooted tree and 0 for all other rooted trees. If the multiplication is also restricted to $a(\bullet) \neq 0, b(\bullet) \neq 0$, the resulting system $G_S^\star$ is a group. If the multiplication is further restricted to $a(\bullet) = b(\bullet) = 1$, then this defines a subgroup G_S^1. The group G_S^1 is the character group of the rooted tree Hopf algebra of Calaque, Ebrahimi-Fard, and Manchon.

There is also a functional representation of quotient rooted tree convolution. This deals with formal functions of the form $g \mapsto \alpha(g) = \sum_{n=0}^{\infty} (1/n!) a_n g^n$. Such a function is denoted α to indicate that it depends only on the coefficients a_n and not on the input g. The power series depends on the choice of α and is of the form

$$f^c(t, \alpha, g) = \sum_{n=0}^{\infty} \sum_{T \in \mathcal{A}[U_n]} \frac{t^n}{n!} c(T) \prod_{\ell \in U_n} \alpha^{(|T^{-1}(\ell)|)}(g).$$

The zero order term is $c(0)g$. This leads to the following theorem [8].

Theorem 2 (Substitution). *Suppose that* $a(\emptyset) = 0$ *and* $c = a \star b$ *is the quotient rooted tree convolution. Then*

$$f^c(t, \beta, g) = f^b(1, f^a(t, \beta, \cdot), g).$$

Proof. The proof here uses the calculus formulas in the appendix. Let D_t be the partial derivative with respect to t. It is sufficient to show that applying D_t^n and then setting t to zero gives the same result for both sides of the equation. For the left hand side this is

$$D_t^{|U|} f^c(t, \beta, g)|_{t=0} = \sum_{T \in \mathcal{A}[U]} c(T) \prod_{\ell \in U} D_g^{|T^{-1}(\ell)|} \beta(g),$$

where U is a label set with n points.

Product Rule. The computation for the right hand side begins with

$$f^b(1, f^a(t,\beta,\cdot), g) = \sum_{\hat{T}} \frac{1}{|\hat{T}|!} b(\hat{T}) \prod_{i\in[\hat{T}]} D_g^{|\hat{T}^{-1}(i)|} f^a(t,\beta,g)$$

By the product rule

$$D_t^{|U|} \prod_{i\in[\hat{T}]} D_g^{|\hat{T}^{-1}(i)|} f^a(t,\beta,g) = \sum_{\psi:U\to[\hat{T}]} \prod_{i\in[\hat{T}]} D_t^{|\psi^{-1}(i)|} D_g^{|\hat{T}^{-1}(i)|} f^a(t,\beta,g).$$

Set $t = 0$. Since $f^a(0,\beta,g) = a(\emptyset)g = 0$, the contributions from i with $\psi^{-1}(i) = \emptyset$ are zero. So $\psi : U \to [\hat{T}]$ induces a set partition Γ of U and a bijection from Γ to $[\hat{T}]$. There are $|\hat{T}|!$ such bijections. This leads to

$$D_t^{|U|} f^b(1, f^a(t,\beta,\cdot), g)|_{t=0} = \sum_{\Gamma\in\mathrm{Part}[U]} \sum_{\hat{T}\in\mathcal{A}[\Gamma]} c(\Gamma,\hat{T}),$$

where

$$c(\Gamma,\hat{T}) = \prod_{B\in\Gamma} b(\hat{T}) D_g^{|\hat{T}^{-1}(B)|} D_t^{|B|} f^a(t,\beta,g)|_{t=0}.$$

Distributive Law. The next stage is to insert

$$D_t^{|B|} f^a(t,\beta,g)|_{t=0} = \sum_{H\in\mathcal{A}[B]} a(H) \prod_{j\in B} D_g^{|H^{-1}(j)|} \beta(g)$$

and use the distributive law. This gives a forest sum

$$D_t^{|U|} f^b(1, f^a(t,\beta,\cdot), g)|_{t=0} = \sum_{\Gamma\in\mathrm{Part}[U]} \sum_{\hat{T}\in\mathcal{A}[\Gamma]} \sum_{F} c(\Gamma,\hat{T},F),$$

where

$$c(\Gamma,\hat{T},F) = \prod_{B\in\Gamma} b(\hat{T}) a(F(B)) D_g^{|\hat{T}^{-1}(B)|} \prod_{j\in B} D_g^{|F(B)^{-1}(j)|} \beta(g).$$

Product Rule. The product rule for differentiation produces

$$c(\Gamma,\hat{T},F) = \prod_{B\in\Gamma} b(\hat{T}) a(F(B)) \sum_{\phi:\hat{T}^{-1}(B))\to B} \prod_{j\in B} D_g^{|\phi^{-1}(j)|} D_g^{|F(B)^{-1}(j|} \beta(g).$$

Distributive Law. The distributive law gives

$$D_t^{|U|} f^b(1, f^a(t,\beta,\cdot), g)|_{t=0} = \sum_{\Gamma\in\mathrm{Part}[U]} \sum_{\hat{T}\in\mathcal{A}[\Gamma]} \sum_{F} \sum_{\Phi} c(\Gamma,\hat{T},F,\Phi),$$

where

$$c(\Gamma,\hat{T},F,\Phi) = \prod_{B\in\Gamma} \prod_{j\in B} b(\hat{T}) a(F(B)) D_g^{|\Phi(B)^{-1}(j)|} D_g^{|F(B)^{-1}(j|)} \beta(g).$$

The sum is over set partitions Γ of U and over forest functions F that send block B to rooted tree $F(B)$ on vertex set B. It is also over rooted trees $\hat{T}$ on vertex set Γ. Finally, it is over functions Φ that send each block B in Γ to a function $\Phi(B)$ that takes each $\hat{T}$ preimage block B' and sends it to a vertex in B.

Construction of Rooted Tree and Subforest. These data determine a rooted tree on U that is made from the rooted trees $F(B)$ internal to the blocks B and from the rooted tree $\hat{T}$ and the function Φ. If $\hat{T}(B') = B$, then there is an edge from the root of the rooted tree on B' to the $\Phi(B)(B')$ in B.

The corresponding contribution involves the coefficients $b(\hat{T})$ and $a(F(B))$ and a contribution from the rooted tree. At a given vertex j this involves a derivative of $\beta(g)$ of an order equal to the total number of edges impinging on this vertex, both from within the block and from the other blocks.

Giving these data is the same as giving the pair $T \in \mathcal{A}[U]$ together with subforest F. So the final expression is

$$D_t^{|U|} f^b(f^a(t,\beta,\cdot),g)|_{t=0} = \sum_{T\in\mathcal{A}[U]} \sum_{F\sqsubseteq T} a^{\times}(F) b(T/F) f_T(g).$$

This establishes the result. □

The authors [8] give two applications of this result. For both the idea is to consider the coefficients $e(\tau) = 1/\tau!$ that give the exact solution of an ordinary differential equation $dx/dt = \beta(x)$. In backward error analysis the idea is to take c corresponding to some numerical method and solve $c = a \star e$ for a. This produces a modified differential equation that agrees with the numerical method. In the application to modified integrators, start with a numerical method given by b and solve $e = a \star b$ for a. This produces a modified numerical method that agrees with the solution of the differential equation.

Appendix: Algebra and Calculus in Combinatorics

Here are some basic results from algebra and calculus in forms that are useful for combinatorics. These are stated in the setting of functions of one variable. There are even more illuminating multi-variable results, but they are not needed in the present exposition.

The Distributive Law. A version of the distributive law of algebra is the following. Suppose that B is a set and for each $b \in B$ there is a corresponding set F_b. Then the product over $b \in B$ of sums indexed by F_b is a sum of products:

$$\prod_{b\in B} \sum_{t\in F_b} a_b(t) = \sum_s \prod_{b\in B} a_b(s(b)).$$

The sum on the right is over all functions $s : B \to \bigcup_b F_b$ such that for each b the value $s(b) \in F_b$. The set of all such functions is the product space $\prod_b F_b$.

In the special case when the $F_b = F$ are all the same, the sum is over all functions $s : B \to F$. In this case the set of all such functions is the Cartesian power space F^B.

The Product Rule. A version of the product rule for differentiation is the following. Let U be a set with $|U|$ elements. Then the $|U|$ order derivative of a product function is given by

$$D^{|U|} \prod_{b \in B} F_b = \sum_{\phi \in B^U} \prod_{b \in B} D^{|\phi^{-1}(b)|} F_b.$$

Here B^U consists of all functions $\phi : U \to B$. Sometimes a function ϕ is described by its inverse images $U_b = \phi^{-1}(b)$, so one can think of this as a sum over the corresponding maps $b \mapsto U_b$.

The Chain Rule. A version of the chain rule is the following. Let U be a set with $|U|$ elements. Then the $|U|$ order derivative of a composite function is given by

$$D^{|U|}(F \circ G) = \sum_{\Gamma \in \mathrm{Part}[U]} (D^{|\Gamma|} F) \circ G \cdot \prod_{B \in \Gamma} D^{|B|} G.$$

Here Part$[U]$ consists of all set partitions of U into disjoint non-empty subsets with union U.

References

1. Bergeron, F., Labelle, G., Leroux, P.: Combinatorial Species and Tree-like Structures. Cambridge University Press, Cambridge (1998)
2. Flajolet, P., Sedgewick, R.: Analytic Combinatorics. Cambridge University Press, Cambridge (2009)
3. MacLane, S., Birkhoff, G.: Algebra. Chelsea, New York (1988)
4. Bogfjellmo, G., Schmeding, A.: The Lie group structure of the Butcher group. Found. Comput. Math. **17**(1), 127–159 (2017)
5. Brouder, Ch.: Runge-Kutta methods and renormalization. Eur. Phys. J. C **12**, 521–534 (2000)
6. Butcher, J.C.: An algebraic theory of integration methods. Math. Comput. **26**, 79–106 (1972)
7. Calaque, D., Ebrahimi-Fard, K., Manchon, D.: Two interacting Hopf algebras of trees. Adv. Appl. Math. **47**, 282–308 (2011)
8. Chartier, P., Hairer, E., Vilmart, G.: Algebraic structures of B-series. Found. Comput. Math. **10**, 407–427 (2010)
9. Connes, A., Kreimer, D.: Hopf algebras, renormalization and noncommutative geometry. Commun. Math. Phys. **199**, 203–242 (1998)
10. Faris, W.G.: Combinatorics and cluster expansions. Probab. Surv. **7**, 157–206 (2010)
11. Faris, W.G.: Combinatorial species and Feynman diagrams. Sèminaire Lotharingien de Combinatoire **61A**, Article B61An (2011)
12. Girelli, F., Krajewski, T., Martinetti, P.: An algebraic Birkhoff decomposition for the continuous renormalization group. J. Math. Phys. **45**, 4679–4697 (2004)

13. Hairer, E., Wanner, G.: Multistep-multistage-multiderivative methods for ordinary differential equations. Computing **11**, 287–303 (1973)
14. Hairer, E., Wanner, G.: On the Butcher group and general multi-value methods. Computing **13**, 1–15 (1974)
15. Kerber, A.: Applied Finite Group Actions, 2nd edn. Springer, Berlin (1999)
16. Lundervold, A., Munthe-Kaas, H.: Hopf algebras of formal diffeomorphisms and numerical integration on manifolds. Comb. Phys. Contemp. Math. **539**, 295–324 (2011)

A Stronger Topology for the Brownian Web

Luiz Renato Fontes(✉)

IME-USP, Rua do Matão 1010, São Paulo, SP 05508-090, Brazil
lrfontes@usp.br

In celebration of Chuck Newman's 70th birthday.

Abstract. We propose a metric space of *coalescing* pairs of paths on which we are able to prove in a fairly direct way convergence of objects such as the *persistence probability* in the (one dimensional, nearest neighbor, symmetric) voter model or the diffusively rescaled *weight distribution* in a silo model (as well as the equivalent *output distribution* in a river basin model), interpreted in terms of (dual) diffusively rescaled coalescing random walks, to corresponding objects defined in terms of the Brownian web.

Keywords: Brownian web · Coalescing random walks · Metric spaces · Convergence · Voter model · Persistence

1 Introduction

Chuck Newman first noticed that the convergence results of [2] were not by themselves enough to show convergence of the *persistence probability* in the voter model to a corresponding probability in the Brownian web. As is well known, the voter model may be described in terms of a system of coalescing random walks, which in turn has in [2] been shown to converge weakly to the Brownian web (under diffusive scaling). As pointed out by Chuck, the issue has to do with the topology in the space of trajectories adopted in [2]. Let us elaborate these points, and introduce other instances where they come up.

Persistence in the Voter Model. Let us suppose that individuals are placed on the sites of $\mathbb{Z}$, and that initially all the individuals on $2\mathbb{Z}$ have each a different opinion on a certain matter. At time $n \geq 1$, each individual of $2\mathbb{Z} + (n \bmod 2)$[1] adopts the opinion at time $n-1$ of a nearest neighbor individual (on $\mathbb{Z}$) chosen uniformly at random, independently from other individuals and choices.

[1] That is, the even or odd sublattices of $\mathbb{Z}$, depending on the parity of n, respectively.

Partially supported by CNPq grant 311257/2014-3 and FAPESP grant 2017/10555-0.

V. Sidoravicius (Ed.): *Sojourns in Probability Theory and Statistical Physics - II*, PROMS 299, pp. 167–185, 2019.
https://doi.org/10.1007/978-981-15-0298-9_7

At time $n = 2k$, the opinion of the individual at the origin may be traced back to the opinion of an individual at time 0. Let the choice of neighbor by individual i at time ℓ, as described above, be graphically represented as an edge from (i, ℓ) to either $(i-1, \ell-1)$ or $(i+1, \ell-1)$, depending on whether the chosen neighbor was the one to the left or the one on the right, respectively. Let us orient each such edge *backwardly* from ℓ to $\ell - 1$. We thus obtain a system of *backward* paths, one from each point of $\mathbb{Z}^2_{e,+}$, the even sublattice of $\mathbb{Z}^2$ in the upper half plane down to a point of $2\mathbb{Z} \times \{0\}$, given by a simple random walk path from the initial point to the final point. We say that the path model thus obtained is *dual* to the voter model. The paths of the path model are independent before meeting, and when they meet, they coalesce into the single path starting at the meeting space-time point (with time running backwards).

The collection of diffusively rescaled such coalescing backward paths converges as the scaling parameter vanishes to the Brownian web. This is the convergence result of [2] mentioned above.

Let us now define the persistence probability of the voter model as the probability that the individual at the origin does not change opinion from time $m = 2j$ to time n. This object was considered in [5] for the equivalent zero temperature Metropolis dynamics for the one dimensional nearest neighbor Ising model. In terms of the dual path model, it is equivalent to the probability that all backward paths starting from $\{(0, 2\ell);\, \ell = j, \dots, k\}$ coalesce before or at time 0.

Rescaling the model diffusively, with m and n also rescaled accordingly, and in such a way that $m/n \sim \alpha \in (0, 1)$, it is natural to expect that the persistence probability for the voter model will converge as the scaling parameter vanishes to the probability that all the (forward) paths from the Brownian web starting from the vertical segment $\{0\} \times [0, \alpha]$ coalesce before time 1. But this does not follow from the convergence result of [2] by itself.

Before proceeding with this particular point, let us discuss another model, where a similar issue comes up.

Weight Distribution in a Silo Model. In [1] a silo model is proposed as follows. A unit weight bead is located at every point of $\mathbb{Z}^2_{e,+}$. Each bead, located at say (i, ℓ) supports the full weight of either the one at $(i-1, \ell+1)$ or the one at $(i+1, \ell+1)$, which one is determined by a uniformly random choice, independent from bead to bead. The *total weight* supported by a bead is the sum of the weights of all the beads it supports from all the rows of beads above it, including its own weight.

In order to understand this quantity, let us start by putting downward directed edges from the supported to the supporting beads, and consider the downward paths starting from each bead to a bead at $2\mathbb{Z} \times \{0\}$ it connects to via those downward edges. Each pair of these paths are independent random walk paths till they meet, when they coalesce into a single random walk path.

Let us now consider the following set of dual upward directed edges in $\mathbb{Z}^2_{o,+}$, the *odd* sublattice of $\mathbb{Z}^2$ in the upper half plane. An upward edge from (j, k) in $\mathbb{Z}^2_{o,+}$ to either $(j-1, k+1)$ or $(j+1, k+1)$ is present if and only if it does not cross any of the downward edges described in the previous paragraph. Let us now consider the upward paths starting from the points of $\mathbb{Z}^2_{o,+}$ and going up along the upward edges of $\mathbb{Z}^2_{o,+}$ just described. One may readily check that any two distinct such upward paths are independent until they meet, after which time they coalesce. We say that the family of upward paths is *dual* to the system of downward paths[2].

It may be also checked that the total weight supported by a bead located at (i, ℓ) is given by the number of beads enclosed by the upward paths starting from $(i-1, \ell)$ and $(i+1, \ell)$.

We want to study the total weight distribution at the bottom of the silo, which may be formulated as the measure on $2\mathbb{Z}$ assigning to $\{i\}$ the total weight supported by the bead at $(i, 0)$.

By properly rescaling this measure, we may expect that it converges weakly to the measure on $\mathbb{R}$ that assigns to a finite interval $[a, b]$ the area encompassed by the following paths from the Brownian web: the leftmost path from $(a, 0)$ and the rightmost path from $(b, 0)$[3]. Again this does not follow from the convergence result of [2].

Output Distribution in a River Basin Model. The river basin/drainage system proposed in [8] (see also [7] for a broader discussion of such models) may be described as follows. From each site (i, ℓ) of $\mathbb{Z}^2_{e,+}$ there flows, along the appropriate edge, a unit volume of water to either $(i-1, \ell+1)$ or $(i+1, \ell+1)$, which one is determined by a uniformly random choice, independently from one originating site to the other. The downward flows to $2\mathbb{Z} \times \{0\}$ are thus described as downward random walk paths from the sites of $\mathbb{Z}^2_{e,+}$ to sites of $2\mathbb{Z} \times \{0\}$, abstractly identical to the ones from the silo model discussed above. We are interested in the distribution of total water flow on the sites of $2\mathbb{Z} \times \{0\}$. This is abstractly identical to the weight distribution measure introduced in the previous example, and thus should converge also as before to the Brownian web measure introduced above.

Let us next discuss why the convergence result of [2] is not enough to establish the convergences in the above examples.

The Trouble with the Topology of [2]. It comes up already when considering two coalescing paths, say starting from different space-time points (each in the vertical segment $\{0\} \times [0, \alpha]$, as in the first example; or both on $\mathbb{R} \times \{0\}$, as in the other examples); these paths should not touch till a *coalescence time*, up from which they are identical. Precisely the coalescence time should be continuous in the sup kind of metric of [2] in order for the convergences in the examples to follow from the convergence result of [2]. But it is not: two pairs of coalescing paths may be close in the sup metric without their coalescence times being close. Indeed, it is not continuous at any pair of coalescing paths; even the *coalescence property* (as implicitly given above) is not continuous in the metric space where the Brownian web is defined in [2].

---

[2] This is a different notion of duality than the one we described for the voter model above.

[3] In the Brownian web we may find multiple paths starting from a single space-time point.

A New Metric Space. This paper is an effort to address the issue raised above by introducing a new metric space for the Brownian web which takes into consideration the coalescence property and its continuity (in an appropriate sense). We discuss the new space in Sect. 2, and in Sect. 3 show that an appropriate description of the Brownian web belongs to it. There is *not* an effort at generality, so this space will accommodate the simple coalescing random walks of the examples, but not non simple ones. It is possible that a mild variation of this space and metric will include the latter models, but we do not discuss the matter further here.

We then show in Sect. 4 a weak convergence result of rescaled random walk paths starting from all space locations *at fixed times* to the corresponding Brownian web paths in this new space. This is enough to establish the convergences of the examples fairly directly in Sect. 5. Unfortunately, we could not show weak convergence of the full family of paths starting from *all* space-time points to the full Brownian web in this new space; more on this point in the last section of the paper.

2 Topological Set-Up

For simplicity, we will consider paths starting from a bounded region of the space-time plane—which we take to be the rectangle $\mathcal{R} := [-1, 1] \times [0, 1]$—, rather than the compactified $\mathbb{R}^2$ as in [2]. We will actually consider pairs of paths starting from $\mathcal{R}$: we start with a space of pairs of coalescing paths; and then a (Hausdorff) space of compact sets of the former space. The latter space will be the sample space of the (restricted) Brownian web.

Pairs of Coalescing Paths. The first space contains pairs of paths from a single path space which we define now. Let

$$\Pi' = \Pi'_{\mathcal{R}} = \bigcup_{t_0 \in [0,1]} C[t_0] \times \{t_0\},$$

where $C[t_0]$ denotes the set of continuous functions $f : [t_0, \infty) \to \mathbb{R}$ such that $f(t_0) \in [-1, 1]$. The elements of Π' should be seen as continuous paths in the space-time plane starting in $\mathcal{R}$, with time running upwards, each such path starting at a given time t_0 from $f(t_0)$; such an element is denoted (f, t_0).

To describe the first space, we need a notion of coalescence of a pair of paths from Π', to be defined now.

Definition 1. *For $i = 1, 2$ let $t_i \in [-1, 1]$, and suppose $(f_i, t_i) \in \Pi'$. Let*

$$t^c = \inf\{t > t^+ := t_1 \vee t_2 : f_1(t) = f_2(t)\},$$

with the usual convention that $\inf \emptyset = \infty$. *We will say that (f_1, t_1) and (f_2, t_2)* coalesce, *or are* coalescing, *or is a* coalescing pair (of paths) *if either $t^c = \infty$ or $f_1(t) = f_2(t)$ for $t \geq t^c$.*

Remark 1

1. The *coalescence time* t^c may equal t^+—a case that happens only if $f_1(t^+) = f_2(t^+)$. In particular, two identical paths of Π' are coalescing. But it may happen that $f_1(t^+) = f_2(t^+)$ and $t^c > t^+$; this indeed takes place in the Brownian web.
2. In each coalescing pair of paths, we may distinguish a *left path* and a *right path*, such that the left path is always to the left of (or coincides with) the right path (for $t \geq t^+$)—that is, $f_1(t) \leq f_2(t)$ for $t \geq t^+$. We will denote a coalescing pair by $[(f_1, t_1), (f_2, t_2)]$, where the first path, namely (f_1, t_1), will always denote the left path; occasionally we will also use the notation $[f_1, f_2]$ for short. In this context, we will occasionally write about an *ordered* pair of paths

The first space is finally defined as

$$\mathcal{C} = \{(f_1, t_1), (f_2, t_2) \in \Pi' : (f_1, t_1) \text{ and } (f_2, t_2) \text{ coalesce}\}.$$

We now want to put a metric in $\mathcal{C}$ under which it is complete and separable. Completeness is the tricky thing. Simple variants of the sup metric do not prevent Cauchy sequences which, so to say, want to converge to pairs of paths of Π' which touch and later coalesce, and are thus not in $\mathcal{C}$. In order to have such sequences not be Cauchy, we add to (a suitable variant of) the sup metric a term tailored to this end. We discuss that term next.

Sup Metrics in Π' *and* $\mathcal{C}$. Let us first consider the sup metric in Π' defined as follows: given (f, t_0), $(g, s_0) \in \Pi'$, let

$$d'((f, t_0), (g, s_0)) = |t_0 - s_0| \vee \sum_{n \geq 1} 2^{-n} \left(\sup_{t \in [0,n]} |\hat{f}(t) - \hat{g}(t)| \wedge 1 \right),$$

where, given $(f, t_0) \in \Pi'$, $\hat{f} : [0, \infty] \to \mathbb{R}$ is such that $\hat{f}(t) = f(t)$ for $t \geq t_0$, and $\hat{f}(t) = f(t_0)$ for $t \in [0, t_0]$.

Let $\bar{d}$ denote the Hausdorff-type metric induced in $\mathcal{C}$ by the metric d' in Π', namely, given two pairs of coalescing paths $[(f_1, t_1), (f_2, t_2)]$ and $[(g_1, s_1), (g_2, s_2)]$ (each pair in $\mathcal{C}$), let

$$\bar{d}([(f_1, t_1), (f_2, t_2)], [(g_1, s_1), (g_2, s_2)]) = \max_{i=1,2} \min_{j=1,2} d'((f_i, t_i), (g_j, s_j)).$$

It may be readily checked that $\bar{d}$ is a metric in $\mathcal{C}$, but *not* a complete one: there are Cauchy sequences $[(f_1^{(n)}, t_1^{(n)}), (f_2^{(n)}, t_2^{(n)})]$ in $(\mathcal{C}, \bar{d})$ such that as $n \to \infty$ $(f_1^{(n)}, t_1^{(n)}) \to (g_1, s_1)$ and $(f_2^{(n)}, t_2^{(n)}) \to (g_2, s_2)$ in (Π', d'), with (g_1, s_1) and (g_2, s_2) such that there exist t', s^* with $t' \in (s^+, s^*)$ satisfying that $g_1(t) \neq g_2(t)$ for $t \in (s^+, t') \cup (t', s^*)$, and $g_1(t) = g_2(t)$ for $t = t', s^*$; thus $[(g_1, s_1), (g_2, s_2)] \notin \mathcal{C}$. (Another way to put it is that $\mathcal{C}$ is not closed in $(\Pi' \times \Pi', \bar{d})$.)

We will then add a term to $\bar{d}$ in order to prevent this situation and have a resulting complete (and separable) metric in $\mathcal{C}$, as follows.

Pods. We rely on the following concept of a *pod* of a pair of coalescing paths in $\mathcal{C}$, defined roughly as the portion of the pair of paths between the beginning of the later path and the coalescence point. We find it necessary to separate the portions of the paths close to the beginning of the pod (the beginning of the later path) from the portions close to the end of the pod (the coalescence time) thus producing a *standard pod.*

For convenience, things will be defined in terms of the difference between the paths, as follows. Given a pair $[(f_1,t_1),(f_2,t_2)] \in \mathcal{C}$, recall that $t^+ = t_1 \vee t_2$ and t_c is the coalescence point. Let $\Psi(\cdot) = \tanh(\cdot)$ and make

$$\tilde{t}^+ = \Psi(t^+),\ \tilde{t}^c = \Psi(t^c),$$

If $t^+ < t^c$, then let $\tilde{p} : [\tilde{t}^+, \tilde{t}^c) \to (0,\infty)$ be such that

$$\tilde{p}(x) = f_2(\Psi^{-1}(x)) - f_1(\Psi^{-1}(x)) \tag{1}$$

$x \in [\tilde{t}^+, \tilde{t}^c)$, where Ψ^{-1} is the inverse function of Ψ. We recall that, according to the convention set above, $f_2 \geq f_1$.

If $t^+ = t^c$, then let $\tilde{p} : \{\tilde{t}^+\} \to \{0\}$.

We note that if $t^c < \infty$, then $\tilde{p}(\tilde{t}^c) := \lim_{x \uparrow \tilde{t}^c} \tilde{p}(x) = 0$ in all cases.

Definition 2. *We call $\tilde{p}$ the* pod (function) *of $[(f_1,t_1),(f_2,t_2)]$. We will refer below to $\tilde{t}^+$ and $\tilde{t}^c$ as the beginning and end of the pod, respectively. We will also refer to the* length *of the pod, meaning $\tilde{t}^c - \tilde{t}^+$, and also to the* diameter *of the pod, meaning $\sup_{x \in (\tilde{t}^+, \tilde{t}^c)} \tilde{p}(x)$. We finally define the* dimension *of a pod as the max of its length and diameter.*

Now let $\tilde{t}^m$ be the midpoint of $[\tilde{t}^+, \tilde{t}^c]$ and for $x \in [0,1)$ make

$$p(x) = \begin{cases} \tilde{p}(x + \tilde{t}^+), & x \in [0, \tilde{t}^m - \tilde{t}^+]; \\ \tilde{p}(x - (1 - \tilde{t}^c)), & x \in [1 - (\tilde{t}^c - \tilde{t}^m), 1); \\ \tilde{p}(\tilde{t}^m) + \frac{1}{2}[1 - (\tilde{t}^c - \tilde{t}^m)], & x = 1/2. \end{cases} \tag{2}$$

If $t^+ = t^c$, we define $p(1) = 0$. We then complete the definition of p in $[0,1)$ by making it linear in $[\tilde{t}^m - \tilde{t}^+, 1/2]$ and in $[1/2, 1 - (\tilde{t}^c - \tilde{t}^m)]$.

We call p the *standard pod (function)* of $[(f_1,t_1),(f_2,t_2)]$, and note that p satisfies

$$p : (0,1) \to (0,\infty); \quad \lim_{x \uparrow 1} p(x) = 0, \text{ if } t^c < \infty; \quad p(0) \geq 0, \tag{3}$$

with equality if and only if $f_1(t^+) = f_2(t^+)$. Additionally, p has derivative 1 and -1 in $(\tilde{t}^m - \tilde{t}^+, 1/2)$ and $(1/2, 1 - (\tilde{t}^c - \tilde{t}^m))$, respectively.

For p, q with the above properties (namely, (3) and continuity) let

$$\Delta(p,q) = \sum_{n \geq 2} 2^{-n} \left(\sup_{x \in [\frac{1}{n}, 1 - \frac{1}{n}]} \left| \frac{1}{p(x)} - \frac{1}{q(x)} \right| \wedge 1 \right).$$

The following result will be needed in Sect. 4 below.

Lemma 1. *If $\tilde{p}$ and $\tilde{q}$ are pods with dimensions at most $\sigma \leq 1/8$, then $\Delta(p, q) \leq \varphi(\sigma)$, with $\lim_{\sigma \to 0} \varphi(\sigma) = 0$, where p and q are the respective standard pods of $\tilde{p}$ and $\tilde{q}$.*

Proof. It follows from (2) that, for all $\sigma \leq 1/8$, p and q are both linear separately in the intervals $[\sigma, 1/2]$ and $[1/2, 1-\sigma]$, with inclinations 1 and -1 in each interval, respectively. This implies that

$$\sup_{x \in [\sigma^{1/3}, 1-\sigma^{1/3}]} \left| \frac{1}{p(x)} - \frac{1}{q(x)} \right| \leq \frac{2\sigma}{\sigma^{2/3} - \sigma^2} =: \varphi'(\sigma).$$

It readily follows that $\varphi(\sigma) = \varphi'(\sigma) + \sum_{n \geq \sigma^{-1/3}} 2^{-n}$ verifies the claim. □

A Stronger Metric in $\mathcal{C}$. We discuss now the term to be added to $\bar{d}$ in order to have a metric in $\mathcal{C}$ making it complete.

Let $[(f_1, t_1), (f_2, t_2)]$ and $[(g_1, s_1), (g_2, s_2)]$ be two coalescing pairs of $\mathcal{C}$, and let p, q be the standard pods respectively of $[(f_1, t_1), (f_2, t_2)], [(g_1, s_1), (g_2, s_2)]$.

Let us define

$$\tilde{d}([(f_1, t_1), (f_2, t_2)], [(g_1, s_1), (g_2, s_2)]) \\ = \bar{d}([(f_1, t_1), (f_2, t_2)], [(g_1, s_1), (g_2, s_2)]) + \Delta(p, q)$$

for $[(f_1, t_1), (f_2, t_2)]$, $[(g_1, s_1), (g_2, s_2)] \in \mathcal{C}$.

$(\mathcal{C}, \tilde{d})$ Is a Complete and Separable Metric Space. That $\tilde{d}$ is a metric in $\mathcal{C}$ is quite clear; separability is also fairly clear after a moment's thought.

To argue completeness (sketchily), let $[(f_1^{(n)}, t_1^{(n)}), (f_2^{(n)}, t_2^{(n)})]$, $n \geq 1$, be a Cauchy sequence in $(\mathcal{C}, \tilde{d})$. Let t_n^+ and t_n^c be the beginning and end of the respective pods.

From the $\bar{d}$ part of $\tilde{d}$, we have that, for $i = 1, 2$, there exist $(f_i, t_i) \in \Pi'$ such that, as $n \to \infty$,

$$(f_i^{(n)}, t_i^{(n)}) \to (f_i, t_i) \text{ in } (\Pi', d'); \tag{4}$$

$$t_i^{(n)} \to t_i,\ t_n^+ \to t^+ = t_1 \vee t_2 \text{ in } \mathbb{R}. \tag{5}$$

And combining this with the Δ part of $\tilde{d}$, we find that

$$t_n^c \to t^c \text{ in } \bar{\mathbb{R}} \tag{6}$$

for some $t^c \in \bar{\mathbb{R}} := \mathbb{R} \cup \{\infty\}$. It may be readily checked that if $t^c < \infty$, then $f_1(t) = f_2(t)$ for $t \geq t^c$.

If $t^+ = t^c$, then we immediately have that $(f_1, t_1) = (f_2, t_2)$, and then of course $[(f_1, t_1), (f_2, t_2)] \in \mathcal{C}$ in this case.

Let us now verify that if $t^+ < t^c$, then

$$f_1(t) \neq f_2(t) \text{ for } t \in (t^+, t^c), \tag{7}$$

and thus we have that $[(f_1, t_1), (f_2, t_2)] \in \mathcal{C}$ also in this case.

From (5) and (6), we have that $\tilde{t}_n^+ \to \tilde{t}^+$ and $\tilde{t}_n^c \to \tilde{t}^c$, and $t^+ < t^c$ implies that $\tilde{t}^+ < \tilde{t}^c$. Given now $[s, s'] \subset (t^+, t^c)$, we have that $\tilde{s} = \Psi(s), \tilde{s}' = \Psi(s')$ are such that $[\tilde{s}, \tilde{s}'] \subset (\tilde{t}^+, \tilde{t}^c)$.

We will at this point assume for simplicity, but without compromising the generality of conclusion of the argument, that $\tilde{t}^m \in (\tilde{s}, \tilde{s}')$.

Let p_n, $n \geq 1$, be the sequence of standard pods of $[(f_1^{(n)}, t_1^{(n)}), (f_2^{(n)}, t_2^{(n)})]$. Since by hypothesis $\Delta(p_m, p_n) \to 0$ as $m, n \to \infty$, it follows that there exists $p : (0,1) \to (0, \infty)$ such that $\frac{1}{p_n} \to \frac{1}{p}$ uniformly over closed intervals of $(0,1)$. Notice also that from the $\bar{d}$ part of the metric, p must be of the form (1)–(2). It follows that $p > 0$ in $[\tilde{s} - \tilde{t}^+, 1 - \tilde{s}']$; in turn, $\tilde{p} > 0$ in $[\tilde{s}, \tilde{s}']$, from which it immediately follows that $f_1 \neq f_2$ in $[s, s']$, and (7) is established.

To finish the completeness argument, it remains to show that

$$[(f_1^{(n)}, t_1^{(n)}), (f_2^{(n)}, t_2^{(n)})] \to [(f_1, t_1), (f_2, t_2)]$$

in $(\mathcal{C}, \tilde{d})$ as $n \to \infty$, but this follows readily from (4)–(6) and the arguments of the previous paragraph.

Remark 2. If $[(f_1, t_1), (f_2, t_2)]$ and $[(f_1^{(n)}, t_1^{(n)}), (f_2^{(n)}, t_2^{(n)})]$, $n \geq 1$, are in $\mathcal{C}$ and (4), (6) hold with t^c the coalescence time of $[(f_1, t_1), (f_2, t_2)]$, then, as may be readily checked,

$$[(f_1^{(n)}, t_1^{(n)}), (f_2^{(n)}, t_2^{(n)})] \to [(f_1, t_1), (f_2, t_2)] \text{ in } (\mathcal{C}, \tilde{d}).$$

Hausdorff Metric Space. We may now define $(\mathcal{H}_\mathcal{C}, d_{\mathcal{H}_\mathcal{C}})$ as the Hausdorff metric space associated to $(\mathcal{C}, \tilde{d})$, namely

$$\mathcal{H}_\mathcal{C} = \{K \subset \mathcal{C} : K \text{ is non empty and compact}\}$$

and, given $K, K' \in \mathcal{H}_\mathcal{C}$,

$$d_{\mathcal{H}_\mathcal{C}}(K, K') = \sup_{[f_1, f_2] \in K} \inf_{[g_1, g_2] \in K'} \tilde{d}([f_1, f_2], [g_1, g_2]) \vee \sup_{[g_1, g_2] \in K'} \inf_{[f_1, f_2] \in K} \tilde{d}([f_1, f_2], [g_1, g_2])$$

(see the last sentence of the 4th point of Remark 1 above).

Remark 3. Given the properties of completeness and separability of $(\mathcal{C}, \tilde{d})$, established above, it follows that $(\mathcal{H}_\mathcal{C}, d_{\mathcal{H}_\mathcal{C}})$ has the same properties.

3 The Brownian Web as a Compact Set of Coalescing Pairs of Paths

The Brownian web was defined in [2] as a closed set of coalescing Brownian paths from (Π, d), a single path space similar to (Π', d'), with the difference that paths live in a compactified $\mathbb{R}^2$—see Section 3 of [2]. It was then showed to be compact. In this section, we will define a *restricted* Brownian web as a set of pairs of paths from $(\mathcal{C}, \tilde{d})$, and then show that it is compact, and thus belongs to $\mathcal{H}_\mathcal{C}$. This will set up the stage for proving a (somewhat restricted) convergence of rescaled coalescing simple random walks to the Brownian web in $(\mathcal{H}_\mathcal{C}, d_{\mathcal{H}_\mathcal{C}})$ in the next section.

Let us briefly make a definition based on a definition from [2]. Let $\mathcal{D}$ be a dense countable subset of $\mathcal{R}$ and let $\mathcal{W} = \mathcal{W}(\mathcal{D})$ be collection of coalescing Brownian paths (with unit diffusion coefficient), one starting from each point of $\mathcal{D}$. See [2], Section 3, for details of a construction. Let now $\bar{\mathcal{W}} = \bar{\mathcal{W}}(\mathcal{D})$ be the closure of $\mathcal{W}$ as a subset of (Π', d'). $\bar{\mathcal{W}}$ is what we call the *restricted Brownian web*, and $\mathcal{W}$ its *skeleton*.

It is quite clear that almost surely every pair of paths in $\mathcal{W}$ is coalescing according to Definition 1 above. $\bar{\mathcal{W}}$ has the same property, as follows readily from Proposition 4.2 of [2]. So the collection of coalescing (ordered) pairs

$$\mathcal{P} := \{((f_1,t_1),(f_2,t_2)) \in \bar{\mathcal{W}} \times \bar{\mathcal{W}} : f_2(t) \geq f_1(t) \text{ for } t \geq t_1 \vee t_2\}$$

is a almost surely subset of $\mathcal{C}$.

$\mathcal{P}$ Is a Compact Set of $(\mathcal{C}, \tilde{d})$. We will argue now the almost sure compactness of $\mathcal{P}$ in $(\mathcal{C}, \tilde{d})$.

We know from Proposition 3.2 in [2] that $\bar{\mathcal{W}}$ is compact in (Π, d), so given a sequence $[(f_1^{(n)}, t_1^{(n)}), (f_2^{(n)}, t_2^{(n)})]$, $n \geq 1$, from $\mathcal{P}$, we have that there exists a subsequence (n') of (n) and $[(f_1,t_1),(f_2,t_2)] \in \mathcal{P}$ such that

$$(f_i^{(n')}, t_i^{(n')}) \to (f_i, t_i) \text{ in } (\Pi, d) \text{ as } n' \to \infty. \tag{8}$$

Since Brownian paths do not go to $\pm\infty$ in finite time, we readily conclude that

$$(f_i^{(n')}, t_i^{(n')}) \to (f_i, t_i) \text{ in } (\Pi', d') \text{ as } n' \to \infty,$$

and, as pointed out in Remark 2, we will have that

$$[(f_1^{(n')}, t_1^{(n')}), (f_2^{(n')}, t_2^{(n')})] \to [(f_1,t_1),(f_2,t_2)] \text{ in } (\mathcal{C}, \tilde{d}) \text{ as } n' \to \infty$$

once (6) holds. Let us argue the latter point. It follows from the proof of Proposition 4.3 of [3] and (8) that given $\varepsilon > 0$, there exist a pair of paths $(f_1', t_1'), (f_2', t_2') \in \mathcal{W}$ and n_0 such that for every $n' \geq n_0$, we have that

$$\begin{aligned} f_1^{(n')}(t) &= f_1'(t), \quad \text{for } t \geq t_1 + \varepsilon \\ f_2^{(n')}(t) &= f_2'(t), \quad \text{for } t \geq t_2 + \varepsilon, \end{aligned}$$

and (6) follows.

Definition 3. *We will call $\mathcal{P}$ the* Brownian pair web (BPW).

Remark 4. As argued above, the BPW is almost surely compact, and as such, it belongs to $(\mathcal{H}_\mathcal{C}, d_{\mathcal{H}_\mathcal{C}})$.

Definition 4. *For $\tau \in [0,1]$, let*

$$\mathcal{P}_\tau = \{[(f_1,t_1),(f_2,t_2)] \in \mathcal{P} : t_1 = t_2 = \tau\} \tag{9}$$

be the pairs in $\mathcal{P}$ both paths of which start from time τ.

It may be readily checked that $\mathcal{P}_\tau$ is compact for every $\tau \in [0,1]$ almost surely.

4 Coalescing Simple Random Walks

Let Y and $Y^{(\delta)}$ be respectively the collection of coalescing simple random walk paths, and the diffusively rescaled collection of the same set of paths, with scale parameter $\delta > 0$, both collections consisting of paths starting from points in the space-time plane. See Introduction and Section 6 of [2]. We recall that the starting space-time points of Y consist of all of $\mathbb{Z}^2$. We will refer below to the starting space-time points of $Y^{(\delta)}$ as *(rescaled) discrete space-time points.* We will consider the subcollection of paths from $Y^{(\delta)}$ starting from $\mathcal{R}$. Let us denote that collection by $\bar{Y}^{(\delta)}$, and their starting points by $\mathcal{R}^{(\delta)}$. Let us now consider the set of (ordered) pairs from $\bar{Y}^{(\delta)}$:

$$\mathcal{P}^{(\delta)} := \{((f_1,t_1),(f_2,t_2)) \in \bar{Y}^{(\delta)} \times \bar{Y}^{(\delta)} : f_2(t) \geq f_1(t) \text{ for } t \geq t_1 \vee t_2\}.$$

We call $\mathcal{P}^{(\delta)}$ the *Discrete Pair Web* (DPW). We note that, for every fixed $\delta > 0$, $\mathcal{P}^{(\delta)}$ is a finite subset of $\mathcal{C}$, so it belongs to $(\mathcal{H}_\mathcal{C}, d_{\mathcal{H}_\mathcal{C}})$.

Definition 5. *For $\tau \in [0,1]$, let*

$$\mathcal{P}_\tau^{(\delta)} = \{[(f_1,t_1),(f_2,t_2)] \in \mathcal{P}^{(\delta)} : t_1 = t_2 = \delta^2\lfloor\delta^{-2}\tau\rfloor\}$$

be the pairs in $\mathcal{P}^{(\delta)}$ both paths of which start from time $\delta^2\lfloor\delta^{-2}\tau\rfloor$.

We will now argue that as $\delta \to 0$, the distribution of $\mathcal{P}_\tau^{(\delta)}$ in $(\mathcal{H}_\mathcal{C}, d_{\mathcal{H}_\mathcal{C}})$ converges to $\mathcal{P}_\tau$ defined in (9) above.

$\mathcal{P}_\tau^{(\delta)}$ *Converges to $\mathcal{P}_\tau$.* Due to the obvious time invariance of the distributions of $\mathcal{P}_\tau^{(\delta)}$ and $\mathcal{P}_\tau$, it suffices to take $\tau = 0$.

Let us recall from [2] that $(\mathcal{H}, d_\mathcal{H})$ is the Hausdorff metric space generated by (Π, d), and that $Y^{(\delta)}$ converges to $\bar{\mathcal{W}}$ in distribution in $(\mathcal{H}, d_\mathcal{H})$. Resorting to Skorohod representation, we may take this convergence to be almost sure. From that it follows fairly readily that

$$\bar{d}_\mathcal{H}(\mathcal{P}^{(\delta)}, \mathcal{P}) \to 0 \tag{10}$$

almost surely as $\delta \to 0$, where $\bar{d}_\mathcal{H}$ is the Hausdorff metric associated to $(\mathcal{N}, \bar{d})$, where

$$\mathcal{N} := \{((f_1,t_1),(f_2,t_2)) \in \Pi' \times \Pi' : f_2(t) \geq f_1(t) \text{ for } t \geq t_1 \vee t_2\},$$

the subset of ordered non crossing pairs of paths from $\Pi' \times \Pi'$. In particular

$$\bar{d}_{\mathcal{H}}(\mathcal{P}_0^{(\delta)}, \mathcal{P}_0) \to 0 \tag{11}$$

almost surely as $\delta \to 0$. It may be also readily checked that $\mathcal{P}^{(\delta)}$ and $\mathcal{P}$ are compact subsets of $(\mathcal{N}, \bar{d})$.

We will strengthen (11) to

$$d_{\mathcal{H}c}(\mathcal{P}_0^{(\delta)}, \mathcal{P}_0) \to 0 \tag{12}$$

in probability as $\delta \to 0$.

Below we will occasionally write $\mathcal{P}_0^{(0)}$ for $\mathcal{P}_0$.

4.1 Argument for (12)

For $\varepsilon > 0$, let $\mathcal{T}^{(\delta)}$ denote the set of pairs of neighboring points of $\delta\mathbb{Z} \times \{0\}$ such that the pair of paths of $\mathcal{P}_0^{(\delta)}$ starting from those points take longer than ε to coalesce. Let us call such pairs of points *double pairs* (of points), and denote the collection of such paths by $\mathcal{P}_{0,\varepsilon}^{(\delta)}$.

Let us recall that a $(0, 2)$-point, or a *(Brownian) double point*, from $\bar{\mathcal{W}}$ is a space-time point (x, t) touched by no path of $\bar{\mathcal{W}}$ starting before t, and from which there start (exactly) 2 paths from $\bar{\mathcal{W}}$. See Definition 3.8 in [4]. Let $\mathcal{T}$ denote the set of Brownian double points in $\mathbb{R} \times \{0\}$ such that the pairs of paths of $\mathcal{P}$ starting from those points take longer than ε to coalesce. Let us denote the collection of such paths by $\mathcal{P}_{0,\varepsilon}$, and also occasionally by $\mathcal{P}_{0,\varepsilon}^{(0)}$.

Let $T^{(\delta)}$, T denote the cardinalities of $\mathcal{T}^{(\delta)}$, $\mathcal{T}$, respectively. It follows from what we know about $\bar{\mathcal{W}}$—see Proposition 4.3 in [2]—that, for each $\varepsilon > 0$, T is a finite random variable, and $T^{(\delta)}$, $\delta > 0$, is a tight family of random variables.

We will call the pairs in $\mathcal{P}_{0,\varepsilon}$ and $\mathcal{P}_{0,\varepsilon}^{(\delta)}$, $\delta \geq 0$, as *(Brownian)* and *(discrete) double pairs* of paths (of $\mathcal{P}_0$ and $\mathcal{P}_0^{(\delta)}$, respectively), or *(Brownian)* and *(discrete) double paths*, for short.

Convergence of Double Paths. It follows readily from (11) that there almost surely exists a 1 to 1 correspondence for each δ small enough between discrete and Brownian double paths so that to each Brownian double path there converges in $\bar{d}$ distance the corresponding discrete double path as $\delta \to 0$.

Now let us fix a Brownian double path, and argue that the corresponding discrete double path converges to that Brownian path also in the $\tilde{d}$ distance. For that, it remains to show that the discrete standard pod associated to the discrete double path converges to that for the Brownian one.

It is enough to show that the coalescence times converge suitably. One may readily check that (11) implies that the liminf as $\delta \to 0$ of the discrete coalescence time (the coalescence time of the discrete double path) must equal or exceed the Brownian one. Let us show that the latter possibility is a null event.

If the limsup as $\delta \to 0$ of the discrete coalescence time exceeds the Brownian one, then given the constituting paths of the Brownian double path till their coalescence time and place, say (x_0, t_0), then given $M > 0$, we have that for small enough $\delta > 0$, we may find (macroscopic) coalescing random walk paths starting from a (macroscopic) interval around $\delta^{-1}x_0$ of length $\frac{1}{M}\delta^{-1}$ at time $\lceil\delta^{-2}t_0\rceil$ which stay within distance $\frac{1}{M}\delta^{-1}$ from each other without coalescing for a time (counting from $\lceil\delta^{-2}t_0\rceil$) of length $\delta^{-2}T$, with $T > 0$ not depending on M. Consider now the locations of the intersections of those coalescing random walk paths with $\mathbb{R} \times \{\lceil\delta^{-2}(t_0 + T/2)\rceil\}$. We must then be able to find a pair of such points at distance at most $\frac{1}{M}\delta^{-1}$ and such that the coalescing random walk paths starting from them stay within distance $\frac{1}{M}\delta^{-1}$ from each other without coalescing for a time (counting from $\lceil\delta^{-2}(t_0 + T/2)\rceil$) of length $\delta^{-2}T/2$.

Resorting again to Proposition 4.3 in [2], we may claim that the size of the set of intersection locations is tight, and we will then readily conclude the argument if we show that for a single fixed pair of locations at distance at most $\frac{1}{M}\delta^{-1}$, the probability that the coalescing random walk paths starting from that pair of locations stay within distance $\frac{1}{M}\delta^{-1}$ from each other without coalescing for a time of length $\delta^{-2}T/2$ vanishes as $M \to \infty$. But this follows from standard results about the tail of the hitting time of the exterior of an interval by a single random walk starting from the interior of that interval.

Argument for the Other Pairs of Paths. A similar argument as that for double paths above shows that pairs of paths from $\mathcal{P}_0^{(\delta)}$ starting from points from different double pairs also converge to the respective Brownian pair of paths from $\bar{\mathcal{W}}$ in the $\tilde{d}$ distance.

Let us enumerate the double pairs of $\mathcal{T}^{(\delta)} = \{(x_j, y_j), j = 1, \dots, k\}$ in increasing order. Let us include in $\mathcal{T}^{(\delta)}$, in case they are not already present, the double pairs (x_0, y_0) and (x_{k+1}, y_{k+1}) such that $x_0 \leq -1$ and $y_{k+1} \geq 1$[4]. Let λ_j and ρ_j be the portions inside $\mathbb{R} \times [0, \varepsilon]$ of the paths of $Y^{(\delta)}$ starting from x_j and y_j, respectively, and, for $j = 0, \dots, k$, let Γ_j be the open subset of $\mathbb{R} \times (0, \varepsilon)$ bounded by ρ_j, and λ_{j+1}, respectively. We will refer below to ρ_j and λ_{j+1}, $j \geq 0$, as the *bounding subpaths* of Γ_j, $j \geq 0$. For $j \geq 0$, let $\bar{\Gamma}_j$ denote the closure of Γ_j and make $\mathcal{I}_j = (\mathbb{R} \times \{0\}) \cap \bar{\Gamma}_j$.

Remark 5
We note that each path starting from $\mathcal{I}_j$, $j \geq 1$, coalesces within time ε with a path from a double path of $\mathcal{P}_{0,\varepsilon}^{(\delta)}$, or with ρ_0, or with λ_{k+1}—indeed it coalesces with one of the paths bounding Γ_j.

We next derive a bound on the *(horizontal) diameters* of the regions $\Gamma_j^{(\delta)}$, $j \geq 0$, meaning the sup distance between the bounding subpaths of $\Gamma_j^{(\delta)}$. We will show that those diameters are bounded above by $\varepsilon^{1/3}$ uniformly in j *with high probability*.

[4] The paths starting from either x_0, y_0, x_{k+1} or y_{k+1} may not belong to $\bar{Y}^{(\delta)}$.

Bound on the Diameter of Γ *Regions.* Let us start by partitioning $[-2,2] \times \{0\}$ into intervals of equal length ε^2, and consider the event that such an interval contains more than 1 double pair. By the version of Proposition 4.1 in [2] which holds for $Y^{(\delta)}$ (see argument in [3]), together with standard facts about hitting times of the simple symmetric random walk on $\mathbb{Z}^5$, this event has probability bounded above by ε^3 uniformly for all δ sufficiently small. So the probability of the event of not finding such an interval in all of $[-2,2] \times \{0\}$ is bounded below by $1-\varepsilon$ (uniformly for all δ sufficiently small). We will assume from now on that we are in such an event.

Let us now enumerate the intervals in this partition which contain a double pair, from left to right: J_1, $J_2, \ldots$. Since the right path from the double path starting in J_ℓ has to coalesce with the left path from the one in $J_{\ell+1}$ before time ε, the probability that we find a pair $J_\ell, J_{\ell+1}$ such that the respective double pairs are further than $\varepsilon^{1/3}$ apart is bounded above by $16\varepsilon^{-4}$ times the probability that the a simple random walk on $\mathbb{Z}^6$ starting from $\varepsilon^{1/3}\delta^{-1}$ hits the origin before $\varepsilon\delta^{-2}$ steps. The latter probability is bounded above by $e^{-\frac{1}{3}\varepsilon^{-1/3}}$ for ε small enough, uniformly in δ sufficiently small. We thus conclude that outside an event of probability at most $e^{-\frac{1}{4}\varepsilon^{-1/3}}$, neighboring double pairs are at most $\varepsilon^{1/3}$ apart (for ε small enough, uniformly in δ sufficiently small). A similar argument shows that outside an event of probability at most constant times ε for ε small enough, uniformly in δ sufficiently small, any $\Gamma_j^{(\delta)}$ region has diameter at most $\varepsilon^{1/3}$ (here including the event whose probability was estimated in the previous paragraph).

The Brownian Case. The claims argued above for $\mathcal{P}_0^{(\delta)}$ go through with no essential modification for the Brownian case of $\mathcal{P}_0$, namely, outside an event of vanishing probability as $\varepsilon \to 0$, the Γ regions have diameters at most $\varepsilon^{1/3}$.

Conclusion. In order to conclude our argument for the validity of (12), it is enough to show that for all fixed $\eta > 0$, we have that outside an event of vanishing probability as $\varepsilon \to 0$ uniformly in δ sufficiently small[7], given a pair of paths in $\mathcal{P}_0^{(\delta)}$ with standard pod p, we may find a pair of paths in $\mathcal{P}_0$ with standard pod q such that

$$\Delta(p,q) \leq \eta, \tag{13}$$

and conversely, exchanging the roles of $\mathcal{P}_0$ and $\mathcal{P}_0^{(\delta)}$. The converse case is similar to the first case, so let us argue only the first case.

It readily follows from our arguments above that we have the bound (13) for pairs of paths from $\mathcal{P}_0^{(\delta)}$ starting from space-time points belonging to (not

[5] Indeed, the process that comes up is the difference of two independent simple random walks on $\mathbb{Z}$, but this behaves the same as the simple symmetric case as far as this issue is concerned.

[6] Idem.

[7] A $\limsup_{\delta\to 0}$ is taken first; and then a $\limsup_{\varepsilon\to 0}$.

necessarily the same) double pairs; since the number of such double pairs is tight in δ as $\delta \to 0$, we get (13) for all these pairs of paths simultaneously with high probability.

In order to extend this to all paths of $\mathcal{P}_0^{(\delta)}$, we argue as follows. Given a pair of paths $[(f_1^{(\delta)}, 0), (f_2^{(\delta)}, 0)]$ from $\mathcal{P}_0^{(\delta)}$, with standard pod $p^{(\delta)}$, let $[(\bar{f}_1^{(\delta)}, 0), (\bar{f}_2^{(\delta)}, 0)]$ be the pair from $\mathcal{P}_0^{(\delta)}$ such that $(\bar{f}_m^{(\delta)}, 0)$ and $(f_m^{(\delta)}, 0)$ coalesce as described in Remark 5 for $m = 1, 2$.

Let $q^{(\delta)}$ be the standard pod of $[(\bar{f}_1^{(\delta)}, 0), (\bar{f}_2^{(\delta)}, 0)]$. From what we just argued in the previous paragraph, there exists a pair $[(\bar{f}_1, 0), (\bar{f}_2, 0)]$ in $\mathcal{P}_0$ with standard pod q such that with high probability the bound (13) holds for the pair of pods $q^{(\delta)}, q$, with $\eta/2$ replacing η.

It should be quite clear from the above considerations on the diameter of the Γ regions that

$$\bar{d}\left(\left[\left(f_1^{(\delta)}, 0\right), \left(f_2^{(\delta)}, 0\right)\right], \left[\left(\bar{f}_1^{(\delta)}, 0\right), \left(\bar{f}_2^{(\delta)}, 0\right)\right]\right) \leq 2\varepsilon^{1/3}$$

with high probability.

Let us now analyse $\Delta(p^{(\delta)}, q^{(\delta)})$. Let $\tilde{p}^{(\delta)}, \tilde{q}^{(\delta)}$ be the pods associated respectively to $p^{(\delta)}, q^{(\delta)}$, so that the latter standard pods are obtained from the former pods according to (2). Notice that $\tilde{p}^{(\delta)}$ and $\tilde{q}^{(\delta)}$ have the same beginning, namely 0. Let us denote their respective ends by t^c and s^c.

Remark 6

1. If each path determining $p^{(\delta)}$ is such that its portion from times 0 to ε is a bounding path of Γ_j for some j, then $p^{(\delta)} = q^{(\delta)}$ and $\Delta(p^{(\delta)}, q^{(\delta)})$ vanishes.
2. Excluding the situation in item 1,
   (a) if the starting locations of the paths determining $p^{(\delta)}$ both belong to the same $\mathcal{I}_j$ for some j, then either $t^c = s^s$, or $t^c \leq \varepsilon$ and $s^c = 0$;
   (b) excluding the situation in item 2a, if the starting locations of the paths determining $p^{(\delta)}$ belong to different $\mathcal{I}_j$'s, then $t^c = s^s$.
3. Whenever $t^c = s^s \geq 2\varepsilon$, we have that $\tilde{p}^{(\delta)}(x) = \tilde{q}^{(\delta)}(x)$ for $x \geq \tilde{\varepsilon} := \Psi(\varepsilon)$, and thus $p^{(\delta)}(x) = q^{(\delta)}(x)$ for $x \geq \tilde{\varepsilon}$.

It follows from the above considerations that given $\eta > 0$, with high probability, the following events take place simultaneously.

1. If $t^c \geq 2\varepsilon$, then $\Delta(p^{(\delta)}, q^{(\delta)}) \leq \sum_{n \geq \tilde{\varepsilon}^{-1}} 2^{-n} \leq \eta/2$;
2. If $t^c < 2\varepsilon$, then replacing ε by 2ε in the analysis above (started with the bounds on diameters of the Γ regions), we find that $s^c \leq 2\varepsilon$ and that the diameters of both $\tilde{p}^{(\delta)}$ and $\tilde{q}^{(\delta)}$ are bounded above by $\varepsilon^{1/3}$. Lemma 1 now implies the same conclusion as in the previous item, namely, $\Delta(p^{(\delta)}, q^{(\delta)}) \leq \eta/2$ (for all ε small enough).

The upshot is that $\Delta(p^{(\delta)}, q^{(\delta)}) \leq \eta/2$ in all cases with high probability, and (12) is established.

5 Examples

In this section, we discuss the examples mentioned in the introduction, namely persistence in the voter model and the weight distribution at the bottom of a silo/water output distribution at a level set of a river basin (modeled by coalescing random walks, as discussed in the introduction).

5.1 Weight/Output Distribution in a Silo/River Basin Model

As discussed in the introduction, the (properly rescaled) weight distribution at a section of the bottom of the silo model there described is the random measure $\mu^{(\delta)}$ on $2\delta\mathbb{Z} \cap [-1, 1]$ such that $\mu^{(\delta)}(\{(2\delta k, 0)\})$ is the area between the rescaled (upward, in this context—recall the description at the introduction) random walk paths from $\bar{Y}^{(\delta)}$ issuing from $((2k-1)\delta, 0)$ and $((2k+1)\delta, 0)$. The output of the river basin model along a section of a level line can be described by the same measure.

Let μ be the random measure on $[-1, 1]$ such that given $a, b \in [-1, 1]$, $a \le b$, $\mu([a, b])$ is the area between the leftmost and rightmost paths from $\bar{\mathcal{W}}$ issuing from $(a, 0)$ and $(b, 0)$, respectively. We want to show that

$$\mu^{(\delta)} \to \mu \tag{14}$$

in distribution, with the vague topology in the space of Radon measures on $[-1, 1]$. We argue (14) next.

$\mu^{(\delta)}$ and μ as Images of a Continuous Map

A Restricted Space. Let us start by considering the following subspace of $\mathcal{H}_C$.

$$\begin{aligned}\mathcal{H}_C^0 = \{K \in \mathcal{H}_C :\ & \text{all paths from } K \text{ start at time } 0;\\ & \text{for every point of } [-1,1] \times \{0\} \text{ there is}\\ & \text{a path of } K \text{ starting from that point}\}\end{aligned}$$

One readily checks that $\mathcal{H}_C^0$ is a closed subset of $(\mathcal{H}_C, d_{\mathcal{H}_C})$. It then follows that $(\mathcal{H}_C^0, d_{\mathcal{H}_C})$ is complete and separable.

In order to have the convergence (12) take place in $(\mathcal{H}_C^0, d_{\mathcal{H}_C})$ as well, we must extend the definition of $\bar{Y}^{(\delta)}$ suitably. A simple way to do so is by incorporating paths starting from space-time points in $\{((2j-1)\delta, (2j+1)\delta) \cap [-1, 1]\} \times \{0\}$, $j \in \mathbb{Z}$, so that their portion in $\mathbb{R} \times [0, \delta^2]$ coalesces linearly at time δ^2 with the path from $\bar{Y}^{(\delta)}$ starting from $(2j\delta, 0)$; from $((2j+1)\delta, 0)$, we incorporate two paths, one coalescing linearly at time δ^2 with the path from $\bar{Y}^{(\delta)}$ starting from $(2j\delta, 0)$; the other coalescing linearly at time δ^2 with the path from $\bar{Y}^{(\delta)}$ starting from $(2(j+1)\delta, 0)$. Let us denote by $\bar{\mathcal{P}}_0^{(\delta)}$ the collection of pairs of paths thus obtained. It is quite clear that $\bar{\mathcal{P}}_0^{(\delta)} \in \mathcal{H}_C^0$ and that the convergence (12) with $\bar{\mathcal{P}}_0^{(\delta)}$ replacing $\mathcal{P}_0^{(\delta)}$ takes place in $(\mathcal{H}_C^0, d_{\mathcal{H}_C})$ as well—to verify the latter claim it is enough to check that $d_{\mathcal{H}_\mathcal{C}}(\bar{\mathcal{P}}_0^{(\delta)}, \mathcal{P}_0^{(\delta)}) \to 0$ as $\delta \to 0$, which is a straightforward matter.

Let us now consider the following subset of $\mathcal{H}_\mathcal{C}^0$.

$$\hat{\mathcal{H}}_\mathcal{C}^0 = \big\{K \in \mathcal{H}_\mathcal{C}^0 : \text{ all pairs of paths of } K \text{ coalesce in finite time}\big\} \tag{15}$$

$\hat{\mathcal{H}}_\mathcal{C}^0$ is open in $(\mathcal{H}_\mathcal{C}^0, d_{\mathcal{H}_\mathcal{C}})$.

Let now $\mathcal{M} : \hat{\mathcal{H}}_\mathcal{C}^0 \to [0,\infty)$ be such that $\mathcal{M}(K)$ is the Radon measure on $[-1,1]$ such that given $a, b \in [-1,1]$, $a \le b$, $\mathcal{M}(K)([a,b])$ is the area between the leftmost and rightmost paths of K[8] issuing from $(a,0)$ and $(b,0)$, respectively. The existence of the leftmost and rightmost paths of K is ensured by the compactness of K and the non crossing property of its paths.

Let us argue that if K, $K_n, n \ge 1$, are in $\hat{\mathcal{H}}_\mathcal{C}^0$ and

$$d_{\mathcal{H}_\mathcal{C}}(K_n, K) \to 0 \text{ as } n \to \infty, \tag{16}$$

then

$$\mathcal{M}(K_n) \to \mathcal{M}(K) \text{ as } n \to \infty \tag{17}$$

in the vague topology, as mentioned above (see paragraph of (14)).

Continuity of $\mathcal{M}$. Let us first argue that

$$\mathcal{M}(K_n)([-1,1]) \to \mathcal{M}(K)([-1,1]). \tag{18}$$

Let $[(f_1^{(n)},0),(f_2^{(n)},0)]$ and $[(f_1,0),(f_2,0)]$ denote the pairs of leftmost and rightmost paths in K_n and K, respectively. (16) implies the existence of pairs of paths $[(\tilde{f}_1^{(n)},0),(\tilde{f}_2^{(n)},0)]$ in K such that

$$\tilde{d}([(f_1^{(n)},0),(f_2^{(n)},0)],[(\tilde{f}_1^{(n)},0),(\tilde{f}_2^{(n)},0)]) \to 0.$$

Since K is compact, there exists $[(\tilde{f}_1,0),(\tilde{f}_2,0)]$ in K and a subsequence (n') such that

$$\tilde{d}([(\tilde{f}_1^{(n')},0),(\tilde{f}_2^{(n')},0)],[(\tilde{f}_1,0),(\tilde{f}_2,0)]) \to 0.$$

It is clear that $(\tilde{f}_1,0),(\tilde{f}_2,0)$ start respectively from $(-1,0),(1,0)$. If we have that either $(\tilde{f}_1,0) \ne (f_1,0)$ or $(\tilde{f}_2,0) \ne (f_2,0)$, we get a contradiction with either the extremity of $(\tilde{f}_1^{(n')},0)$ or $(\tilde{f}_2^{(n')},0)$, respectively, for n' large enough. If follows that $(\tilde{f}_1,0) = (f_1,0)$ and $(\tilde{f}_2,0) = (f_2,0)$, and thus

$$\tilde{d}([(f_1^{(n)},0),(f_2^{(n)},0)],[(f_1,0),(f_2,0)]) \to 0,$$

and (18) follows.

To conclude, we must argue that

$$\mathcal{M}(K_n)([x,y]) \to \mathcal{M}(K)([x,y]), \tag{19}$$

[8] Out of all the pairs of paths in K.

where $x, y \in [-1,1]$ are continuity points of $\mathcal{M}(K)$ such that $x < y$. This means that the paths from K out of $(x,0)$ and $(y,0)$ are unique. Now an argument similar to the one to establish (18) in the previous paragraph (however dispensing with the contradiction part, since we have the uniqueness just alluded to), we have that

$$\tilde{d}([(g_1^{(n)},0),(g_2^{(n)},0)],[(g_1,0),(g_2,0)]) \to 0,$$

where $(g_1^{(n)},0)$ and $(g_2^{(n)},0)$ are the leftmost and rightmost paths from K_n issuing from $(x,0)$ and $(y,0)$, respectively; $(g_1,0)$ and $(g_2,0)$ are the unique paths from K issuing from $(x,0)$ and $(y,0)$, respectively. (19) follows.

Conclusion. The convergence (12) with $\bar{\mathcal{P}}_0^{(\delta)}$ replacing $\mathcal{P}_0^{(\delta)}$ takes place in $(\mathcal{H}_{\mathcal{C}}^0, d_{\mathcal{H}_{\mathcal{C}}})$, as argued at the beginning of the previous subsubsection—see paragraph right above (15)—and (17) then it immediately follows that

$$\bar{\mu}^{(\delta)} := \mathcal{M}(\bar{\mathcal{P}}_0^{(\delta)}) \to \mathcal{M}(\mathcal{P}_0) = \mu$$

as $\delta \to 0$. To get (14) it is then enough to check that for any continuous function $h : [-1,1] \to \mathbb{R}$, we have that

$$\left| \int_{-1}^{1} h\, d\bar{\mu}^{(\delta)} - \int_{-1}^{1} h\, d\mu^{(\delta)} \right| \to 0$$

as $\delta \to 0$, which is a straightforward matter.

5.2 Persistence in the Voter Model

As discussed at the introduction, the persistence probability may be expressed as follows. For $\alpha \in (0,1)$, let $L = \{0\} \times [0,\alpha]$[9], and

$$\mathcal{H}_{\mathcal{C}}^{L} = \{K \in \mathcal{H}_{\mathcal{C}} : \text{ all paths from } K \text{ start from } L\}.$$

$\mathcal{H}_{\mathcal{C}}^{L}$ is a closed subset of $(\mathcal{H}_{\mathcal{C}}, d_{\mathcal{H}_{\mathcal{C}}})$, and so $(\mathcal{H}_{\mathcal{C}}^{L}, d_{\mathcal{H}_{\mathcal{C}}})$ is complete and separable.

Let now $\mathcal{S} : \mathcal{H}_{\mathcal{C}}^{L} \to [0,\infty)$ be such that for $K \in \mathcal{H}_{\mathcal{C}}^{L}$

$$\mathcal{S}(K) = \sup\{t^c([(f_1,t_1),(f_2,t_2)]) : [(f_1,t_1),(f_2,t_2)] \in K\},$$

that is, $\mathcal{S}(K)$ is the sup of the coalescence times of pairs of paths from K.

$\mathcal{S}$ is continuous in $(\mathcal{H}_{\mathcal{C}}^{L}, d_{\mathcal{H}_{\mathcal{C}}})$.

Let us show next that

$$\mathcal{S}(\mathcal{P}_L^{(\delta)}) \to \mathcal{S}(\mathcal{P}_L) \tag{20}$$

in distribution as $\delta \to 0$, where $\mathcal{P}_L^{(\delta)}$ and $\mathcal{P}_L$ are the sets of pairs of paths from $\mathcal{P}^{(\delta)}$ and $\mathcal{P}$, respectively, that start both from L.

[9] For convenience, we assume α to be a multiple of δ^2.

Argument for (20). We again assume (10), from which it follows that

$$\bar{d}_{\mathcal{H}}(\mathcal{P}_L^{(\delta)}, \mathcal{P}_L) \to 0 \tag{21}$$

almost surely as $\delta \to 0$.

From (12), we may assume that

$$d_{\mathcal{H}_C}(\mathcal{P}_\alpha^{(\delta)}, \mathcal{P}_\alpha) \to 0 \tag{22}$$

almost surely as $\delta \to 0$.

Let $x_\ell^{(\delta)}$ and $x_r^{(\delta)}$, resp. x_ℓ and x_r, be the leftmost and rightmost points of $\mathbb{R} \times \{\alpha\}$ touched by paths from $\mathcal{P}_L^{(\delta)}$, resp. $\mathcal{P}_L$. Theorem 3.14 from [4] ensures that (x_ℓ, α) and (x_r, α) have almost surely both exactly one path from $\bar{\mathcal{W}}$ issuing from each of them. Let $(f_\ell^{(\delta)}, \alpha), (f_r^{(\delta)}, \alpha)$ be the paths from $\bar{Y}^{(\delta)}$ issuing from $(x_\ell^{(\delta)}, \alpha), (x_r^{(\delta)}, \alpha)$, respectively; and let $(f_\ell, \alpha), (f_r, \alpha)$ be the paths from $\bar{\mathcal{W}}$ issuing from $(x_\ell, \alpha), (x_r, \alpha)$, respectively

We note that

$$\mathcal{S}(\mathcal{P}_L^{(\delta)}) = t^c([(f_\ell^{(\delta)}, \alpha), (f_r^{(\delta)}, \alpha)]), \;\; \mathcal{S}(\mathcal{P}_L) = t^c([(f_\ell, \alpha), (f_r, \alpha)]).$$

(21) now implies that

$$x_\ell^{(\delta)} \to x_\ell, \quad x_r^{(\delta)} \to x_r, \tag{23}$$

and (20) follows readily from (22)[10] and (23) via an argument similar to one used for the continuity of $\mathcal{M}$ in the previous subsection (see paragraph of (19) above).

6 Final Comments

A glaring issue not addressed in Sect. 4 is the weak convergence of the full family of random walk paths $\mathcal{P}^{(\delta)}$ to $\mathcal{P}$. The difficulty here comes in trying to obtain a uniform bound for $\Delta(p, q)$ over pairs of rescaled random walk paths with different pod lengths. This difficulty does not occur for pairs of paths which start at the same time and coalesce at the same time (and thus have the same pod lengths), and this facilitated the argument for (12) above. The point is that in the former case we do not have the validity of something like the last point of the third item in Remark 6; this led in one of our attempts to the need of controlling the modulus of continuity of the *inverse* of a random walk path (well) after time 0 and before hitting the origin, which we did not find a way of accomplishing.

Even if we were able to apply (12) to the motivating examples, there are other interesting examples which conceivably do not follow readily from that result, and would require the full (or a fuller) convergence result, possibly under another,

[10] We should in this application have (22) for paths starting from $[-r, r] \times \{0\}$ for arbitrary fixed r, but this of course follows as in the special case $r = 1$.

more flexible metric, in a suitable space[11]. One such example is the curve forming the boundary between the downward and upward families of coalescing random walk paths (those families were described in the discussion of the silo model). The convergence to the appropriately defined curve defined for the Brownian web (and its dual web) was undertaken in [6] by a direct approach.

Another example is the full weight distribution in the silo (the collection of weights supported by all beads in the silo, not just the ones at the bottom). An intriguing point in this respect would be how to best describe this distribution (it is *not* a measure in the space-time plane), and in which metric space.

Acknowledgements. At various times in the last years, besides Chuck Newman, I discussed the issues of this article with a number of colleagues and collaborators, among whom I would like to thankfully mention Rafael Grisi, who participated in early stages of this project, Pablo Ferrari, Glauco Valle, and Antônio Luiz Pereira.

References

1. Coppersmith, S.N., Liu, C.-h., Majumdar, S., Narayan, O., Witten, T.A.: Model for force fluctuations in bead packs. Phys. Rev. E **53**(5), 4673–4685 (1996)
2. Fontes, L.R.G., Isopi, M., Newman, C.M., Ravishankar, K.: The Brownian web: characterization and convergence. Ann. Probab. **32**, 2857–2883 (2004)
3. Fontes, L.R.G., Isopi, M., Newman, C.M., Ravishankar, K.: The Brownian web: characterization and convergence (long version). arXiv:math/0304119 (2003)
4. Fontes, L.R.G., Isopi, M., Newman, C.M., Ravishankar, K.: Coarsening, nucleation, and the marked Brownian web. Annales de l'Institut Henri Poincaré, Probabilités et Statistiques **42**, 37–60 (2006)
5. Fontes, L.R.G., Isopi, M., Newman, C.M., Stein, D.L.: Aging in $1D$ discrete spin models and equivalent systems. Phys. Rev. Lett. **87**, 110201-1–110201-4 (2001)
6. Newman, C.M., Ravishankar, K.: Convergence of the Tóth lattice filling curve to the Tóth–Werner plane filling curve. ALEA Lat. Am. J. Probab. Math. Stat. **1**, 333–345 (2006)
7. Rodriguez-Iturbe, I., Rinaldo, A.: Fractal River Basins: Chance and Self-Organization. Cambridge Univ. Press, New York (1997)
8. Scheidegger, A.E.: A stochastic model for drainage patterns into an intramontane trench. Bull. Ass. Sci. Hydrol. **12**, 15–20 (1967)

[11] In a fuller, more flexible setting, one might conceivably also be able to treat the motivating examples more directly.

FKG (and Other Inequalities) from (Generalized and Approximate) FK Random Cluster Representation (and Iterated Folding)

Alberto Gandolfi(✉)

New York University Abu Dhabi, Abu Dhabi, United Arab Emirates
albertogandolfi@nyu.edu

Abstract. In this paper we prove several inequalities by means of diagrammatic expansions, a technique already used in [1]. This time we show that iterations of the folding of a probability leads to the proof of some inequalities by means of a generalized and approximate random cluster representation of the iterated foldings. One of the inequalities is the well known FKG inequality, which ends up being proven, quite unexpectedly, by means of the (generalized) FK representation.

Although most of the results are not new, we hope that the techniques will find applications in other contexts.

Keywords: FK · Random cluster representation · FKG · Positive association · Negative association · Approximate random cluster representation · Folding · Tree

1 Introduction

[1] introduces a technique which allows to prove some inequalities of BK type concerning disjoint occurrences of events by means of diagrammatic expansions. The main tool there is to consider foldings of a probability and then a generalized FK random cluster representation of the folded measures. Depending on the features of the random cluster representation, one could prove several different inequalities.

We consider here a further step, in which the folding operation is iterated and a tree of possible foldings is generated. We develop the terminology and study the main properties.

Next, we show that it is possible to consider the random cluster representation of the distribution obtained after infinitely many foldings, and that specific representations can be obtained depending on the properties of the starting probability. Such specific representations can then be used to prove inequalities. The first examples involve the FKG inequality.

The FKG inequality was originally proven as a tool to study models in rigorous statistical mechanics, such as the Ising model [12]. Almost the same group

V. Sidoravicius (Ed.): *Sojourns in Probability Theory and Statistical Physics - II*, PROMS 299, pp. 186–207, 2019.
https://doi.org/10.1007/978-981-15-0298-9_8

of authors, and at about the same time, developed, in an independent work, the FK random cluster representation, as an alternative parametrization of the Ising model [11]. Both the FKG inequality and the FK representations have been generalized, and then used in a great number of situations (see [13], for example). Note, in particular, that several FK random cluster models, the representation of the ferromagnetic Ising model for instance, satisfy the FKG inequality. In spite of these superficial connections, no close relation has ever been found between the FK representation and the FKG inequality. Using the methods introduced in this paper, however, we give here an alternative proof of the FKG inequality based on (a generalized version of) the FK random cluster representation, a quite unexpected connection between the two concepts. Other inequalities are also derived.

The strategy of the paper is as follows.

First, we recap from [1] how folding and generalized random cluster representation of the folded probabilities allows to prove BK type inequalities. We then develop an approximate version of the main results of [1].

Next, we consider iteration of the folding operation giving rise to a tree of foldings, and note that one can take the limit of infinitely many foldings. In addition, we observe that one can restrict to the case in which essential foldings, as defined below, are all done at the beginning, thus effectively reducing the study of the limit distribution to a finite number of possible limits.

We then observe that the FKG theory, which states that the FKG condition implies positive association, can be translated into the foldings where it takes an even more natural form: a probability satisfies the FKG condition if and only if it does the same in each folded version, and hence in the limit of infinitely many foldings. In the opposite direction, if the folded versions satisfy a certain inequality labelled (1) below, then the initial probability has positive association. The above mentioned inequality (1) does indeed climb back the tree of foldings (see Lemma 2); the price to pay in this is an exponential error factor, which is however controlled by a super exponential convergence of the distributions to their limits down the tree of foldings.

Starting from a probability P which satisfies the FKG condition, the FKG condition descends thus down through the tree of foldings; it is there, in the infinite limit on the tree of foldings, that the FKG condition meets a simple FK random cluster representation able to easily justify inequality (1) which, in turn, climbs back to prove positive association of P (see Theorem 4).

When all of this is applied to negative association it gives some sufficient conditions for negative association, including Pemantle's result [20], see Section 6.

We hope that all this machinery, which is used here to give an alternative proof to largely known results, may turn useful for novel results as well. For simplicity, we consider only finite sets here, leaving questions about infinite volume limits to further researches.

2 Preliminaries

2.1 Definitions

Let Λ be a finite set, and $\mathcal{F} = \{F_i\}_{i\in\Lambda}$ be a collection of finite sets F_i. For each subset $M \subseteq \Lambda$ we denote $\Omega_M = \prod_{i\in M} F_i$. We define the usual concatenation of configurations: if $\omega(1)$ is defined on $B(1)$ and $\omega(2)$ in $B(2)$, with $B(1)\cap B(2) = \emptyset$, then $\omega(1)\omega(2)$ is defined on $B(1)\cup B(2)$ such that $(\omega(1)\omega(2))_i = \omega(1)_i$ if $i \in B(1)$ and $(\omega(1)\omega(2))_i = \omega(2)_i$ if $i \in B(2)$.

For each set S, let $\mathcal{P}(S)$ be the family of all subsets of S; we are interested in studying probabilities P on $\mathcal{P}(\Omega_\Lambda)$.

Throughout the paper Z_i are normalizing constants implicitly defined by the context. When J, K etc. represent entities like indices or sets, boldfaced characters $\mathbf{J}, \mathbf{K}$ etc. represent finite ordered collections of such entities indexed in some form, i.e. vectors. In the sequel $\mathbb{I}_A$ represents the indicator function of A.

For $K \subseteq \Lambda$ and $\omega \in \Omega_\Lambda$, $[\omega]_K$ is the cylinder $\{\omega' : \omega'_i = \omega_i \text{ for all } i \in K\}$.

2.2 Generalized FK-RCR

A generalized FK random cluster representation, hereafter called RCR, has been developed in [1], Sect. 2.1, as follows.

Definition 1. *Given finite sets M and $F_i, i \in M, \Omega_M = \prod_{i\in M} F_i$, a family $\mathcal{B} \subseteq \mathcal{P}(M)$; $H = \prod_{b\in\mathcal{B}} \mathcal{P}(\Omega_b)$; and a probability P on $\mathcal{P}(\Omega_M)$, a $\mathcal{B}$-RCR of P is a probability ν on $\mathcal{P}(H)$ such that for every $\omega \in \Omega_M$*

$$P(\omega) = \frac{1}{Z_1} \sum_{\eta\in H} \nu(\eta) \prod_{b\in\mathcal{B}} \mathbb{I}_{\omega_b\in\eta_b} = \frac{1}{Z_1} \sum_{\eta\in H, \eta\sim\omega} \nu(\eta),$$

where Z_1 is a normalizing factor, and $\eta \sim \omega$ means that $\omega_b \in \eta_b$ for all $b \in \mathcal{B}$.

We call $\mathcal{B}$ the collection of hyperbonds of the RCR, and ν its base probability or simply base.

As discussed in [1], our terminology is slightly different from that used in [11] and subsequent literature [13]. In the first place, we consider set variables η instead of the usual real valued ones. Then we focus on the base probability ν of the RCR; the works preceding [1], instead, were focusing on the joint distribution $Q(\omega,\eta) = \frac{1}{Z_2}\nu(\eta)\prod_{b\in\mathcal{B}} \mathbb{I}_{\omega\sim\eta}$, and, in particular, on the projection on the η variables $\phi(\eta) = \sum_\omega Q(\omega,\eta)$, in which the well-known expression $2^{Cl(\eta)}$ appears. The use of set variables and the focus on the base probability ν help streamlining the theory.

To illustrate the connection between the two notations, we now review the FK representation for the Ising model without magnetic field. The collection of hyperbonds consists of the edges $b \in E$ of some graph, and hence $|b| = 2$. For each $b = \{i, j\}$, in the standard FK we have the real valued random variable $\mathbb{I}_{\omega_i=\omega_j}$, which corresponds to the set variable η_b taking values either $\{(1,1),(-1,-1)\} = \{\omega_{i,j} : \mathbb{I}_{\omega_i=\omega_j} \geq 1\}$, or $\Omega_b = \{\omega_{i,j} : \mathbb{I}_{\omega_i=\omega_j} \geq 0\}$; hence ν

is concentrated on $\prod_{b\in\mathcal{B}} H_b \subseteq H$ with $H_b = \{\{(1,1),(-1,-1)\},\Omega_b\}$. Our ν is Bernoulli: $\nu = \prod_{b\in\mathcal{B}} \nu_b$, where ν_b is defined on H_b by $\nu_b(\{(1,1),(-1,-1)\}) = p = 1-\nu_b(\Omega_b)$. We then get

$$\nu(\eta) = \frac{1}{Z_1} p^{|\{b:\eta_b=\{(1,1),(-1,-1)\}\}|}(1-p)^{|\{b:\eta_b=\Omega\}|} = \frac{1}{Z_1} p^{n_1(\eta)}(1-p)^{n_0(\eta)}$$

where n_i indicates the number of bonds b with $|\eta_b| = 4-2i$. The joint representation is then $Q(\omega,\eta) = \frac{1}{Z_2} p^{n_1(\eta)}(1-p)^{n_0(\eta)}\mathbb{I}_{\omega\sim\eta}$; the marginal on ω is the Ising model with interaction J such that $p = 1-e^{-2J}$, and the marginal on η is the original FK model

$$\phi(\eta) = \frac{1}{Z_2} p^{n_1(\eta)}(1-p)^{n_0(\eta)} 2^{Cl(\eta)}$$

where $Cl(\eta)$ is the number of vertex clusters determined by the configuration η by stating that two vertices i,j are connected if the $\mathbb{I}_{\omega_i=\omega_j} = 1$, i.e. $\eta_{i,j} \neq \Omega_b$ One pleasant feature of the general RCR is that a probability P is Gibbs iff it has a Bernoulli RCR (see [1]).

All other extensions of the FK representation, such as the one in [8], can be easily translated in our framework.

A notion of connectedness, similar to the one in the original work on random cluster representations, is developed in [1] for a general RCR: a hyperbond b is called active in η if $\eta_b \neq \Omega_b$; and two sets of vertices in Λ are connected if there is a sequence of active hyperbonds connecting them. This notion is central in proving the main inequalities in [1].

2.3 Foldings of a Probability

In [1,7,21] the operation of folding of a probability measure P is introduced; it amounts to the following: fix one region K, one configuration α on K, two pointwise different configurations $\boldsymbol{\beta} = (\beta(1),\beta(2))$ of K^c, and consider a random configuration of K^c which pointwise takes the value of either $\beta(1)$ or $\beta(2)$, with the probability induced by $P\times P$ given that both configurations coincide with α on K. This determines a collection of probabilities as $K,\alpha,\boldsymbol{\beta}$ vary. As each of the resulting probabilities is symmetric (see below), the entire process can be seen as a dependent coupling of P with itself, which separates an asymmetric nonrandom "drift" and a symmetric randomness.

Definition 2. *Given a probability P on $\mathcal{P}(\Omega_\Lambda)$, a subset $K\subseteq\Lambda$, a configuration $\alpha\in\Omega_K$, and two configurations $\boldsymbol{\beta} = (\beta(1),\beta(2)) \in \Omega_{K^c}$, with $\beta_i(1)\neq\beta_i(2)$ for all $i\in K^c$, we define the $(K,\alpha,\boldsymbol{\beta})$-folded version of P as the probability $P^{K,\alpha,\boldsymbol{\beta}}$ on Ω_{K^c} given by*

$$P^{K,\alpha,\boldsymbol{\beta}}(\omega_{K^c}) = \frac{1}{Z_{K,\alpha,\boldsymbol{\beta}}} P(\alpha\omega_{K^c})P(\alpha\overline{\omega}^{\boldsymbol{\beta}}_{K^c}),$$

if there is at least one ω_{K^c} for which the r.h.s. is not zero; in the definition we used concatenation of configurations, and the reverse operation $\overline{\omega}^{\boldsymbol{\beta}}$ which *consists of exchanging ω_i from $\beta_i(1)$ to $\beta_i(2)$ or viceversa, if $\omega_i \in \{\beta_i(1), \beta_i(2)\}$, and inserting a symbol with zero P probability otherwise. Clearly, $P^{K,\alpha,\beta}(\omega_{K^c})$ is concentrated on the $\omega_{K^c} \in \prod_{i\in K^c}\{\beta_i(1), \beta_i(2)\}$.*

An equivalent definition is obtained by considering

$$W_{K,\alpha,\beta} = \{(\omega, \omega') \in \Omega \times \Omega : \omega_K = \omega'_K = \alpha;\\ \text{and for all } i \in K^c, \omega_i, \omega'_i \in \{\beta_i(1), \beta_i(2)\} \text{ and } \omega_i \neq \omega'_i\},$$

then taking $P^{K,\alpha,\beta}(\omega_{K^c}) = (P \times P)(\cdot|W_{K,\alpha,\beta})$, and finally projecting on the first configuration. This entails that $Z_{K,\alpha,\beta} = (P \times P)(W_{K,\alpha,\beta})$.

When there are only two symbols, i.e. $|F_i| = 2$, then $\boldsymbol{\beta}$ are irrelevant and can be omitted from the notation.

Example 1. The folding of an Ising model $\mu_{\mathbf{J},\mathbf{h}}(\omega) = \frac{1}{Z}\exp(\sum_{i,j} J_{i,j}\omega_i\omega_j + \sum_i h_i\omega_i)$, $\omega \in \{-1,1\}^\Lambda$, is an Ising model $P^{K,\alpha,\beta} = P^{K,\alpha} = \mu_{2\mathbf{J},\mathbf{0}}$ on Ω_{K^c}, with twice the interaction and zero external field.

2.4 Disjoint Occurrences of Events

[2] introduces the concept of disjoint occurrence of events, proving the first version of the BK inequality, shown then in full generality in [21]. [1] introduces several different versions of disjoint occurrence, using the RCR of the folding of a probability P to show that such inequalities hold for certain P's.

In all these works, an important role is played by a function which indicates which pairs of sets of indices can be used to identify two events in a given configuration. More precisely, for $A, B \subseteq \Omega_\Lambda$, the set of disjoint occurrence pairs is defined by

$$\mathcal{D}(A, B, \omega) = \{(K, L) : K, L \subseteq \Lambda, K \cap L = \emptyset, [\omega]_K \subseteq A, [\omega]_L \subseteq B\},$$

where we recall that $[\omega]_K$ indicates the cylinder of ω in K. A selection rule is a function Ψ which assigns to each (A, B, ω) a (possibly empty) subset $\Psi(A, B, \omega)$ of $\mathcal{D}(A, B, \omega)$, i.e. a set of disjoint pairs of subsets of Λ from which A and B can be disjointly recognized. By means of a selection rule we can define the generalized box operation [1]

$$A \boxed{\Psi} B = \{\omega \in \Omega_\Lambda : \Psi(A, B, \omega) \neq \emptyset\}.$$

If $\Psi(A, B, \omega) = \mathcal{D}(A, B, \omega)$ we get the usual box operation as $A\boxed{\mathcal{D}}B = A\Box B$.

Example 2. If $F_i = \{0, 1\}$ for all $i \in \Lambda$ and A and B are increasing, then we can require that $(K, L) \in \Psi(A, B, \omega)$ implies $K, L \subseteq \omega^{-1}(1)$ to disallow the possibility of recognizing events in sets which uselessly contain vertices i in which $\omega_i = 0$.

Example 3. Similarly, if $F_i = \{0, 1\}$ for all $i \in \Lambda$, A is increasing and B is decreasing, then we can take $K \subseteq \omega^{-1}(1)$ and $L \subseteq \omega^{-1}(0)$.

In this paper we will eventually be dealing only with these last two cases, but we state the results for a general Ψ as it does not require much more efforts.

2.5 Inequalities from Approximate RCR and Folding

We start by extending the main result of [1] to the case in which the RCR of the folding is only approximate.

Definition 3. *Given finite sets M and $\Lambda, M \subseteq \Lambda, F_i, i \in M, \Omega_M = \prod_{i\in M} F_i$, a family $\mathcal{B} \subseteq \mathcal{P}(\Lambda)$, constraints $H \subseteq \prod_{b\in\mathcal{B}'} \mathcal{P}(\Omega_b)$ on hyperbonds configurations, with $\mathcal{B}' = \{b' = b \cap M, b \in \mathcal{B}\}$, a probability P on $\mathcal{P}(\Omega_M)$, and $\epsilon > 0$, a ϵ-$\mathcal{B}$-**RCR** of P is a probability ν on $\mathcal{P}(H)$ such that for every $\omega \in \Omega_M$*

$$P(\omega) = \frac{1}{Z_1} \sum_{\eta \in H, \eta \sim \omega} \nu(\eta) \pm \epsilon,$$

where $x = y \pm \epsilon$ stands for $|x - y| \leq \epsilon$.

Combining connectedness in a RCR, as defined at the end of Sect. 2.2, with a selection rule, we say that given two events A and B, a RCR ν of a probability P, and a selection rule Ψ, Ψ uses disjoint clusters of ν for A, B if for all ω such that $\Psi(A, B, \omega) \neq \emptyset$ there is a pair $(M, N) \in \Psi(A, B, \omega)$ such that

$$\nu(\eta : \eta \sim \omega \text{ and } M \text{ is connected to } N \text{ using active hyperedges of } \eta) = 0.$$

Lemma 1. *Let Λ be a finite set, $\Omega_\Lambda = \prod_{i\in\Lambda} F_i$, and suppose $|F_i| = 2$ for all $i \in \Lambda$; let P be a probability on Ω_Λ, $A, B \subseteq \Omega_\Lambda$, Ψ a selection rule. Suppose that, for $\epsilon > 0$, P has a symmetric $\frac{\epsilon}{2^{|\Lambda|+1}}$-$\mathcal{B}$-RCR ν such that Ψ uses disjoint random clusters of ν for A, B, then*

$$P(A \boxed{\Psi} B) \leq P(A \cap \overline{B}) + \epsilon, \tag{1}$$

where we recall that $\overline{B}$ is the event obtained from B by changing all the ω_i's into $\overline{\omega}_i = 1 - \omega_i$. Here symmetric means that $\nu(\prod_{b\in\mathcal{B}} \eta_b) = \nu(\prod_{b\in\mathcal{B}} \overline{\eta}_b)$.

Proof. The proof of a similar statement is given in [1], so we omit some tedious details here. Given a configuration η, the set of its active hyperbonds b falls apart into $t(\eta)$ connected components $C_1(\eta), \ldots, C_{t(j)}(\eta)$ called clusters. Fix now $\ell \in \{1, \ldots, t(\eta)\}$ and a configuration ω such that $\omega_b \in \eta_b$ for each $b \subseteq \cup_{I=1}^{t(\eta)} C_\ell(\eta)$. By the symmetry of ν and the definition of active hyperbonds, with ν-probability one the configuration $\overline{\omega}^{C_\ell(\eta)}$, which is the reversed $1 - \omega_i$ for all $i \in C_\ell(\eta)$ and coincides with ω_i for all other i's, also satisfies $\overline{\omega}_b^{C_\ell(\eta)} \in \eta_b$ for all b's. Let then $\Omega_{\eta,\omega}$ be the set of configurations $\xi \in \Omega_\Lambda$ which in $C_\ell(\eta)$ coincide with either ω or $\overline{\omega}^{C_\ell(\eta)}$ for each $\ell = 1, \ldots, t(\eta)$; we have $|\Omega_{\eta,\omega}| = 2^{t(\eta)}$. Consider the joint probability distribution $Q(\rho, \xi) = \frac{1}{Z_2}\nu(\rho) \prod_{b\in\mathcal{B}} \mathbb{I}_{\xi\sim\rho}$; then for the given η and ω, $\mu_{\eta,\omega}(\xi) = Q((\rho, \xi)|\rho = \eta, \xi \in \Omega_{\eta,\omega}) = \frac{\mathbb{I}_{\xi\in\Omega_{\eta,\omega}}}{2^{t(\eta)}}$ is concentrated on $\Omega_{\eta,\omega}$. This distribution can then be encoded into i.i.d. two valued symmetric Bernoulli variables by a map $T_\eta : \Omega_{\eta,\omega} \to \{0, 1\}^{t(\eta)}$ with $(T_\eta(\xi))_\ell = \mathbb{I}_{\xi|_{C_\ell(\eta)} = \omega|_{C_\ell(\eta)}}$. Reimer's [21] results apply to $T_\eta(\mu_\eta)$ showing that $T_\eta(\mu_\eta)(A' \Box B') \leq T_\eta(\mu_\eta)(A' \cap \overline{B}')$ for all events $A', B' \subseteq \{0, 1\}^{t(\eta)}$.

In addition, one can verify, as it is explicitly proven in [1] – see the claims on Page 171 – that $T_\eta(A\boxed{\Psi}B) \subseteq T_\eta(A)\Box T_\eta(B)$, and $T_\eta(A\cap B) = T_\eta(A)\cap T_\eta(B)$. The relation $R(\xi,\omega)$ iff $\xi \in \Omega_{\eta,\omega}$ partitions $\{\omega \in \Omega_\Lambda : \omega \sim \eta\}$ into equivalence classes: two η compatible configurations are in the same class if they coincide on each cluster of η except possibly for cluster dependent spin flips. Indicating $\Omega(\eta) = \{\omega \in \Omega_\Lambda : \omega \sim \eta\}/R$, we have

$$\begin{aligned}
P(A\boxed{\Psi}B) &= \sum_{\omega\in A\boxed{\Psi}B} P(\omega) \\
&= \sum_{\omega\in A\boxed{\Psi}B} \frac{1}{Z_1}\sum_{\eta\in H}\nu(\eta)\mathbb{I}_{\eta\sim\omega} \pm 2^{|\Lambda|}\frac{\epsilon}{2^{|\Lambda|+1}} \\
&= \sum_{\eta\in H} Q(\eta, A\boxed{\Psi}B) \pm \frac{\epsilon}{2} \\
&= \sum_{\eta\in H}\sum_{\omega\in\Omega(\eta)} \mu_{\eta,\omega}(A\boxed{\Psi}B)Q(\Omega_{\eta,\omega}) \pm \frac{\epsilon}{2} \\
&\leq \sum_{\eta\in H}\sum_{\omega\in\Omega(\eta)} T_\eta(\mu_{\eta,\omega})(T_\eta(A)\Box T_\eta(B))Q(\Omega_{\eta,\omega}) + \epsilon \\
&\leq \sum_{\eta\in H}\sum_{\omega\in\Omega(\eta)} T_\eta(\mu_{\eta,\omega})(T_\eta(A)\cap \overline{T_\eta(B)})Q(\Omega_{\eta,\omega}) + \epsilon \\
&= \sum_{\eta\in H}\sum_{\omega\in\Omega(\eta)} T_\eta(\mu_{\eta,\omega})(T_\eta(A\cap \overline{B}))Q(\Omega_{\eta,\omega}) + \epsilon \\
&= \sum_{\eta\in H}\sum_{\omega\in\Omega(\eta)} \mu_{\eta,\omega}(A\cap \overline{B})Q(\Omega_{\eta,\omega}) + \epsilon \\
&= P(A\cap\overline{B}) + \epsilon
\end{aligned}$$

□

Let $R_{K,\alpha,\beta}(\omega)$ be either the restriction ω_{K^c} of ω to K^c if the restriction ω_K equals α and $\omega_i \in \{\beta(1)_i, \beta(2)_i\}$ for all $i \in K^c$, or else the empty set; in other words, let

$$R_{K,\alpha,\beta}(\omega) = \{\omega_{K^c} : \alpha\omega_{K^c} = \omega \text{ and } (\omega_{K^c})_i \in \{\beta_i(1), \beta_i(2)\} \text{ for all } i \in K^c\}.$$

If Ψ is a selection rule on Λ, $K \subseteq \Lambda$ and $\alpha \in \Omega_K$, we let $\Psi_{K,\alpha}$ be defined by $\Psi_{K,\alpha}(A', B', \omega_{K^c}) = \Psi(\alpha A', \alpha B', \alpha\omega_{K^c}) \cap K^c$ for all $A', B' \subseteq \Omega_{K^c}$, where $\alpha A'$ is the concatenation of the configurations of A' with α, and where for a collection of pairs of sets (M, N)'s the intersection with K^c is simply the collection of intersections $(M \cap K^c, N \cap K^c)$. $\Psi_{K,\alpha}$ is a selection rule in $\mathcal{P}(\Omega_{K^c}) \times \mathcal{P}(\Omega_{K^c}) \times \Omega_{K^c}$, and one can see that

$$R_{K,\alpha,\beta}(A\boxed{\Psi}B) \subseteq R_{K,\alpha,\beta}(A)\boxed{\Psi_{K,\alpha}}R_{K,\alpha,\beta}(B).$$

Lemma 2. *For $\Lambda, F_i, \Omega_\Lambda, P, \Psi$ $A, B \in \Omega_\Lambda$, as in Lemma 1 before, we have that if for each $K, \alpha, \boldsymbol{\beta}$*

$$P^{K,\alpha,\boldsymbol{\beta}}(R_{K,\alpha,\boldsymbol{\beta}}(A)\boxed{\Psi_{K,\alpha}}R_{K,\alpha,\boldsymbol{\beta}}(B)) \leq P^{K,\alpha,\boldsymbol{\beta}}(R_{K,\alpha,\boldsymbol{\beta}}(A) \cap \overline{R_{K,\alpha,\boldsymbol{\beta}}(B)}) + \epsilon$$

then

$$P(A\boxed{\Psi}B) \leq P(A)P(B) + \epsilon.$$

Proof. Recall that the $W_{K,\alpha,\boldsymbol{\beta}}$'s, with varying $K \subseteq \Lambda, \alpha \in \Omega_K, \boldsymbol{\beta} \in \Omega_{K^c}$ form a partition of $\Omega_\Lambda \times \Omega_\Lambda$. Furthermore, $P^{K,\alpha,\boldsymbol{\beta}}(\omega_{K^c}) = (P \times P)(\alpha\omega_{K^c} \times \Omega_\Lambda | W_{K,\alpha,\boldsymbol{\beta}})$ so that for $A \subseteq \Omega_\Lambda$,

$$(P \times P)(A \times \Omega_\Lambda | W_{K,\alpha,\boldsymbol{\beta}}) = P^{K,\alpha,\boldsymbol{\beta}}(R_{K,\alpha,\boldsymbol{\beta}}(A)).$$

Thus

$$\begin{aligned} P(A\boxed{\Psi}B) &= (P \times P)(A\boxed{\Psi}B \times \Omega_\Lambda) \\ &= \sum_{K \subseteq \Lambda} \sum_{\alpha \in \Omega_K} \sum_{\boldsymbol{\beta}} (P \times P)(A\boxed{\Psi}B \times \Omega_\Lambda | W_{K,\alpha,\boldsymbol{\beta}}) \\ &\quad \times (P \times P)(W_{K,\alpha,\boldsymbol{\beta}}) \\ &\leq \sum_{K \subseteq \Lambda} \sum_{\alpha \in \Omega_K} \sum_{\boldsymbol{\beta}} ((P \times P)(A \times B | W_{K,\alpha,\boldsymbol{\beta}}) + \epsilon) \\ &\quad \times (P \times P)(W_{K,\alpha,\boldsymbol{\beta}}) \\ &= P(A)P(B) + \epsilon \end{aligned}$$

as for each $K \subseteq \Lambda, \alpha \in \Omega_K, \boldsymbol{\beta} \in \Omega_{K^c}$ we have, using Lemma 1 in the second inequality,

$$\begin{aligned} &(P \times P)(A\boxed{\Psi}B \times \Omega_\Lambda | W_{K,\alpha,\boldsymbol{\beta}}) \\ &= P^{K,\alpha,\boldsymbol{\beta}}(R_{K,\alpha,\boldsymbol{\beta}}(A\boxed{\Psi}B)) \\ &\leq P^{K,\alpha,\boldsymbol{\beta}}(R_{K,\alpha,\boldsymbol{\beta}}(A)\boxed{\Psi_{K,\alpha}}R_{K,\alpha,\boldsymbol{\beta}}(B)) \\ &\leq P^{K,\alpha,\boldsymbol{\beta}}(R_{K,\alpha,\boldsymbol{\beta}}(A) \cap \overline{R_{K,\alpha,\boldsymbol{\beta}}(B)}) + \epsilon \\ &= (P \times P)(R_{K,\alpha,\boldsymbol{\beta}}(A) \times R_{K,\alpha,\boldsymbol{\beta}}(B) | W_{K,\alpha,\boldsymbol{\beta}}) + \epsilon \\ &= (P \times P)(A \times B | W_{K,\alpha,\boldsymbol{\beta}}) + \epsilon \end{aligned}$$

since the last equality holds conditioned to $W_{K,\alpha,\boldsymbol{\beta}}$. □

Using the results of Lemmas 1 and 2 for $\epsilon = 0$, it is proven in [1] that if Ψ used disjoint clusters of each RCR of the foldings of a probability P, then $P(A\boxed{\Psi}B) \leq P(A)P(B)$.

3 Iterated Foldings

3.1 Iterated Foldings

We now introduce the notation for dealing with iterated foldings. After the first folding the distributions are binary and symmetric, so the first folding is often singled out in the notation, and results for subsequent foldings assume symmetry.

We use the usual symbols $\Lambda, \mathcal{B}(\Lambda), F_i, P$ on Ω_Λ.

Definition 4. *For* $i = 1, \ldots, n$, *let* $K_i \subseteq \left(\cup_{j=1}^{i-1} K_j\right)^c$, $K_j \subseteq \Lambda$, *be a sequence of disjoint sets, and indicate* $\mathbf{K}_i = (K_1, \ldots, K_i)$; *let* $\alpha_1 \in \Omega_{K_1}$, *and for* $i \geq 2$, $\alpha_i \in \Omega_{K_i}(\boldsymbol{\beta})$ *and* $\boldsymbol{\alpha}_i = (\alpha_1, \ldots, \alpha_i)$; *and, finally, let* $\boldsymbol{\beta} = (\beta(1), \beta(2))$ *with* $\beta(1), \beta(2) \in \Omega_{K_1^c}$.

Given $\mathcal{B}(\Lambda)$, *the* $(\mathbf{K}_n, \boldsymbol{\alpha}_n, \boldsymbol{\beta})$*-*$\mathcal{B}$*-folding* $P^{\mathbf{K}_n, \boldsymbol{\alpha}_n, \boldsymbol{\beta}}$ *of* P *is defined recursively on* $\Omega_{(\cup_{K \in \mathbf{K}_i} K)^c}$ *by* $P^{\mathbf{K}_i, \boldsymbol{\alpha}_i, \boldsymbol{\beta}} := \left(P^{\mathbf{K}_{i-1}, \boldsymbol{\alpha}_{i-1}, \boldsymbol{\beta}}\right)^{k_i, \alpha_i}$ *for* $i \geq 2$, *starting from* $P^{\mathbf{K}_1, \boldsymbol{\alpha}_1, \boldsymbol{\beta}} := P^{K_1, \alpha_1, \boldsymbol{\beta}}$. *It is natural to denote* $\mathbf{P}^{\mathbf{0}} = P$, *with* $(\mathbf{K}_0, \boldsymbol{\alpha}_0, \boldsymbol{\beta}) = \mathbf{0}$.

The set of iterated foldings has the structure of an infinite tree, indexed by $(\mathbf{K}_n, \boldsymbol{\alpha}_n, \boldsymbol{\beta})$, $n = 0, 1, \ldots$. The direct descendants of $(\mathbf{K}_i, \boldsymbol{\alpha}_i, \boldsymbol{\beta})$ are of the form $((\mathbf{K}_i, K_{i+1}), (\boldsymbol{\alpha}_i, \alpha_{i+1}), \boldsymbol{\beta})$ for some $K_{i+1} \subseteq (\cup_{j=1}^{i-1} K_j)^c, \alpha_{i+1} \in \Omega_{K_{i+1}}$; notice that $\boldsymbol{\beta}$ is not relevant after the first folding. Moreover, if $K_i = \emptyset$ then $\alpha_i \in \Omega_{K_i}$ is just a formal symbol (sometimes we use a generic symbol α for this).

An **infinite branch** in a tree of foldings is a sequence $(\mathbf{K}_i, \boldsymbol{\alpha}_i, \boldsymbol{\beta})$ of vertices of the tree. The sequence $(\cup_{K \in \mathbf{K}_i} K)^c$ is non increasing, and in an infinite branch of the tree of foldings it must be asymptotically constant. Thus, for any infinite branch there exists L such that $K_i = \emptyset$ for all $i \geq L$.

The **limiting distribution in an infinite branch** of the tree of foldings such that $K_i = \emptyset$ for all $i \geq L$ is the probability P^∞ defined on $\overline{\Omega}_\Lambda := \Omega_{(\cup_{K \in \mathbf{K}_L} K)^c}$ by $P^\infty(\omega) = \lim_{i \to \infty} P^{\mathbf{K}_i, \boldsymbol{\alpha}_i, \boldsymbol{\beta}}(\omega)$.

Lemma 3. *For* $L \geq 1$ *the limiting distribution in an infinite branch of the tree of foldings such that* $K_i = \emptyset$ *for all* $i \geq L$ *always exists, and equals*

$$P^\infty(\omega) = \frac{1}{Z_\infty} \mathbb{I}_{\{\omega : P^{\mathbf{K}_L, \boldsymbol{\alpha}_i, \boldsymbol{\beta}}(\omega) \geq P^{\mathbf{K}_L, \boldsymbol{\alpha}_i, \boldsymbol{\beta}}(\omega') \text{ for all } \omega' \in \Omega_{(\cup_{K \in \mathbf{K}_L} K)^c}\}},$$

i.e. it is the uniform distribution on the maxima of $P^{\mathbf{K}_L, \boldsymbol{\alpha}_L, \boldsymbol{\beta}}$, *where* Z_∞ *is the number of such maxima. Moreover, for* $i > L$, *and any* ω *such that* $P^\infty(\omega) > 0$

$$\sup_{\omega \in \Omega_L} |P^{\mathbf{K}_i, \boldsymbol{\alpha}_i, \boldsymbol{\beta}}(\omega) - P^\infty(\omega)| \leq |\Omega_\Lambda| \frac{\max_{\omega'' : P^\infty(\omega'') = 0} (P^{\mathbf{K}_L, \boldsymbol{\alpha}_L, \boldsymbol{\beta}}(\omega''))^{2^{i-L}}}{(\max_{\omega'} P^{\mathbf{K}_L, \boldsymbol{\alpha}_L, \boldsymbol{\beta}}(\omega'))^{2^{i-L}}} =: a^{2^i}$$

with $a < 1$.

Proof. Let $i > L$; since $K_i = \emptyset$ all $P^{\mathbf{K}_i, \boldsymbol{\alpha}_i, \boldsymbol{\beta}}$'s are defined on the same $\Omega_{(\cup_{i=1}^L K_i)^c}$; each direct descendant is a folding of the parent distribution, which is already symmetric since $L \geq 1$. Thus $P^{\mathbf{K}_{i+1},\boldsymbol{\alpha}_{i+1},\beta}(\omega) = \frac{(P^{\mathbf{K}_i,\boldsymbol{\alpha}_i,\beta}(\omega))^2}{\sum_{\omega'} P^{\mathbf{K}_i,\boldsymbol{\alpha}_i,\beta}(\omega'))^2}$. Hence, if ω is a maximizer of $P^{\mathbf{K}_L,\boldsymbol{\alpha}_L,\beta}$, then

$$\lim_{i\to\infty} P^{\mathbf{K}_i,\boldsymbol{\alpha}_i,\beta}(\omega) = \lim_{i\to\infty} \frac{(P^{\mathbf{K}_L,\boldsymbol{\alpha}_L,\beta}(\omega))^{2^{i-L}}}{\sum_{\omega'}(P^{\mathbf{K}_L,\boldsymbol{\alpha}_L,\beta}(\omega'))^{2^{i-L}}} = \frac{1}{Z_\infty},$$

and the limit equals 0 otherwise. Moreover, for all ω such that $P^\infty(\omega) > 0$, as $Z_\infty \geq 1$ and $P^{\mathbf{K}_L,\boldsymbol{\alpha}_L,\beta}(\omega) = \max_{\omega'\in\Omega_\Lambda} P^{\mathbf{K}_L,\boldsymbol{\alpha}_L,\beta}(\omega')$, we have

$$\begin{aligned}
&\left|P^{\mathbf{K}_i,\boldsymbol{\alpha}_i,\beta}(\omega) - \frac{1}{Z_\infty}\right| \\
&= \left|\frac{(P^{\mathbf{K}_L,\boldsymbol{\alpha}_L,\beta}(\omega))^{2^{i-L}}}{Z_\infty(P^{\mathbf{K}_L,\boldsymbol{\alpha}_L,\beta}(\omega))^{2^{i-L}} + \sum_{\omega'':P^\infty(\omega'')=0}(P^{\mathbf{K}_L,\boldsymbol{\alpha}_L,\beta}(\omega''))^{2^{i-L}}} - \frac{1}{Z_\infty}\right| \\
&\leq \frac{\sum_{\omega'':P^\infty(\omega'')=0}(P^{\mathbf{K}_L,\boldsymbol{\alpha}_L,\beta}(\omega''))^{2^{i-L}}}{(P^{\mathbf{K}_L,\boldsymbol{\alpha}_L,\beta}(\omega))^{2^{i-L}}} \\
&\leq |\Omega_\Lambda| \frac{\max_{\omega'':P^\infty(\omega'')=0}(P^{\mathbf{K}_L,\boldsymbol{\alpha}_L,\beta}(\omega''))^{2^{i-L}}}{(P^{\mathbf{K}_L,\boldsymbol{\alpha}_L,\beta}(\omega))^{2^{i-L}}} \\
&= |\Omega_\Lambda| \left(\frac{\max_{\omega'':P^\infty(\omega'')=0} P^{\mathbf{K}_L,\boldsymbol{\alpha}_L,\beta}(\omega'')}{\max_{\omega'} P^{\mathbf{K}_L,\boldsymbol{\alpha}_L,\beta}(\omega'))}\right)^{2^{i-L}} < 1,
\end{aligned}$$

so that we can take $a < 1$. □

3.2 Essential Tree of Foldings

An **essential folding** is either a folding in which $K_i \neq \emptyset$ or the first folding (which is "essential" by itself as it simmetrizes the distribution); the non essential foldings are called inessential. The **essential tree of foldings** is a subtree of the tree of foldings such that in each branch the first foldings are all essential and the remaining ones are all inessential; in other words, it does not alternate between the two types of foldings. In this paper, by **subtree** of a tree we always mean a connected subtree containing the root.

We now see that the essential tree of foldings contains all the information of the tree of foldings.

Lemma 4. *If $P^{\mathbf{K}_n,\boldsymbol{\alpha}_n,\beta}$ is a folding then there is a folding in the essential tree of foldings which is equal to it.*

Proof. If $\mathbf{K}_n = (K_1, K_2, \ldots, K_n)$ then let $K_{i_1}, K_{i_2}, \ldots, K_{i_r}$ be the nonempty K_i's among those with $i \geq 2$. We see that for

$$\mathbf{K}'_n = (K_1, K_{i_1}, K_{i_2}, \ldots, K_{i_r}, \emptyset, \ldots, \emptyset),$$

and $\boldsymbol{\alpha}'_n$ similarly rearranged, $P^{\mathbf{K}_n,\boldsymbol{\alpha}_n,\beta} = P^{\mathbf{K}'_n,\boldsymbol{\alpha}'_n,\beta}$. In fact, all we have to show is that, after the first folding, exchanging an essential and an inessential folding gives the same probability. After the first folding the probabilities are symmetric; therefore, assuming that P is already a folding, we have

$$\begin{aligned}
P^{(K_1,\emptyset),(\alpha_1,\alpha),\beta}(\omega) &= \frac{1}{Z_2} P^{K_1,\alpha_1,\beta}(\omega) P^{K_1,\alpha_1,\beta}(\overline{\omega}) \\
&= \frac{1}{Z_2} P^{K_1,\alpha_1,\beta}(\omega)^2 \\
&= \frac{1}{Z_2}(\frac{1}{Z_1} P(\alpha_1\omega)P(\alpha_1\overline{\omega}))^2 \\
&= \frac{(P(\alpha_1\omega))^2 (P(\alpha_1\overline{\omega}))^2}{\sum_{\omega'\in\Omega_{K_1^c}} (P(\alpha_1\omega'))^2 (P(\alpha_1\overline{\omega}'))^2} \\
&= \frac{\frac{(P(\alpha_1\omega))^2}{\sum_{\omega''}(P(\alpha_1\omega''))^2} \frac{(P(\alpha_1\overline{\omega}))^2}{\sum_{\omega''}(P(\alpha_1\omega''))^2}}{\sum_{\omega'\in\Omega_{K_1^c}} \frac{P(\alpha_1\omega'))^2}{\sum_{\omega''}(P(\alpha_1\omega''))^2} \frac{(P(\alpha_1\overline{\omega}'))^2}{\sum_{\omega''}(P(\alpha_1\omega''))^2}} \\
&= \frac{P^{\emptyset,\alpha}(\alpha_1\omega) P^{\emptyset,\alpha}(\alpha_1\overline{\omega})}{\sum_{\omega'\in\Omega_{K_1^c}} P^{\emptyset,\alpha}(\alpha_1\omega') P^{\emptyset,\alpha}(\alpha_1\overline{\omega}')} \\
&= P^{(\emptyset,K_1),(\alpha,\alpha_1),\beta}(\omega)
\end{aligned}$$

□

We restrict from now on to the essential tree of foldings. Note, in particular, that the essential tree of foldings has at most $(2^{|\Lambda|}|\Omega_\Lambda|)^{|\Lambda|} < \infty$ infinite branches. In addition, in each branch there are at most $L = |\Lambda|$ initial essential foldings, and then all the others are inessential.

3.3 RCR's in the Tree of Foldings

Given $\Lambda, \mathcal{B} \subseteq \mathcal{P}(\Lambda)$, and a probability P on $\mathcal{P}(\Lambda)$, various collections of foldings of P can admit a $\mathcal{B}$-RCR. Some relevant cases are as follows.

Definition 5. *We say that P has a* **finite tree $\mathcal{B}$-RCR**, *if there exists a finite subtree T of the tree of foldings such that for each leaf $(\mathbf{K}, \boldsymbol{\alpha}, \boldsymbol{\beta})$ of T there exists a $\mathcal{B}$-RCR $\nu_{\mathbf{K},\boldsymbol{\alpha},\boldsymbol{\beta}}$ of $P^{\mathbf{K},\boldsymbol{\alpha},\boldsymbol{\beta}}$.*

For $\epsilon > 0$, we say that P has a **finite tree ϵ-$\mathcal{B}$-RCR** , *if there exists a finite subtree T of the tree of foldings such that for each leaf $(\mathbf{K}, \boldsymbol{\alpha}, \boldsymbol{\beta})$ of T there exists a ϵ-$\mathcal{B}$-RCR $\nu_{\mathbf{K},\boldsymbol{\alpha},\boldsymbol{\beta}}$ of $P^{\mathbf{K},\boldsymbol{\alpha},\boldsymbol{\beta}}$.*

We say that P has an **$\mathcal{B}$-RCR at infinity** *if for each infinite branch in the essential tree of branches the limiting distribution P^∞ in that branch has a $\mathcal{B}$-RCR.*

We say that P has an **asymptotic $\mathcal{B}$-RCR** *if there exists an integer valued function n_ϵ such that $\lim_{\epsilon\to 0} 2^{n_\epsilon}\epsilon = 0$ for which the following happens: for every $\epsilon > 0$, P has a finite tree ϵ-$\mathcal{B}$-RCR on the subtree T_ϵ formed by the essential tree of foldings truncated after n_ϵ generations.*

Lemma 5. *If P has a $\mathcal{B}$-RCR ν_∞ at infinity, then ν_∞ is also an asymptotic $\mathcal{B}$-RCR.*

Proof. Let P have a $\mathcal{B}$-RCR at infinity, and consider $\epsilon > 0$. Recall that in each infinite branch there are at most L essential foldings at the beginning, and choose $n = n_\epsilon$ the smallest integer such that

$$a^{2^n} = |\Omega_\Lambda| \frac{\max_{\omega'':P^\infty(\omega'')=0}(P^{\mathbf{K}_L,\boldsymbol{\alpha}_L,\boldsymbol{\beta}}(\omega''))^{2^{n-L}}}{\max_{\omega'}(P^{\mathbf{K}_L,\boldsymbol{\alpha}_L,\boldsymbol{\beta}}(\omega'))^{2^{n-L}}} \leq \epsilon \tag{2}$$

where a is taken as in Lemma 3. Since P^∞ has a $\mathcal{B}$-RCR by assumption, and it approximates $P^{\mathbf{K}_L,\boldsymbol{\alpha}_L,\boldsymbol{\beta}}$ by less than $a^{2^n} < \epsilon$ by Lemma 3, it follows that $P^{\mathbf{K}_L,\boldsymbol{\alpha}_L,\boldsymbol{\beta}}$ has a ϵ-$\mathcal{B}$-RCR using $T = T_\epsilon$ as finite subtree. In addition, as n_ϵ is chosen to be the smallest integer such that (2) holds, $\epsilon \approx a^{2^{n_\epsilon}}$ and $\lim_{\epsilon\to 0} 2^{n_\epsilon}\epsilon = \lim_{\epsilon\to 0} 2^{n_\epsilon} a^{2^{n_\epsilon}} = 0$ as $a < 1$ from Lemma 3; hence P has an asymptotic $\mathcal{B}$-RCR. □

4 FKG Theory

4.1 Positive Association

A probability P is positively associated (PA) if for all increasing $A, B \subseteq \Omega_\Lambda$, $P(A \cap B) \geq P(A)P(B)$. Equivalently, for all increasing A and decreasing B, $P(A \cap B) \leq P(A)P(B)$.

We are going to find a very simple sufficient condition for positive association in terms of RCR. To this purpose, given a finite set Λ and ordered sets F_i, we denote by $\hat{\omega}_\Lambda$ the configuration such that $(\hat{\omega}_\Lambda)_i = \max\{s : s \in F_i\}$.

If $|b| \leq 2$ for all $b \in \mathcal{B}(\Lambda)$ and the F_i's are all ordered sets, then a $\mathcal{B}$-RCR ν is **ferromagnetic** if for $b = \{i, j\}$, $\hat{\omega}_b \in \eta_b$ for all $\eta_b \neq \emptyset$. Recall that if $|F_i| = 2$ we say that ν is symmetric if $\omega_b \in \eta_b$ implies $\overline{\omega}_b \in \eta_b$. We can assume that $F_i = \{0, 1\}$, in case after relabeling of the elements of F_i; in such case, a symmetric and ferromagnetic $\mathcal{B}$-RCR with $|b| \leq 2$ for all b is such that either $\eta_b = \Omega_b$ or $\eta_b = \{\{0, 0\}, \{1, 1\}\}$, analogously to what happens in the original FK representation (see Sect. 2.2).

Theorem 1. *Let F_i be ordered and P be a probability on Ω_Λ. Suppose that there is a finite subtree T of the tree of foldings such that each leaf of T has a symmetric ferromagnetic $\mathcal{B}$-RCR with $|b| \leq 2$ for all $b \in \mathcal{B}(\Lambda)$; then P is PA.*

Proof. For each leaf $(\mathbf{K}, \boldsymbol{\alpha}, \boldsymbol{\beta})$, $P^{\mathbf{K},\boldsymbol{\alpha},\boldsymbol{\beta}}$ is defined on $\prod_i\{\beta_i(1), \beta_i(2)\}$, which can be relabeled into $F_i = \{0, 1\}$. The proof is divided in various steps.

(I) For each leaf $(\mathbf{K}, \boldsymbol{\alpha}, \boldsymbol{\beta})$ there is a $\mathcal{B}$-RCR $\nu_{\mathbf{K},\boldsymbol{\alpha},\boldsymbol{\beta}}$ which is symmetric and ferromagnetic. If $\nu(\eta) > 0$, $C(\eta)$ is one of the clusters formed by active hyperbonds in η, and $\omega \in \Omega_{(\cup_{i=1}^n K_i)^c}$, then either $C(\eta) \subseteq \omega^{-1}(1)$ or $C(\eta) \subseteq \omega^{-1}(0)$ as a bond $b = \{i, j\}$ with $\omega_i \neq \omega_j$ can never be active.

Let Ψ be a selection rule which only selects pairs of sets (M, N) with $M \subseteq \omega^{-1}(1)$ and $N \subseteq \omega^{-1}(0)$; under this selection rule the event A can only be recognized by 1's and the event B by 0's. Therefore, at most one of $C(\eta) \cap M \neq \emptyset$

or $C(\eta) \cap N \neq \emptyset$ is possible, hence Ψ uses disjoint clusters of η. Lemma 1 with $\epsilon = 0$ gives that for all increasing A and decreasing B we have that for each leaf

$$P^{\mathbf{K},\boldsymbol{\alpha},\boldsymbol{\beta}}(A \cap B) = P^{\mathbf{K},\boldsymbol{\alpha},\boldsymbol{\beta}}(A\boxed{\Psi}B) \leq P^{\mathbf{K},\boldsymbol{\alpha},\boldsymbol{\beta}}(A \cap \overline{B}). \tag{3}$$

(II) (a false step) Let $(\mathbf{K}, \boldsymbol{\alpha}, \boldsymbol{\beta})$ be a node of the tree of foldings, and assume that for its direct descendant (3) holds with the appropriate symbol replacements to indicate the descendant. Its descendants are also obtained from a folding, so Lemma 2 applies with $\Lambda = (\cup_{i=1}^n K_i)^c$ and $\epsilon = 0$ to see that for A increasing, B decreasing in $\Omega_{(\cup_{i=1}^n K_i)^c}$, for each node of the tree of foldings whose direct descendants satisfy (3) we have

$$P^{\mathbf{K},\boldsymbol{\alpha},\boldsymbol{\beta}}(A \cap B) = P^{\mathbf{K},\boldsymbol{\alpha},\boldsymbol{\beta}}(A\boxed{\Psi}B) \leq P^{\mathbf{K},\boldsymbol{\alpha},\boldsymbol{\beta}}(A)P^{\mathbf{K},\boldsymbol{\alpha},\boldsymbol{\beta}}(B). \tag{4}$$

This gives the required inequality, but for the node $(\mathbf{K}, \boldsymbol{\alpha}, \boldsymbol{\beta})$. To get the result for P we have to backtrack a little and bootstrap things.

(III) Suppose that for a node $(\mathbf{K}, \boldsymbol{\alpha}, \boldsymbol{\beta})$ of the tree of foldings all of its direct descendant satisfy (3). If B is decreasing then $\overline{B}$ is increasing, and $(\overline{B})^c$ is again decreasing. Then it follows from (3) and (4) that

$$P^{\mathbf{K},\boldsymbol{\alpha},\boldsymbol{\beta}}(A \cap (\overline{B})^c) \leq P^{\mathbf{K},\boldsymbol{\alpha},\boldsymbol{\beta}}(A)P^{\mathbf{K},\boldsymbol{\alpha},\boldsymbol{\beta}}((\overline{B})^c),$$

which yields

$$P^{\mathbf{K},\boldsymbol{\alpha},\boldsymbol{\beta}}(A \cap \overline{B}) \geq P^{\mathbf{K},\boldsymbol{\alpha},\boldsymbol{\beta}}(A)P^{\mathbf{K},\boldsymbol{\alpha},\boldsymbol{\beta}}(\overline{B}). \tag{5}$$

By symmetry of each folding after the first one, for all B,

$$P^{\mathbf{K},\boldsymbol{\alpha},\boldsymbol{\beta}}(\overline{B}) = P^{\mathbf{K},\boldsymbol{\alpha},\boldsymbol{\beta}}(B).$$

Then, for all increasing A and decreasing B we have that for a vertex $(\mathbf{K}, \boldsymbol{\alpha}, \boldsymbol{\beta})$ of the tree of foldings other than the root, all of whose direct descendant satisfy (3) it holds

$$\begin{aligned} P^{\mathbf{K},\boldsymbol{\alpha},\boldsymbol{\beta}}(A \cap B) &\leq P^{\mathbf{K},\boldsymbol{\alpha},\boldsymbol{\beta}}(A)P^{\mathbf{K},\boldsymbol{\alpha},\boldsymbol{\beta}}(B) \\ &= P^{\mathbf{K},\boldsymbol{\alpha},\boldsymbol{\beta}}(A)P^{\mathbf{K},\boldsymbol{\alpha},\boldsymbol{\beta}}(\overline{B}) \leq P^{\mathbf{K},\boldsymbol{\alpha},\boldsymbol{\beta}}(A \cap \overline{B}), \end{aligned} \tag{6}$$

which is again (3), but now brought up one step.

(IV) As (3) holds for all the leaves, and if it holds for all descendants it holds also for the parent node, provided this is different from the root, then (3) holds for all nodes, including thus all descendants of the root $\mathbf{0}$.

(V) The final step is like Step (II), but performed on the original P. By (IV), we can apply Lemma 2 to get that for increasing A and decreasing B, $P(A \cap B) = P(A\boxed{\Psi}B) \leq P(A)P(B)$ so that P is PA as required. □

Lemma 6. *Let F_i be ordered and P be a probability on Ω_Λ. Suppose P has an asymptotic $\mathcal{B}$-RCR such that each leaf of T_ϵ has a symmetric ferromagnetic ϵ-$\mathcal{B}$-RCR with $|b| \leq 2$ for all $b \in \mathcal{B}(\Lambda)$; then P is PA.*

Proof. The proof is a repetition of the proof of Theorem 1 with some modifications. Applying Lemma 1 with the error term, (3) is replaced by

$$P^{\mathbf{K},\alpha,\beta}(A \cap B) = P^{\mathbf{K},\alpha,\beta}(A \boxed{\Psi} B) \leq P^{\mathbf{K},\alpha,\beta}(A \cap \overline{B}) + \epsilon 2^{|\Lambda|+1}.$$

Then (4) adds $\epsilon 2^{|\Lambda|+1}$, (5) is replaced by

$$P^{\mathbf{K},\alpha,\beta}(A \cap \overline{B}) \geq P^{\mathbf{K},\alpha,\beta}(A) P^{\mathbf{K},\alpha,\beta}(\overline{B}) - \epsilon 2^{|\Lambda|+1} \tag{7}$$

and (6) is replaced by

$$P^{\mathbf{K},\alpha,\beta}(A \cap B) \leq P^{\mathbf{K},\alpha,\beta}(A \cap \overline{B}) + 2\epsilon 2^{|\Lambda|+1}. \tag{8}$$

Notice the extra factor of 2 due to the possible summation of the errors in (7) and (4).

Following Part (IV) above, since T_ϵ has n_ϵ generations we have that the direct descendants of the root $\mathbf{0}$ satisfy

$$P^{\mathbf{K},\alpha,\beta}(A \cap B) \leq P^{\mathbf{K},\alpha,\beta}(A \cap \overline{B}) + 2^{n_\epsilon} \epsilon 2^{|\Lambda|+1},$$

and following Part (V) $P(A \cap B) \leq P(A)P(B) + 2^{n_\epsilon} \epsilon 2^{|\Lambda|+1}$. By definition of asymptotic representability, $\lim_{\epsilon \to 0} 2^{n_\epsilon} \epsilon = 0$, so that $P(A \cap B) \leq P(A)P(B)$. □

Combining Lemmas 5 and 6 we get the final result of this section:

Theorem 2. *Let F_i be ordered and P be a probability on Ω_Λ. If P has a $\mathcal{B}$-RCR at infinity which is ferromagnetic and symmetric, with $|b| \leq 2$ for all $b \in \mathcal{B}(\Lambda)$, then P is PA.*

4.2 FKG Theorem

The FKG theorem, which we are going to show at the end of this section, states that if a probability on a lattice satisfies the FKG condition then it is PA. We say that P **satisfies the FKG condition** if for all $\omega, \omega' \in \Omega_\Lambda = \prod_{i \in \Lambda} F_i$, F_i ordered sets, $|F_i| \leq 2$

$$P(\omega \vee \omega') P(\omega \wedge \omega') \geq P(\omega) P(\omega') \tag{9}$$

Recall that $\hat{\omega}_i = \max_{x \in F_i} x$. The FKG condition takes a nice form in the foldings.

Lemma 7. *P satisfies the FKG condition if and only if for every $K \subseteq \Lambda, \alpha \in \Omega_K, \beta(1), \beta(2) \in \Omega_{K^c}, \beta_i(1) \neq \beta_i(2), \omega_{K^c} \in \Omega_{K^c}$*

$$P^{K,\alpha,\beta}(\omega_{K^c}) \leq P^{K,\alpha,\beta}(\hat{\omega}_{K^c}). \tag{10}$$

Proof. If (9) is satisfied, then for all $K, \alpha, \beta(1), \beta(2), \omega_{K^c}$ let $\omega = \alpha \hat{\omega}_{K^c}$ and $\omega' = \alpha \overline{\omega_{K^c}}$. We have

$$\omega = \alpha \omega_{K^c}, \quad \omega' = \alpha \overline{\omega_{K^c}}, \quad \omega \vee \omega' = \alpha \hat{\omega}_{K^c}, \quad \omega \wedge \omega' = \alpha \overline{\hat{\omega}_{K^c}}. \tag{11}$$

Thus

$$\begin{aligned} P^{K,\alpha,\beta}(\omega_{K^c}) &= P(\alpha\omega_{K^c})P(\alpha\overline{\omega_{K^c}}) = P(\omega)P(\omega') \\ &\leq P(\omega \vee \omega')P(\omega \wedge \omega') \\ &= P(\alpha\hat{\omega}_{K^c})P(\alpha\overline{\hat{\omega}_{K^c}}) = P^{K,\alpha,\beta}(\hat{\omega}_{K^c}). \end{aligned} \tag{12}$$

Viceversa, if P satisfies (10) and $\omega, \omega' \in \Omega_\Lambda$, let K be the set of $i \in \Lambda$ in which the two configurations ω, ω' are equal; let also $\alpha_i = \omega_i$ for $i \in K$, $\beta_i(1) = \omega_i$, $\beta_i(2) = \omega'_i$ for $i \in K^c$, $(\omega_{K^c})_i = \omega_i$ for $i \in K^c$. Then (11) still holds and the (12) can be used in reversed order to show that (9) holds. □

Lemma 8. *P satisfies the FKG condition if and only if for all foldings $(K, \alpha, \boldsymbol{\beta})$, $P^{K,\alpha,\beta}$ satisfies the FKG condition as well.*

Proof. For all $(K, \alpha, \boldsymbol{\beta})$, and for all $u \in K^c$, $\omega(1), \omega(2) \in \Omega_{K^c}$

$$\begin{aligned} \overline{\omega_u(1) \vee \omega_u(2)} &= \overline{\omega_u(1)} \wedge \overline{\omega_u(2)} \\ \overline{\omega_u(1) \wedge \omega_u(2)} &= \overline{\omega_u(1)} \vee \overline{\omega_u(2)}. \end{aligned}$$

If P is FKG then

$$\begin{aligned} &P^{K,\alpha,\beta}(\omega(1) \vee \omega(2))P^{K,\alpha,\beta}(\omega(1) \wedge \omega(2)) \\ &= P(\alpha(\omega(1) \vee \omega(2)))P(\alpha\overline{(\omega(1) \vee \omega(2))})P(\alpha(\omega(1) \wedge \omega(2)))P(\alpha\overline{(\omega(1) \wedge \omega(2))}) \\ &= P(\alpha(\omega(1) \vee \omega(2)))P(\alpha(\overline{\omega(1)} \wedge \overline{\omega(2)}))P(\alpha(\omega(1) \wedge \omega(2)))P(\alpha(\overline{\omega(1)} \vee \overline{\omega(2)})) \\ &\geq P(\alpha\omega(1))P(\alpha\omega(2))P(\alpha\overline{\omega(1)})P(\alpha\overline{\omega(2)}) \\ &= P^{K,\alpha,\beta}(\omega(1))P^{K,\alpha,\beta}(\omega(2)). \end{aligned}$$

Viceversa, if for all $(K, \alpha, \boldsymbol{\beta})$, $P^{K,\alpha,\beta}$ satisfies the FKG condition, then for $\omega \in \Omega_{K^c}$, and, by symmetry of the foldings,

$$\begin{aligned} (P^{K,\alpha,\beta}(\hat{\omega}))^2 &= P^{K,\alpha,\beta}(\omega \vee \overline{\omega})P^{K,\alpha,\beta}(\omega \wedge \overline{\omega}) \\ &\geq P^{K,\alpha,\beta}(\omega)P^{K,\alpha,\beta}(\overline{\omega}) = (P^{K,\alpha,\beta}(\omega))^2. \end{aligned}$$

Hence P is FKG by Lemma 7. □

It follows that if P satisfies the FKG conditions then so do all its iterated foldings, and the probability P^∞ in the limit of each infinite branch. Each of the P^∞ is a two-valued, symmetric distribution uniform on some subset $D \subseteq \Omega_{(\cup_{n\in\mathbb{N}} K_n)^c}$; we see now that all probabilities of this form which satisfy the FKG condition are products of independent clusters, so that they have a ferromagnetic and symmetric $\mathcal{B}$-RCR made of pairs.

Theorem 3. *If P is two-valued, symmetric, uniform on $D \subseteq \Omega_\Lambda = \prod_{u\in\Lambda}\{0,1\}$ for some set Λ, and satisfies the FKG condition, then P has a ferromagnetic and symmetric $\mathcal{B}$-RCR such that $|b| \leq 2$ for all $b \in \mathcal{B}$.*

Note that the reverse also holds, but we omit the proof for brevity,

Proof. If P is as in the thesis, then consider $\mathcal{B} = \{b : b = \{u, v\}, u, v \in \Lambda\}$, and let $\bar{\eta} \in \prod_{b \in \mathcal{B}} \Omega_b$ be defined by

$$\bar{\eta}_{\{u,v\}} = \begin{cases} \{(1,1),(0,0)\} & \text{if for all } \omega \in D, \omega_u = \omega_v \\ \Omega_b & \text{if there exists } \omega \in D \text{ such that } \omega_u \neq \omega_v \end{cases}$$

Clearly, if $\omega \in D$ and $\bar{\eta}_{\{u,v\}} = \{(1,1),(0,0)\}$ then $\omega_u = \omega_v$; thus, $\omega \sim \bar{\eta}$. We then let $\nu := \delta_{\bar{\eta}}$; clearly, ν is ferromagnetic and symmetric. We claim that it is a $\mathcal{B}$-RCR of P.

For all $\omega \in D$, $P(\omega) = \frac{1}{Z} \sum_{\eta \sim \omega} \nu(\eta) = \frac{\nu(\bar{\eta})}{Z} = \frac{1}{Z}$; if we show that for $\omega \notin D$, $\sum_{\eta \sim \omega} \nu(\eta) = 0$, which is to say $\eta \not\sim \omega$, then indeed $Z = |D|$ and P would be uniform on D, which is the thesis. It is then sufficient to prove that if ω is constant on the $\bar{\eta}$ clusters, i.e. the clusters formed by the active edges in $\bar{\eta}$, then $\omega \in D$.

Let $\Omega(\bar{\eta}) = \{\omega : \omega \text{ is constant on the } \bar{\eta} \text{ clusters }\}$; clearly, $D \subseteq \Omega(\bar{\eta})$ and we want to show that equality holds. Let $C_1(\bar{\eta}), \ldots, C_m(\bar{\eta})$ be the cluster defined by the active bonds of $\bar{\eta}$, and let $t : \Omega(\bar{\eta}) \to \{0,1\}^m$ be defined by $(t(\omega))_j = \omega_u$ for $u \in C_j(\bar{\eta})$. Observe that $t(D) = \{\tau \in \{0,1\}^m : \text{there exists } \omega \in \Omega(\bar{\eta}) \text{ with } t(\omega) = \tau\}$ satisfies

1. $t(D)$ is a sublattice of $\{0,1\}^m$: in fact, for

$$\tau(1), \tau(2) \in t(D),\ \tau(1) = t(\omega(1)),\ \tau(2) = t(\omega(2)),\ \omega(1), \omega(2) \in D,$$

   we have $P(\omega(1)) = P(\omega(2)) \neq 0$, hence, by the FKG property of P, $P(\omega(1) \vee \omega(2)), P(\omega(1) \wedge \omega(2)) \neq 0$; therefore, $\tau(1) \vee \tau(2) = t(\omega(1) \vee \omega(2)) \in t(D)$ and $\tau(1) \wedge \tau(2) = t(\omega(1) \wedge \omega(2)) \in t(D)$;
2. $t(D)$ is symmetric: by the symmetry of P, if $\tau = t(\omega)$ then $P(\omega) > 0$ which implies $P(\overline{\omega}) > 0$, hence $\overline{\omega} \in D$, thus $\overline{\tau} = t(\overline{\omega}) \in t(D)$;
3. $t(D)$ separates points of $M = \{1, \ldots, m\}$ where, as above, m is the number of clusters defined by active bonds in $\bar{\eta}$, in the sense that for all $i, j \in M, i \neq j$, there is $\tau^{i,j} \in t(D)$ such that $\tau^{i,j}_i \neq \tau^{i,j}_j$: in fact, by definition of $\bar{\eta}$, if $i \neq j$ and $u \in C_i(\bar{\eta}), v \in C_j(\bar{\eta})$, then there exists $\omega \in D$ such that $\omega_u \neq \omega_v$; hence, $(t(\omega))_i = \omega_u \neq \omega_v = (t(\omega))_j$. Therefore, $\tau^{i,j} = t(\omega)$ separates i and j. Here we have used that $\mathcal{B} = \{b : b = \{u, v\}, u, v \in \Lambda\}$.

The next lemma shows that a symmetric sublattice which separates points coincides with the whole lattice; hence, $t(D) = \{0,1\}^m$, and every configuration which is constant on the clusters of $\bar{\eta}$ is in D, which concludes the proof. □

Lemma 9. *Consider $T = \{0,1\}^m$. If a sublattice L of T is symmetric and separates points, then $L = T$.*

Proof. Since L separates points, $L \neq \emptyset$; let $\tau \in L$, then $\overline{\tau} \in L$ by symmetry, and $\hat{\tau} = \tau \vee \overline{\tau} \in L$, as it is a lattice, and $\overline{\hat{\tau}} \in L$ by symmetry.

For $\tau \in T$, let $m(\tau) = |\{i = 1, \ldots, m : \tau_i = 0\}|$. If $m(\tau) = 0$ or m, then $\tau = \hat{\tau}$ or $\tau = \overline{\hat{\tau}}$, respectively, so in any case $\tau \in L$. If $m(\tau) = 1$ then let i be such that $\tau_i = 0$; for $j \neq i$ let $\tau^{(j)} \in L$ separate i and j, which is to say $\tau_i^{(j)} \neq \tau_j^{(j)}$; by symmetry, there is also $\tilde{\tau}^{(j)} \in L$ such that $\tilde{\tau}_i^{(j)} = 0$ and $\tilde{\tau}_j^{(j)} = 1$. Then $\tau = \vee_{j \neq i} \tilde{\tau}^{(j)} \in L$.

Finally, let $\overline{m} = \min\{m' :$ there exists $\tau \in T \setminus L$ such that $m(\tau) = m'\}$. For $\tau \in T \setminus L$ such that $m(\tau) = \overline{m}$ and i such that $\tau_i = 0$, let $\tau^{(i)}$ be such that

$$(\tau^{(i)})_j = \begin{cases} \tau_j & \text{if } j \neq i \\ 1 & \text{if } j = i. \end{cases}$$

Then $m(\tau^{(i)}) = m(\tau) - 1 < \overline{m}$ and $\tau^{(i)} \in L$ by induction. Moreover, $\tau = \wedge_{i:\tau_i=0} \tau^{(i)}$. Hence, $\tau \in L$, which is a contradiction. □

Our main conclusion

Theorem 4 *(FKG theorem). If P satisfies the FKG condition then P is PA.*

Proof. If P satisfies the FKG condition, then so does the limit of the foldings down each infinite branch. The limit satisfies then the conditions of Theorem 3 and hence it has a ferromagnetic and symmetric $\mathcal{B}$-RCR such that $|b| \leq 2$ for all $b \in \mathcal{B}$. But then Theorem 2 implies that P is PA. □

5 Some Theory of Negative Association

5.1 Negative Association

We assume that $\Omega_\Lambda = \prod_{i \in \Lambda} F_i$, with F_i finite and ordered. We say that two events $A, B \in \Omega_\Lambda$ have **disjoint support** if there exists $N \subseteq \Lambda$ (not dependent on a configuration) such that for all $\omega \in A \cap B$, $[\omega]_N \subseteq A$ and $[\omega]_{N^c} \subseteq B$.

Definition 6. *P is negatively associated (NA) if for all $A, B \subseteq \Omega_\Lambda$, increasing and with disjoint support, $P(A \cap B) \leq P(A)P(B)$.*

The theory of negative association is more difficult than that of positive association, see [3–5, 9, 10, 14, 15, 20]. We develop a version here, which for the most part reproduces the results in [20], with a strategy that mirrors the one obtained above for positive association, hence completely different from the one used in [20]. The additional difficulty with respect to our FKG theory can be seen from the fact that a RCR sufficient to guarantee NA is the mirror image of the one for PA, but, in addition, needs to be concentrated on isolated edges.

Let $\mathcal{B}(\Lambda)$ be such that $|b| \leq 2$ for all $b \in \mathcal{B}(\Lambda)$, let $F_i = \{0, 1\}$ for all $i \in \Lambda$, then we say that a $\mathcal{B}$-RCR ν is symmetric and **antiferromagnetic** if for all b, $\eta_b = \{\{(0,1), (1,0)\}, \Omega_b\}$.

We say that ν is **concentrated on isolated edges** if ν is a $\mathcal{B}$-RCR with $|b| \leq 2$ for all $b \in \mathcal{B}(\Lambda)$, and, moreover, if for all $b(1), b(2) \in \mathcal{B}$ such that $\eta_{b(1)} = \eta_{b(2)} \neq \Omega_b$ implies $b(1) \cap b(2) = \emptyset$; in other words, active bonds are isolated with ν probability one. In [1] various inequalities of BK type are proven by RCR of foldings of a probability P which are symmetric, antiferromagnetic and concentrated on isolated edges. We now develop a theory a negative association using such representation for the limit probability of the trees of foldings.

Theorem 5. *Let F_i be ordered and P be a probability on Ω_Λ. If P has a $\mathcal{B}$-RCR at infinity which is antiferromagnetic, symmetric and concentrated on isolated edges, then P is NA.*

Proof. The proof can be obtained by following the proofs of Theorems 1 and 2 with suitable adaptations.

In particular, it is proven in [1] that if all foldings of a probability P have a $\mathcal{B}$-RCR which is antiferromagnetic, symmetric and concentrated on isolated edges, with $|b| \leq 2$ for all $b \in \mathcal{B}(\Lambda)$, then P satisfies the following: for all A, B increasing and with disjoint support, $P(A \cap B) \leq P(A \cap \overline{B})$. This inequality is approximate if the representation is only approximate, as one would get from just having a $\mathcal{B}$-RCR at infinity.

Part (III) of Theorem 1 can be followed without modifications, by just noting that if A and B have disjoint support, then also B^c and $\overline{B}$ have disjoint support from A, as seen by just using the same set N.

All the remaining parts of the proofs work without relevant modifications. □

5.2 Negative FKG Theory

For a configuration $\omega \in \Omega_\Lambda = \{0, 1\}$ we indicate $|\omega| = \sum_{i \in \Lambda} \omega_1$ the number of 1's in ω.

Definition 7. *We say that a probability P satisfies the negative FKG condition, or it is NFKG, if for every folding (K, α) it happens that if $\omega_{K^c} \in \Omega_{K^c}$ satisfies*

$$\left| |\omega_{K^c}| - \frac{|K^c|}{2} \right| \leq 1/2 \tag{13}$$

then $P^{K,\alpha}(\omega_{K^c}) \geq P^{K,\alpha}(\omega'_{K^c})$ for all $\omega'_{K^c} \in \Omega_{B_i \cap K^c}$.

Note that if ω_{K^c} satisfies (13), then also $\overline{\omega_{K^c}}$ does. With these definitions we want to mirror Theorem 4. But this is actually easier done if we start from a more restricted notion of negative FKG:

Definition 8. *We say that a probability P satisfies the strict negative FKG condition, or that it is SNFKG, if for every folding (K, α) it happens that if $\omega_{K^c} \in \Omega_{K^c}$ satisfies (13) then*

1. *$P^{K,\alpha}(\omega_{K^c}) = P^{K,\alpha}(\omega'_{K^c})$ for all $\omega'_{K^c} \in \Omega_{B_i \cap K^c}$ which also satisfy (13);*
2. *$P^{K,\alpha}(\omega_{K^c}) > P^{K,\alpha}(\omega'_{K^c})$ if ω'_{K^c} does not satisfy (13).*

Of course, SNFKG implies NFKG.

Lemma 10. *If P is SNFKG then $P^{K,\alpha}$ is SNFKG for each folding (K, α).*

Proof. Let P be SNFKG, and consider $K_1, K_2 \subseteq K_1^c, \alpha_1 \in \Omega_{K_1^c}, \alpha_2 \in \Omega_{(K_1 \cup K_2)^c}$. Then consider the folding $P^{(K_1,K_2),(\alpha_1,\alpha_2)}$ of P^{K_1,α_1}, and two configurations

$\omega', \tilde{\omega} \in \Omega_{(K_1 \cup K_2)^c}$ such that ω' satisfies (13) for $K = K_1 \cup K_2$; we have

$$\begin{aligned} P^{(K_1,K_2),(\alpha_1,\alpha_2)}(\omega') &= \frac{1}{Z_2 Z_1^2} P(\alpha_1\alpha_2\omega')P(\alpha_1\overline{\alpha_2}\overline{\omega'})P(\alpha_1\alpha_2\overline{\omega'})P(\alpha_1\overline{\alpha_2}\omega') \\ &\geq \frac{1}{Z_2 Z_1^2} P(\alpha_1\alpha_2\tilde{\omega})P(\alpha_1\overline{\alpha_2}\overline{\tilde{\omega}})P(\alpha_1\alpha_2\overline{\tilde{\omega}})P(\alpha_1\overline{\alpha_2}\tilde{\omega}) \\ &= P^{(K_1,K_2),(\alpha_1,\alpha_2)}(\tilde{\omega}) \end{aligned}$$

as for the folding $(K_1 \cup K_2, \alpha_1\alpha_2)$

$$\begin{aligned} P(\alpha_1\alpha_2\omega')P(\alpha_1\alpha_2\overline{\omega'}) &= P^{(K_1\cup K_2),(\alpha_1\alpha_2)}(\omega') \\ &\geq P^{(K_1\cup K_2),(\alpha_1\alpha_2)}(\tilde{\omega}) \\ &= P(\alpha_1\alpha_2\tilde{\omega})P(\alpha_1\alpha_2\overline{\tilde{\omega}}), \end{aligned}$$

and analogously for the folding $(K_1 \cup K_2, \alpha_1\overline{\alpha}_2)$, with equality if also $\tilde{\omega}$ satisfies (13), and strict inequalities otherwise. □

Lemma 11. *If P is SNFKG then P has a an antiferromagnetic and symmetric $\mathcal{B}$-RCR at infinity concentrated on isolated edges.*

Proof. By Definition 8 we consider the uniform probability P^∞ on the maxima of a leaf $(\mathbf{K}, \boldsymbol{\alpha})$ of the essential tree of foldings. As P is SNFKG then each leaf is SNFKG; then, the maxima are configurations ω satisfying (13) with $K = \cup K_i$ with K_i the sets of the essential foldings leading to the leaf.

Let ν be the uniform distribution on the complete pairings of K^c, i.e. configurations η's whose active bonds are disjoint but pair every two vertices (see [1] for more details), with the exception of at most one vertex.

We need to show that each ω satisfying (13) is compatible with the same number of such complete pairings, and that no other ω is compatible.

In fact, recall that the active bonds are antiferromagnetic, so each carries exactly one value 1: a compatible configuration must satisfy (13) with $K = \cup K_i$.

On the other hand, each ω satisfying (13) is compatible with exactly $\lceil \frac{\lfloor K^c \rfloor}{2} \rceil!$ complete pairings, and this finishes the proof. □

Corollary 1. *If P is SNFKG then it is NA.*

Proof. By Lemma 11, P has a $\mathcal{B}$-RCR by an antiferromagnetic and symmetric RCR concentrated on isolated edges. By Theorem 5 this implies that P is NA. □

Theorem 6. *If P is NFKG then it is NA.*

Proof. We prove the result by a perturbation method.

Let P be NFKG, let $A, B \subseteq \Omega_\Lambda$ be increasing events, and suppose there exists $N \subseteq \Lambda$ such that for all $\omega \in A \cap B$, $[\omega]_N \subseteq A$, $[\omega]_{N^c} \subseteq B$.

Given $\epsilon > 0$ consider the probability such that

$$P_\epsilon(\omega) = \frac{P(\omega)}{Z_\epsilon}(1+\epsilon)^{|\{b=\{u,v\}\in\mathcal{B}(\Lambda):\, \omega_u \neq \omega_v\}|}.$$

If P is NFKG then P_ϵ is SNFKG. Suppose, in fact, that $\omega \in \Omega_\Lambda$ satisfies (13).

1. If also $\omega' \in \Omega_\Lambda$ satisfies (13), then $P(\omega) = P(\omega')$, by definition of NFKG, and

$$P_\epsilon(\omega) = \frac{P(\omega)}{Z_\epsilon}(1+\epsilon)^{\lfloor \frac{|\Lambda|}{2} \rfloor} = \frac{P(\omega')}{Z_\epsilon}(1+\epsilon)^{\lfloor \frac{|\Lambda|}{2} \rfloor} = P_\epsilon(\omega')$$

2. If, on the other hand, $\omega' \in \Omega_\Lambda$ does not satisfy (13), then $P(\omega) \geq P(\omega')$, and, by definition, it is enough to show that

$$\left\lfloor \frac{|\Lambda|}{2} \right\rfloor > |\{b = \{u, v\} \in \mathcal{B}(\Lambda), \omega'_u \neq \omega'_v\}|.$$

For this, it is enough that for all m, $\Lambda = \{1, \ldots, m\}, \Omega = \{0,1\}^\Lambda, \epsilon > 0$, the function $\omega \to |\{b = \{u,v\} \in \mathcal{B}(\Lambda), u, v \in \Lambda, \omega_u \neq \omega_v\}|$ has its maximum for the ω's such that $||\omega| - \frac{m}{2}| \leq \frac{1}{2}$ and it is strictly smaller for all ω's such that $||\omega| - \frac{m}{2}| > \frac{1}{2}$. But, in fact, if $|\omega| = k$ then

$$\begin{aligned}
&|\{b = \{u, v\} \in \mathcal{B}(\Lambda), u, v \in \Lambda, \omega_u \neq \omega_v\}| \\
&= k(m-k) - \left(\frac{k(k-1)}{2} + \frac{(m-k)(m-k-1)}{2}\right) \\
&= 3km - 3k^2 - m^2 + m =: a_k.
\end{aligned}$$

Now, $a_{k+1} - a_k = 3m - 3 - 6k$ and $a_{k+1} \geq a_k$ is equivalent to $k \leq \frac{m-1}{2}$ with equality if and only if $|k - \frac{m}{2}| \leq \frac{1}{2}$; in such case $a_k = \lfloor \frac{m}{2} \rfloor$, which is what is required to show that P_ϵ is SNFKG.

Since P_ϵ is SNFKG than it is NA by Corollary 1. Hence, $P_\epsilon(A \cap B) \leq P_\epsilon(A)P_\epsilon(B)$ for every $\epsilon > 0$. As Λ and F_i's are finite, for every A $\lim_{\epsilon \to 0} P_\epsilon(A) = P(A)$, so also P is NA. □

Example 4. If P is exchangeable, then let $p_k = P(\omega)$ if $|\omega| = k$, $k = 0, 1, \ldots, |\Lambda|$. Then if P is NFKG one can deduce that for $s < m/2$

$$p_{r+m-s}p_{r+s} \geq p_{r+m-s+1}p_{r+s-1};$$

from this it is easy to see that P is NFKG if and only if $p_{k+1}p_{k-1} \leq p_k^2$ for $K = 1, \ldots, |\Lambda| - 1$, i.e. P is ultra log concave (ULC). This gives an alternative proof of the result in [20] that an exchangeable ULC is NA (actually, we prove that P is CNA+, in the terminology of [20], as also done there).

Acknowledgments. The author would like to thank J. van den Berg for very valuable discussions and suggestions.

Dedication. This paper is dedicated to Chuck Newman on the occasion of his birthday. The FKG inequality and the FK random cluster representations have been used many times in Chuck's research, from [16–18] to [6]. They appeared as two separate topics, as they have always been perceived so far. They also played a role in many discussions I had with Chuck, during which I gained deep insights on several problems in statistical mechanics, in particular spin glasses, and on many other subjects.

I present here a proof of the FKG inequality which relies on (an extension of) the FK representation (with some additional ingredients, such as foldings). Besides shedding some light on the relation between the two concepts, I believe that the theory developed here can be useful in other directions, in particular towards identifying a role for percolation in the description of the phase transition in Spin Glasses (see [19]). One of the motivations behind this note is the hope that a close relation between percolation and Spin Glasses might, in due time, be found in the foldings.

References

1. van den Berg, J., Gandolfi, A.: BK-type inequalities and generalized random-cluster representations. Probab. Theory Related Fields **157**(1–2), 157–181 (2013)
2. van den Berg, J., Kesten, H.: Inequalities with applications to percolation and reliability. J. Appl. Probab. **22**, 556–569 (1985)
3. Borcea, J., Brändén, P., Liggett, T.M.: Negative dependence and the geometry of polynomials. J. Amer. Math. Soc. **22**, 521–567 (2009)
4. Brändén, P.: Polynomials with the half-plane property and matroid theory. Adv. Math. **216**, 302–320 (2007)
5. Brändén, P., Jonasson, J.: Negative dependence in sampling. Scand. J. Stat. **39**(4), 830–838 (2012)
6. Camia, F., Jiang, J., Newman, C.M.: Exponential decay for the near-critical scaling limit of the planar Ising model. arXiv:1707.02668 (2017)
7. Caputo, P., Sinclair, A.: Entropy production in nonlinear recombination models. Bernoulli J. **24**, 3246–3282 (2018)
8. Chayes, L., Lei, H.K.: Random cluster models on the triangular lattice. J. Stat. Phys. **122**, 647–670 (2006)
9. Dubhashi, D., Jonasson, J., Ranjan, D.: Positive influence and negative dependence. Comb. Probab. Comput. **16**, 29–41 (2007)
10. Dubhashi, D., Ranjan, D.: Balls and bins: a study in negative dependence. Random Struct. Alg. **13**, 99–124 (1998)
11. Fortuin, C.M., Kasteleyn, P.W.: On the random-cluster model. I. Introduction and relation to other models. Physica **57**, 536–564 (1972)
12. Fortuin, C.M., Kasteleyn, P.W., Ginibre, J.: Correlation inequalities on some partially ordered sets. Commun. Math. Phys. **22**(2), 89–103 (1971)
13. Grimmett, G.R.: The Random-Cluster Model. Springer, Berlin (2006)
14. Kahn, J., Neiman, M.: Negative correlation and log-concavity. Random Struct. Alg. **37**, 367–388 (2010)
15. Markström, K.: Closure properties and negatively associated measures violating the van den Berg–Kesten inequality. Elect. Comm. Probab. **15**, 449–456 (2009)
16. Newman, C.: Normal fluctuations and the FKG inequalities. Commun. Math. Phys. **74**, 119–128 (1980)
17. Newman, C.: A general central limit theorem for FKG systems. Commun. Math. Phys. **91**, 75–80 (1983)
18. Newman, C.: Disordered Ising systems and random cluster representations. In: Grimmett, G. (ed.) Probability and Phase Transition, pp. 247–260. Kluwer, Dordrecht (1994)

19. Machta, J., Newman, C.M., Stein, D.L.: The percolation signature of the spin glass transition. J. Stat. Phys. **130**, 113–128 (2008)
20. Pemantle, R.: Towards a theory of negative dependence. J. Math. Phys. **41**, 1371–1390 (2000)
21. Reimer, D.: Proof of the Van den Berg-Kesten conjecture. Comb. Probab. Comput. **9**, 27–32 (2000)

The Rumor Percolation Model and Its Variations

Valdivino V. Junior[1], Fábio P. Machado[2], and Krishnamurthi Ravishankar[3,4](✉)

[1] Federal University of Goias, Campus Samambaia, Goiânia, GO, Brazil
vvjunior@ufg.br
[2] Institute of Mathematics and Statistics, University of São Paulo, Rua do Matão 1010, São Paulo, SP 05508-090, Brazil
fmachado@ime.usp.br
[3] NYU-ECNU Institute of Mathematical Sciences at NYU Shanghai, 3663 Zhongshan Road North, Shanghai 200062, China
[4] NYU-ECNU Institute of Mathematical Sciences at NYU Shanghai, 1555 Century Ave, Pudong, Shanghai, China
kr26@nyu.edu

To Chuck Newman, Friend, Colleague, and Mentor for his 70th Birthday

Abstract. The study of rumor models from a percolation theory point of view has gained a few adepts in the last few years. The persistence of a rumor, which may consistently spread out throughout a population can be associated to the existence of a giant component containing the origin of a graph. That is one of the main interests in percolation theory. In this paper we present a quick review of recent results on rumor models of this type.

Keywords: Epidemic model · Galton–Watson trees · Rumor model · Spherically symmetric trees

1 Introduction and Basic Definitions

We are interested in a long-range percolation model on infinite graphs which we call the *Rumor Percolation Model*. Such models have recently been studied by a few authors in a series of papers. The dynamics of the model describes the spreading of a rumor on a graph in the following way. We assign an independent random *radius of influence* R_v to each vertex v of an infinite, locally finite, connected graph $\mathcal{G}$. Then we define a chain reaction on $\mathcal{G}$ according to the following simple rules: (1) at time zero, only the root (a fixed vertex of $\mathcal{G}$) hears the rumor, (2) at time $n \geq 1$, a new vertex hears the rumor if it is a distance

Research supported by CNPq (303699/2018-3), FAPESP (09/52379-8), PNPD-Capes 536114 and Simons Foundation Collaboration grant 281207.

V. Sidoravicius (Ed.): *Sojourns in Probability Theory and Statistical Physics - II*, PROMS 299, pp. 208–227, 2019.
https://doi.org/10.1007/978-981-15-0298-9_9

at most R_v of some vertex v that previously heard the rumor. We point out that similar models, are of interest in Computer Science, in particular in the area of distributed networks. One of the problems of interest is the broadcasting problem where one node has some information which it wants to pass on to other nodes. Questions of optimal algorithm for achieving this goal are of interest. This question was considered for the case where the nodes are uniformly randomly distributed on an interval $[0, L]$ and the nodes had a transmission radius of one. In [13] asymptotically (in L) optimal algorithm was obtained.

Definition 1. *The Rumor Percolation Model on $\mathcal{G}$.*
Let $\mathcal{G} = (\mathcal{V}, \mathcal{E})$ be an infinite, locally finite, connected graph and let $\{R_v\}_{\{v \in \mathcal{V}\}}$ be a set of independent and identically distributed random variables. Furthermore, for each $u \in \mathcal{V}$, we define the random sets

$$B_u = \{v \in \mathcal{V} : d(u, v) \leq R_u\}. \tag{1}$$

or

$$B_u = \{v \in \mathcal{V}, u \leq v : d(u, v) \leq R_u\}. \tag{2}$$

With these sets we define the Rumor Percolation Model on $\mathcal{G}$, the non-decreasing sequence of random sets $I_0 \subset I_1 \subset \cdots$ defined as $I_0 = \{\mathcal{O}\}$ and inductively $I_{n+1} = \bigcup_{u \in I_n} B_u$ for all $n \geq 0$.

Definition 2. *The Rumor Percolation Model survival.*
Consider $I = \bigcup_{n \geq 0} I_n$ be the connected component of the origin of $\mathcal{G}$. Under the rumor process interpretation, I is the set of vertices which heard the rumor. We say that the process survives *(*dies out*) if $|I| = \infty$ ($|I| < \infty$), referring to the surviving event as V.*

In Sect. 2 we review the paper of Athreya et al. [1]. Instead of considering a graph structure they consider a homogeneous Poisson point process on $\mathbb{R}^d$ and $\mathbb{R}^{d^+}$ with $\{R_v\}$, the *box of influence*, starting from every point v of the point process in the sense of (2). They work with the concept of the coverage of a set $(t, \infty)^d$ for some $t > 0$, the *eventual coverage*. In Sect. 3 we review the paper of Lebenstayn and Rodriguez [11] where authors consider the *Disk Percolation Model*. While the set of *radius of influence*, $\{R_v\}_{\{v \in \mathcal{V}\}}$, has a geometric distribution, the graph $\mathcal{G}$ is quite general. In their version the *radius of influence* of a vertex $v \in \mathcal{G}$ goes in every possible direction as in (1). In Sect. 4 we review the papers of Junior et al. [8] and Gallo et al. [4]. They work with a processes that they made known as Fireworks on $\mathbb{N}$ (direct and reverse). They studied a homogeneous version, where there is one informant per vertex and the *radius of influence* are independent and have identical distribution, and a heterogeneous version, where one of these conditions fail. In their models the *radii of influence* behave as specified in (2). In Sect. 5 the papers of Junior et al. [8,9] and [10] are briefly reviewed. They work with the Cone Percolation model, a Fireworks model on a tree (homogeneous, spherically symmetric, periodic or Galton Watson). In all these models the *radii of influence* behave as specified in (2). In Sect. 6 we review the paper of Bertacchi and Zucca [2]. They consider a type of random environment in the sense that the number of informants in each vertex are random.

2 Random Sets on $\mathbb{R}^d$ and $\mathbb{R}^{d^+}$

The theory of coverage processes was introduced by Hall [7] in 1988. He developed a class of stochastic processes intended to be used as a model for *binary images*, that is, images which partition $\mathbb{R}^d$ into two regions, $\mathcal{C}$ and its complement, representing the "black" and "white" parts of an image. In its basic version the process consists of a point process $P = \{\xi_1, \xi_2, \dots\}$ and a collection of random sets $\{S_1, S_2, \dots\}$. The "black" region $\mathcal{C}$ is then defined to be $\mathcal{C} = \cup_{i=1}^{\infty}(\xi_i + S_i)$. Hall [7] developed probabilistic results on geometrical properties of $\mathcal{C}$, such as the size-distribution of its connected subsets. In that work the main assumptions needed to obtain explicit results is that P is an homogeneous Poisson process with parameter λ and the S_i are independent copies of a random closed set. This version is known as the Poisson Boolean model.

Athreya et al. [1] considered two different models, both related to rumor percolation. For the first model, arising in genome analysis, they consider $\{X_i\}_{i\in\mathbb{N}}$ be a $\{0,1\}$-valued time-homogeneous Markov chain and $\{\rho_i\}_{i\in\mathbb{N}}$ an independent and identically distributed sequence of random variables assuming values on $\mathbb{N}$, independent of the Markov chain. Let $S_i = [i, i+\rho_i]$ whenever $X_i = 1$ ($\emptyset$ otherwise) and $C = \cup_{i=1}^{\infty} S_i$.

Definition 3. *We say that* $\mathbb{N}$ *is eventually covered by* C *(or* C *eventually covers* $\mathbb{N}$*) if there exists a* $t \geq 1$ *such that* $[t, \infty) \subseteq C$.

Theorem 1 (Athreya et al. [1]). *Let* $p_{ij} = \mathbb{P}(X_{n+1} = j | X_n = i)$. *Assume that* $0 < p_{00}, p_{10} < 1$,

(i) If

$$l = \liminf_{j\to\infty} j\mathbb{P}(\rho_1 > j) > 1,$$

then

$$\mathbb{P}(C \textit{ eventually covers } \mathbb{N}) = 1$$

whenever

$$\frac{p_{01}}{p_{10} + p_{01}} > \frac{1}{l}.$$

(ii) If

$$L = \limsup_{j\to\infty} j\mathbb{P}(\rho_1 > j) < \infty,$$

then

$$\mathbb{P}(C \textit{ eventually covers } \mathbb{N}) = 0$$

whenever

$$\frac{p_{01}}{p_{10} + p_{01}} < \frac{1}{L}.$$

Their second model aims to complement known results on complete coverage in stochastic geometry. For $B(0,\rho)$ the closed d-dimensional ball of radius ρ centered at the origin, some important previous results for the random covered region $\cup_{i=1}^{\infty}(\xi_i + B(0,\rho_i))$ are presented in the next two theorems.

Theorem 2 (Hall [7]). *For the Poisson Bolean model on* $\mathbb{R}^d$ *the space is fully covered by* $\cup_{i=1}^{\infty}(\xi_i + B(0,\rho_i))$ *almost surely if and only if* $\mathbf{E}(\rho^d) = \infty$.

If instead of a Poisson point process one considers an arbitrary ergodic process, there is the following result

Theorem 3 (Meester and Roy [12]). *For the Bolean model on* $\mathbb{R}^d$ *the space is fully covered by* $\cup_{i=1}^{\infty}(\xi_i + B(0,\rho_i))$ *almost surely if* $\mathbf{E}(\rho^d) = \infty$.

Athreya et al. [1] take $\mathbb{R}^d_+$ and the random covered region

$$C = \cup_{\{i:\xi\in\mathbb{R}^d_+\}}(\xi_i + [0,\rho_i]^d).$$

Guided by the fact that C will never completely cover $\mathbb{R}^d_+$ because, for any $\epsilon > 0$, $[0,\epsilon]^d$ will not be covered by C with positive probability, they work with the notion of *eventual coverage* for the orthant $\mathbb{R}^d_+$.

Definition 4. *We say that* $\mathbb{R}^d_+$ *is eventually covered by the Poisson Boolean model if there exists a* $t \in (0,\infty)$ *such that* $[t,\infty)^d \subseteq C$.

With this notion Athreya et al. [1] are able to present the following result, considering a Poisson Bolean model on $\mathbb{R}^d_+$. They show that eventual coverage depends on the growth rate of the distribution function of ρ (even when $\mathbf{E}(\rho) = \infty$) as well as on whether $d = 1$ or $d \geq 2$.

Theorem 4 (Athreya et al. [1]). *Assume* $d = 1$.

(i) If

$$0 < l := \liminf_{x\to\infty} x\mathbb{P}(\rho > x) < \infty,$$

then there exists a λ_0 *such that* $0 < \lambda_0 \leq 1/l < \infty$ *and*

$$\mathbb{P}_\lambda(\mathbb{R}_+ \text{ is eventually covered by } C) = \begin{cases} 0 \text{ if } \lambda < \lambda_0, \\ 1 \text{ if } \lambda > \lambda_0; \end{cases}$$

(ii) If

$$0 < L := \limsup_{x\to\infty} x\mathbb{P}(\rho > x) < \infty,$$

then there exists a λ_1 *such that* $0 < 1/L \leq \lambda_1 < \infty$ *and*

$$\mathbb{P}_\lambda(\mathbb{R}_+ \text{ is eventually covered by } C) = \begin{cases} 0 \text{ if } \lambda < \lambda_1, \\ 1 \text{ if } \lambda > \lambda_1; \end{cases}$$

(iii) If

$$\lim_{x\to\infty} x\mathbb{P}(\rho > x) = \infty,$$

then for all $\lambda > 0, \mathbb{R}_+$ *is eventually covered by* C *(*$\mathbb{P}_\lambda$*−a.s.);*

(iv) If

$$\lim_{x\to\infty} x\mathbb{P}(\rho > x) = 0,$$

then for any $\lambda > 0, \mathbb{R}_+$ *is eventually covered by* C *(*$\mathbb{P}_\lambda$*−a.s.).*

Theorem 5 (Athreya et al. [1]**).** *Let* $d \geq 2$*. For all* $\lambda > 0$*,*

(i) If

$$\liminf_{x\to\infty} x\mathbb{P}(\rho > x) > 0,$$

then

$$\mathbb{P}_\lambda(\mathbb{R}^d_+ \textit{ is eventually covered by } C) = 1;$$

(ii) If

$$\lim_{x\to\infty} x\mathbb{P}(\rho > x) = 0,$$

then

$$\mathbb{P}_\lambda(\mathbb{R}^d_+ \textit{ is eventually covered by } C) = 0.$$

It is interesting to observe that while $\mathbf{E}(\rho^d) = \infty$ guarantees complete coverage of $\mathbb{R}^d$ by C, it is not sufficient to guarantee *eventual coverage* for $\mathbb{R}^d_+$. This is due to the fact that a boundary effect is present in the orthant $\mathbb{R}^d_+$ but absent in the whole space $\mathbb{R}^d$.

3 Disk Percolation

Lebensztayn and Rodriguez studied a long-range percolation model on infinite graphs, the *Disk Percolation Model.* They assign a random *radius of influence* R_v to each vertex v of an infinite, locally finite, connected graph G, so that all the assigned radii are independent and identically distributed random variables with geometric distribution with parameter$(1-p)$, which means, satisfying

$$\mathbb{P}(R = k) = (1-p)p^k, k = 0, 1, 2, \ldots$$

Then they defined a growing process on G according to the following rules: (1) at time zero, only the root (a fixed vertex of G) is declared infected, (2) at time $n \geq 1$, a new vertex is infected if it is at graph distance at most R_v of some vertex v previously infected, and (3) infected vertices remain infected forever. They investigated the critical value $p_c(G)$ above which this process spreads indefinitely through the graph with positive probability.

They worked in a few settings including locally finite graphs in the sense that

$$\Delta = \sup_{v\in\mathcal{G}}\{d(v)\} < \infty$$

where $d(v)$ is the number of neighbors (or *degree*) of a vertex v. Let I be the set of vertices which are eventually infected.

An interesting question is whether such a model presents *phase transition* in the sense that for $p_c(\mathcal{G}) := \inf\{p : \mathbb{P}(|I| = \infty\})$ we have that

$$0 < p_c(\mathcal{G}) < 1.$$

They provided an answer which relies on a comparison between the *Disk Percolation Model* and the independent site percolation model. To understand this, consider $p_c^{site}(\mathcal{G})$ the critical probability for the independent site percolation model on $\mathcal{G}$.

Theorem 6 (Lebenstayn and Rodriguez [11]**).** *Let $\mathcal{G}$ be of bounded degree ($\Delta < \infty$) and be such that $p_c^{site}(\mathcal{G}) < 1$. Then*

$$0 < p_c(\mathcal{G}) < 1.$$

The proof they presented relies on the following two propositions, the first one is a comparison which gives an upper bound to $p_c(\mathcal{G})$.

Proposition 1 (Lebenstayn and Rodriguez [11]**)**

$$p_c(\mathcal{G}) \leq p_c^{site}(\mathcal{G})$$

while the second one gives a lower bound for the case when $\mathcal{G}$ is bounded.

Proposition 2 (Lebenstayn and Rodriguez [11]**).** *Suppose that $\mathcal{G}$ is a graph of bounded degree. Then*

$$p_c(\mathcal{G}) \geq -1 + \left(1 + \frac{1}{\Delta - 1}\right)^{1/2}.$$

3.1 Disk Percolation on Trees

Consider a tree $\mathbb{T}$ (a connected graph with no cycles) and its set of vertices $\mathcal{V}(\mathbb{T})$. We say that a tree, $\mathbb{T}_d$, is *homogeneous*, if each one of its vertices has degree (number of neighbours) $d + 1$.

Theorem 7 (Lebenstayn and Rodriguez [11]**).** *For any $d \geq 2$*

$$-1 + \left(1 - \frac{1}{d}\right)^{1/2} \leq p_c(\mathbb{T}_d) \leq 1 - \left(1 - \frac{1}{d}\right)^{1/2}.$$

Corollary 1 (Lebenstayn and Rodriguez [11]**).** *For any $d \geq 2$*

$$p_c(\mathbb{T}_d) = 1/(2d) + \mathcal{O}(1/d^2) \text{ as } d \to \infty.$$

Single out one vertex from $\mathcal{V}(\mathbb{T})$ and call this $\mathcal{O}$, the origin of $\mathcal{V}(\mathbb{T})$. For each two vertices $u, v \in \mathcal{V}(\mathbb{T})$, consider that $u \leq v$ if u belongs to the path connecting $\mathcal{O}$ to v.

For a tree $\mathbb{T}$ and $n \geq 1$ we define

$$\mathbb{T}^u := \{v \in \mathcal{V} : u \leq v\},$$
$$\mathbb{T}^u_n := \{v \in \mathbb{T}^u : d(v, \mathcal{O}) \leq d(u, \mathcal{O}) + n\}$$

and

$$M_n(u) := |\partial \mathbb{T}^u_n| := |\{v \in \mathbb{T}^u : d(v, \mathcal{O}) = d(u, \mathcal{O}) + n\}|.$$

Definition 5. *Let us define for a tree* $\mathbb{T}$

$$\dim\inf \partial \mathbb{T} := \lim_{n\to\infty} \min_{v \in \mathcal{V}} \frac{1}{n} \ln M_n(v).$$

Observe that

$$\dim\inf \partial \mathbb{T}_d = \ln d.$$

Definition 6. *We say that a tree,* $\mathbb{T}_S$*, is spherically symmetric, if any pair of vertices at the same distance from the origin, have the same degree.*

Theorem 8 (Lebenstayn and Rodriguez [11]). *For any spherically symmetric tree* $\mathbb{T}_S$

$$p_c(\mathbb{T}_S) \leq 1 - \left(1 - e^{-\dim\inf \partial \mathbb{T}_S}\right)^{1/2}$$

4 Fireworks on $\mathbb{N}$

The Fireworks processes are another interesting version of the *Rumor Percolation Model.* Junior et al. [8] and Gallo et al. [4] recently studied discrete time stochastic systems on $\mathbb{N}$ modeling processes of rumor spreading. In their models the involved individuals can either have an active role, working as spreaders and transmiting the information within a random distance to their right, or a passive role, hearing the information from spreaders within a random distance to their left. The appetite in spreading or hearing the rumor is represented by a set of random variables whose distributions may depend on the individuals positions on $\mathbb{N}$. Their main goal is to understand - based on the distribution of those random variables - whether the probability of having an infinite set of individuals knowing the rumor is positive or not.

Junior et al. [8] manage to write the survival event as a limit of an increasing sequence of events whose probability can be bounded by a nice use of FKG inequality. The use of a non-standard version of Borel–Cantelli lemma helped in the task of finding conditions for the processes to die out. Gallo et al. [4] based the proofs of their results on a clever relationship between the rumor processes and a specific discrete time renewal process. With this technique they were able to obtain more precise results for homogeneous versions of the processes.

Consider $\{u_i\}_{i\in\mathbb{N}}$ a set of vertices of $\mathbb{N}$ such that $0 < u_1 < u_2 < \cdots$ and a set of independent random variables $\{R_i\}_{i\in\mathbb{N}}$ assuming values in $\mathbb{Z}_+$.

4.1 Fireworks

At time 0, information travels a distance R_0 towards the right side of the origin, in such a way that all vertices $u_i \leq R_0$ get informed. Let V be the event that the number of sites which are eventually informed is infinite. In general, at every discrete time t a vertex u_j informed at time $t-1$ passes the information on (within R_j, its *radius of influence*) and they do this just once, informing the vertices u_i (only those vertices which have not been informed before) $u_j < u_i \leq u_j + R_j$. Observe that, except for the set of vertices $\{u_i\}$, all other vertices are nonactionable, meaning that their *radius of influence* equals 0 almost surely.

Homogeneous Fireworks. Consider all the $R_i \sim R$ (having the same distribution) and $u_i = i$ for all i.

Theorem 9 (Junior et al. [8]). *Consider in the Homogeneous Fireworks Process*

$$a_n = \prod_{i=0}^{n} \mathbb{P}(R \leq i).$$

Then

$$\sum_{n=1}^{\infty} a_n = \infty \text{ if and only if } \mathbb{P}[V] = 0.$$

Theorem 10 (Gallo et al. [4]). *For the Homogeneous Fireworks Process,*

$$\mathbb{P}(V) = \left[1 + \sum_{j=1}^{\infty} \prod_{i=0}^{j-1} \mathbb{P}(R \leq i)\right]^{-1}.$$

Observe that the result presented in Theorem 9 is nicely generalized in Theorem 10.

Example 1. Consider the *Homogeneous Fireworks Process* such that

$$\mathbb{P}(R = k) = \frac{2}{(k+2)(k+3)} \text{ for } k \in \mathbb{N}^*.$$

Then $\mathbb{P}[V] = \frac{1}{2}$.

Corollary 2 (Junior et al. [8]). *For the Homogeneous Fireworks Process, consider*

$$L = \lim_{n \to \infty} n\mathbb{P}(R \geq n).$$

We have that

(i) If $L > 1$ then $\mathbb{P}[V] > 0$.
(ii) If $L < 1$ then $\mathbb{P}[V] = 0$.

(iii) *If* $L = 1$ *and there exists* N *such that for all* $n \geq N$

$$\mathbb{P}(R \geq n) \leq \frac{1}{n-1}, \text{ then } \mathbb{P}[V] = 0.$$

Let M be the final number of spreaders.

Theorem 11 (Gallo et al. [4]). *If* $\mathbf{E}(R) < \infty$ *then the random variable* M *has finite expectation. Besides,* M *has exponential tail distribution when* $\mathbb{P}(R \leq n)$ *increases exponentially fast to* 1.

Under more specific assumptions, it is possible to obtain more precise information on the tail distribution. Items *(i)* and *(iii)* of next proposition follow from Proposition B.2 of Gallo et al. [5], item *(ii)* is due to Remark 5 from Bressaud et al. [3] and item *(iv)* follows from Theorem 1.1 of Garsia and Lamperti [6].

Proposition 3 (Gallo et al. [4]). *We have the following explicit bounds for the tail distributions.*

(i) *If* $\mathbb{P}(R > k) \leq C_r r^k$, $k \geq 1$, *for some* $r \in (0,1)$ *and a constant* $C_r \in (0, \log \frac{1}{r})$ *then*

$$\mathbb{P}(M \geq k) \leq \frac{1}{C_r}(e^{C_r} r)^k.$$

(ii) *If* $\mathbb{P}(R > k) \sim (\log k)^{\beta} k^{-\alpha}$, $\beta \in \mathbb{R}$, $\alpha > 1$, *then there exists* $C > 0$ *such that, for large* k, *we have* $\mathbb{P}(M \geq k) \leq C(\log k)^{\beta} k^{-\alpha}$.

(iii) *If* $\mathbb{P}(R > k) = \frac{r}{k}$, $k \geq 1$ *where* $r \in (0,1)$, *there exists* $C > 0$ *such that, for large* k, *we have*

$$\mathbb{P}(M \geq k) \leq C \frac{(\ln k)^{3+r}}{(k)^{2-(1+r)^2}}.$$

(iv) *If* $\mathbb{P}(R > k) \sim ((k+1)/(k+2))^{\alpha}$, $\alpha \in (1/2, 1)$, *then there exists* $C = C(\alpha) > 0$ *such that, for large* k, *we have*

$$\mathbb{P}(M \geq k) \leq \frac{C}{k^{1-\alpha}}.$$

Heterogeneous Fireworks

Remark 1. Consider the *Heterogeneous Fireworks Process*. One can get a sufficient condition for $\mathbb{P}[V] = 0$ ($\mathbb{P}[V] > 0$) by a coupling argument. Consider $\mathbb{P}(R_i \geq k) \leq \mathbf{P}(R \geq k)$ ($\mathbb{P}(R_i \geq k) \geq \mathbf{P}(R \geq k)$) for some random variable R whose distribution $\mathbf{P}$ satisfies $\lim_{n\to\infty} n\mathbf{P}(R \geq n) < 1$ ($\lim_{n\to\infty} n\mathbf{P}(R \geq n) > 1$). Finally use item (ii) (item (i)) of Corollary 2.

Theorem 12 (Junior et al. [8]). *Consider a Heterogeneous Fireworks Process for which actionable vertices are at integer positions* $u_0 = 0 < u_1 < u_2 < \dots$ *such that* $u_{n+1} - u_n \leq m$, *for* $m \geq 1$. *Besides, let us assume* $\mathbb{P}(R_n < m) \in (0,1)$ *for all* n. *Then*

(i) *If* $\sum_{n=0}^{\infty} [\mathbb{P}(R_n < tm)]^t < \infty$ *for some* $t \geq 1$ *then* $\mathbb{P}[V] > 0$.

(ii) *If for some random variable* R, *with distribution* $\mathbf{P}$, *the following conditions hold*
- $\mathbf{P}(R \geq k) - \mathbb{P}(R_n \geq k) \leq b_k$ *for all* $k \geq 0$ *and all* $n \geq 0$,
- $\lim_{n\to\infty} n[\mathbf{P}(R \geq n) - b_n] > m$,
- $\lim_{n\to\infty} b_n = 0$,

then $\mathbb{P}[V] > 0$.

(iii) $\mathbb{P}(V) \geq \prod_{j=0}^{\infty} \left[1 - \prod_{i=0}^{j} \mathbb{P}(R_{j-i} < (i+1)m)\right]$.

4.2 Reverse Fireworks

At time 0, only the origin has the information. At time 1, individuals placed at vertices u_i such that $u_i \leq R_i$ get the information from the origin. At time $t \in \mathbb{N}$ the set of vertices u_j which can find an informed individual at time $t-1$ within a distance R_j to its left, get the information. Let us call this set A_t. If for some t, A_t is empty the process stops. If the process never stops we say it survives.

Let S be the event "the reverse process survives". Besides, we denote by Z the final number of spreaders.

Homogeneous Reverse Fireworks. Consider all the R_i having the same distribution and $u_i = i$ for all i.

Theorem 13 (Junior et al. [8]). *Consider the Homogeneous Reverse Fireworks Process. We have that*

(i) *If* $\mathbb{E}(R) = \infty$ *then* $\mathbb{P}(S) = 1$.
(ii) *If* $\mathbb{E}(R) < \infty$ *then* $\mathbb{P}(S) = 0$.

Theorem 14 (Gallo et al. [4]). *Consider the Homogeneous Reverse Fireworks Process. If* $\mathbb{E}(R) < \infty$ *then* $Z \sim \mathcal{G}\left(\prod_{k=0}^{\infty} \mathbb{P}(R \leq k)\right)$ *in the sense that for* $p = \prod_{k=0}^{\infty} \mathbb{P}(R \leq k)$ *we have*

$$\mathbb{P}(Z = k) = p(1-p)^k \text{ for all } k.$$

For any $n \geq 1$, let $Z(n)$ be the number of spreaders in $\{1, \dots, n\}$. We will now state limit theorems for the proportion of spreaders within $\{1, \dots, n\}$, $Z(n)/n$, when n tends to ∞.

Let

$$\mu := 1 + \sum_{j=1}^{\infty} \prod_{i=0}^{j-1} \mathbb{P}(R \leq i) \quad \text{and}$$

$$\sigma^2 := \sum_{k=1}^{\infty} k^2 \mathbb{P}(R > k-1) \prod_{i=0}^{k-2} \mathbb{P}(R \leq i) - \mu^2.$$

Notice that $\mu < \infty$ implies that $\prod_{k=0}^{\infty} \mathbb{P}(R \leq k) = 0$ (this implies $\mathbf{E}(R) = \infty$) .

Theorem 15 (Gallo et al. [4]). *If $\mu < \infty$ then*

$$\frac{Z(n)}{n} \xrightarrow{a.s.} \mu^{-1},$$

and thus, with probability one, μ^{-1} is the final proportion of spreaders. Moreover, if $\sigma^2 \in (0, \infty)$, then

$$\sqrt{n}\left(\frac{Z(n)}{n} - \mu^{-1}\right) \xrightarrow{\mathcal{D}} \mathcal{N}\left(0, \frac{\sigma^2}{\mu^3}\right).$$

Otherwise, $Z(n)/n \to 0$.

In particular, observe that if the $\mathbb{P}(R \leq k)$'s satisfy at the same time

$$\prod_{k=0}^{\infty} \mathbb{P}(R \leq k) = 0$$

and $\mu = \infty$ (for instance, if they are as in items (iii) and (iv) of Proposition 3), then the information reaches infinitely many individuals, but the final proportion of informed individuals is zero.

Heterogeneous Reverse Fireworks

Theorem 16 (Junior et al. [8]). *Consider the Heterogeneous Reverse Fireworks Process. It holds that*

(i) $\sum_{k=1}^{\infty} \mathbb{P}(R_{n+k} \geq k) = \infty$ *for all n if and only if* $\mathbb{P}(S) = 1$.
(ii) If $\sum_{n=1}^{\infty} \prod_{k=1}^{\infty} \mathbb{P}(R_{n+k} < k) < \infty$ *then* $\mathbb{P}(S) > 0$.

Remark 2. By a coupling argument and Theorem 13 one can see that if there is a random variable R, whose distribution is $\mathbf{P}$, with $\mathbf{E}[R] < \infty$ ($\mathbf{E}[R] = \infty$), such that $\mathbb{P}(R_n \geq k) \leq \mathbf{P}(R \geq k)$ ($\mathbb{P}(R_n \geq k) \geq \mathbf{P}(R \geq k)$) for all k then $\mathbb{P}(S) = 0$ ($\mathbb{P}(S) = 1$).

Example 2. It is possible to have in the *Heterogeneous Fireworks Process* the expectation of the *radius of influence* infinite for all vertices together and the process dies out almost surely.

Let $\{b_n\}_{n\in\mathbb{N}}$ be a non-increasing sequence convergent to 0 and such that $b_0 < 1$.

(i) $\mathbb{P}(R_n = 0) = 1 - b_n$ and $\mathbb{P}(R_n = k) = b_{n+k-1} - b_{n+k}$ for $k \geq 1$.
(ii) $\sum_{n=0}^{\infty} b_n = \infty$.
(iii) $\lim_{n\to\infty} n b_n = 0$.

Observe that $\mathbb{E}(R_n) = \infty$ for all n from *(ii)*. Besides $\mathbb{P}[V] = 0$ from *(iii)*, because For

$$V_n = \{\text{The individual at vertex } u_n \text{ gets the information}\},$$

$$\mathbb{P}(V_n) \leq \sum_{k=0}^{n-1} \mathbb{P}(R_k \geq n - k) = \sum_{k=0}^{n-1} b_{n-1} = (n-1)b_n. \tag{3}$$

and the fact that $V = \lim_{n\to\infty} V_n$.

Example 3. It is possible to have in the *Heterogeneous Fireworks Process* the expectation of the *radius of influence* finite for all vertices and the process survives with positive probability. Assume that $\sum_{n=0}^{\infty} b_n < \infty$, while

(i) $\mathbb{P}(R_n = 0) = b_n$
(ii) $\mathbb{P}(R_n = 1) = 1 - b_n$

Then $\mathbb{E}(R_n) < 1$ for all n and $\mathbb{P}(V) > 0$ by item (i) of Theorem 12 with $m = t = 1$.

Example 4. Next we present an example where $\mathbb{P}[S] = 1$ for a *Heterogeneous Reverse Fireworks Process* while $\mathbb{P}[V] = 0$ for a *Heterogeneous Fireworks Process.* For this aim consider

(i) $\mathbb{P}(R_n = 0) = 1 - b_n$ and $\mathbb{P}(R_n = n) = b_n$.
(ii) $\sum_{n=0}^{\infty} b_n = \infty$.
(iii) $\lim_{n\to\infty} nb_n = 0$.

Observe that even though $\lim_{n\to\infty} \mathbf{E}[R_n] = 0$ and $\lim_{n\to\infty} \mathbb{P}(R_n = 0) = 1$, from Theorem 16 and *(ii)* it is true for the *Heterogeneous Reverse Fireworks Process* that $\mathbb{P}(S) = 1$. In the opposite direction, by (3) and *(iii)* one have that $\mathbb{P}[V] = 0$ for the *Heterogeneous Fireworks Process.*

5 Cone Percolation on $\mathbb{T}_d$

Junior et al. [9] consider a process which allows us to associate the dynamic activation on the set of vertices to a discrete rumor process. Individuals become spreaders as soon as they hear the rumor. Next time, they propagate the rumor within their *radius of influence* and immediately become stiflers. Junior et al. [9] establish whether the process has positive probability of involving an infinite set of individuals. Besides, they present sharp lower and upper bounds for the probability of that event, depending on the general distribution of the random variables that define the *radius of influence* of each individual. Their proofs are based on comparisons with branching processes.

Pick a $v \in \mathcal{V}(\mathbb{T}_d)$ such that $d(\mathcal{O}, v) = 1$ and consider $\mathbb{T}_d^+ = \mathbb{T}_d \backslash \mathbb{T}_d^+(v)$. Consider $\mathbb{P}_+$ and $\mathbb{P}$ the probability measures associated to the processes on $\mathbb{T}_d^+$ and $\mathbb{T}_d$ (we do not mention the random variable R unless absolutely necessary). By a coupling argument one can see that for a fixed distribution of R

$$\mathbb{P}_+[V] \leq \mathbb{P}[V].$$

Furthermore, by the definition of $\mathbb{T}_d^+$ and its relation with $\mathbb{T}_d$ we have that for a fixed distribution of R

$$\mathbb{P}_+[V] = 0 \text{ if and only if } \mathbb{P}[V] = 0.$$

Let $p_0 = \mathbb{P}(R = 0)$.

Theorem 17 (Junior et al. [9]). *Consider the Cone Percolation Model on $\mathbb{T}_d^+$ with radius of influence R.*

(i) If $(1-p_0)d > 1$, *then* $\mathbb{P}_+[V] > 0$,
(ii) If $(1-p_0)d \leq 1$ *and* $\mathbf{E}(d^R) > 1 + p_0$, *then* $\mathbb{P}_+[V] > 0$,
(iii) If $\mathbf{E}(d^R) \leq 2 - \frac{1}{d}$, *then* $\mathbb{P}_+[V] = 0$.

Theorem 18 (Junior et al. [10]). *Consider a Cone Percolation Model on* $\mathbb{T}_d$. *Then for* $\mathbb{E}(d^R) < 2 - \frac{1}{d}$, *we have*

$$\frac{d + \mathbb{E}\left(d^R\right) - p_0}{d[1 - \mathbb{E}\left(d^R\right) + p_0]} \leq \mathbb{E}(|I|) \leq \frac{\mathbb{E}\left(d^R\right) + d - 2}{2d - 1 - d\mathbb{E}\left(d^R\right)}.$$

Example 5 (Junior et al. [10]). Consider $R \sim \mathcal{G}(1-p)$, a *radius of influence* satisfying

$$\mathbb{P}(R = k) = (1-p)p^k, k = 0, 1, 2, \ldots$$

and assume also $pd < \frac{1}{2}$. So we have

$$\frac{1 - dp + p - p^2}{1 - 2dp + dp^2} \leq \mathbb{E}(|I|) \leq \frac{1 - dp - p}{1 - 2dp}.$$

That gives us a fairly sharp bound even when we pick p and d such that pd is very close to $\frac{1}{2}$ as, for example, $p = 10^{-6}$ and $d = 499,000$. For these parameters we get $250.438 \leq \mathbb{E}(|I|) \leq 250.501$.

Let ρ and ψ be, respectively, the smallest non-negative roots of the equations

$$\mathbf{E}(\rho^{d^R}) + (1-\rho)p_0 = \rho,$$
$$\mathbf{E}(\psi^{\frac{d}{d-1}(d^R - 1)}) = \psi.$$

Theorem 19 (Junior et al. [9]). *Consider the Cone Percolation Model on* $\mathbb{T}_d^+$. *Then*

$$1 - \rho \leq \mathbb{P}_+(V) \leq 1 - \psi.$$

Theorem 20 (Junior et al. [9]). *For the Cone Percolation Model on* $\mathbb{T}_d$ *with radius of influence R, it holds that*

$$1 - \left(1 - \rho^{\frac{d+1}{d}}\right) p_0 - \mathbf{E}\left(\rho^{\frac{(d+1)}{d} d^R}\right) \leq \mathbb{P}[V] \leq 1 - \mathbf{E}\left(\psi^{\frac{(d+1)}{d-1}(d^R - 1)}\right).$$

Consider $d = 2$ and R following a Binomial distribution with parameters 4 and $\frac{1}{2}$ ($R \sim \mathcal{B}(4, \frac{1}{2})$). Therefore ρ and ψ are, respectively, solutions of

$$x^{16} + 4x^8 + 6x^4 + 4x^2 - 16x + 1 = 0,$$
$$x^{30} + 4x^{14} + 6x^6 + 4x^2 - 16x + 1 = 0.$$

So $\rho = 0.0635146$ and $\psi = 0.06350850$, which implies that

$$0.937435919 \leq \mathbb{P}[V] \leq 0.937435962.$$

Theorem 21 (Junior et al. [9]). *The Heterogeneous Cone Percolation Process in $\mathbb{T}_d^+$ has a giant component with positive probability if for some fixed n,*

$$\liminf_{j\to\infty} d^n \prod_{k=0}^{n-1}[1-\prod_{i=0}^{k}\mathbb{P}_+[R_{jn+i}<k+1-i]]>1. \tag{4}$$

A consequence of Theorem 5.4 from Bertacchi and Zucca [2] is the following result

Corollary 3. *Consider a Homogeneous Reverse Fireworks Process on $\mathbb{T}_d$. Then*

$$\mathbb{P}(S)=1 \textit{ if and only if } \sum_{n=1}^{\infty} d^n\mathbb{P}(R\geq n)=\infty.$$

$$\mathbb{P}(S)=0 \textit{ if and only if } \sum_{n=1}^{\infty} d^n\mathbb{P}(R\geq n)\prod_{j=1}^{n-1}[1-\mathbb{P}(R\geq j)]\leq 1$$

Theorem 22 (Junior et al. [10]). *For a Cone Percolation Model in $\mathbb{T}_S$ and R, the radius of influence, $\mathbb{P}(V)>0$ if*

$$\lim_{n\to\infty}\sqrt[n]{\rho_n}>e^{-\dim\inf\partial\mathbb{T}_S}$$

where

$$\rho_n:=\prod_{k=0}^{n-1}[1-\prod_{i=0}^{k}\mathbb{P}(R<i+1)].$$

Corollary 4 (Junior et al. [10]). *For a Cone Percolation Model in $\mathbb{T}_S$ and R, a radius of influence satisfying $\mathbb{P}(R\leq k)=1$ for some $k\in\mathbb{N}$, $\mathbb{P}(V)>0$ if*

$$\dim\inf\partial\mathbb{T}_S>\ln\left[\frac{1}{1-\prod_{j=1}^{k}\mathbb{P}(R<j)}\right].$$

Definition 7. *A k-periodic tree with degree $\tilde{d}=(d_1,\cdots,d_k)$, $d_i\geq 2$ for all $i=1,2,\cdots,k$, is as tree such that for any vertex whose distance to the origin is $nk+i-1$ for some $n\in\mathbb{N}$ has degree d_i+1. We refer to this tree as $\mathbb{T}_{\tilde{d}}$.*

Example 6 (Junior et al. [10]). Consider a *Cone Percolation Model* in $\mathbb{T}_S$ with $R\sim\mathcal{B}(p)$, a *radius of influence* satisfying

$$\mathbb{P}(R=1)=p=1-\mathbb{P}(R=0).$$

(i) If $\dim\inf\partial\mathbb{T}_S>-\ln p$ then $\mathbb{P}(v)>0$,
(ii) If $\mathbb{T}_S=\mathbb{T}_{\tilde{d}}$ and $\sqrt[k]{\prod_{j=1}^{k}d_j}>\frac{1}{p}$ then $\mathbb{P}(V)>0$.

6 Random Environments

In this section we review the Fireworks and the Reverse Fireworks processes, with a random number of stations at each vertex. Bertacchi and Zucca [2] consider an extra source of randomness: the number of individuals sitting on each vertex. They consider two families of random variables $\{N_x\}_{x\in\mathcal{G}}$ and $\{R_{x,i}\}_{x\in\mathcal{G}}^{i\in\mathbb{N}}$ such that $\{N_x, R_{x,i}\}$ are independent and $\{R_{x,i}\}_{i\in\mathbb{N}}$ are identically distributed for all $x \in \mathcal{G}$ that is $R_{x,i} \sim R_x$. In their paper N_x represents the random number of individuals at vertex x (in particular $N_{\mathcal{O}}$ is the number of individuals at the origin) while $\{R_{x,i}\}_{i=1}^{N_x}$ are their *radius of influence*. The main question about this model is to understand under which conditions, the signal, starting from one vertex of a graph ($\mathbb{N}$ or a Galton–Watson tree), will spread indefinitely with positive probability or die out almost surely in a finite number of steps.

Bertacchi and Zucca [2] rely in their analysis on associating the processes with random numbers of stations (fireworks or reverse fireworks), with processes with one station per vertex as in Junior et al. [8]. Indeed, they consider processes with one station on each vertex x and *radius of influence* $\tilde{R}_x = \mathbf{1}_{\{N_x\geq 1\}} \max\{R_{x,j} : j = 1, \ldots, N_x\}$. They call this process, the *deterministic counterpart* or *annealed counterpart* of the original process. They observe that the *annealed counterpart* does not retain any information about the environment, nevertheless the probability of survival for the original process and for its *annealed counterpart* are the same.

6.1 Fireworks

For $x \in \mathcal{G}$, let us define

$$\varphi_{N_x}(t) := \mathbf{E}(t^{N_x}) = \sum_{j=0}^{\infty} \mathbb{P}(N_x = j)t^j$$

Homogeneous Fireworks. Consider $R_i \sim R$ and $N_x \sim N$ for all $x \in \mathcal{G}$. Let us define

$$f_{R,N}(n) := n\{1 - \varphi_N(\mathbb{P}(R < n))\}.$$

Theorem 23 (Bertacchi and Zucca [2])

(i) If $\limsup_{n\to\infty} f_{R,N}(n) < 1$ *then* $\mathbb{P}(V) = 0$.

(ii) If $\liminf_{n\to\infty} f_{R,N}(n) > 1$ *then* $\mathbb{P}(V) > 0$.

(iii) If $\mathbb{E}(N) < \infty$ *and* $\limsup_{n\to\infty} n\mathbb{P}(R \geq n) < \dfrac{1}{\mathbb{E}(N)}$ *then* $\mathbb{P}(V) = 0$.

(iv) If $\mathbb{E}(N) < \infty$ *and* $\mathbb{E}(R) < \infty$ *then* $\mathbb{P}(V) = 0$.

(v) If $\liminf_{n\to\infty} n\mathbb{P}(R \geq n)\varphi'_N(\mathbb{P}(R < n)) > 1$ *then* $\mathbb{P}(V) > 0$.

A consequence of Theorem 1 from Gallo et al. [4] is the following result

Corollary 5

$$\mathbb{P}(V) = \left[1 + \sum_{j=1}^{\infty} \prod_{i=0}^{j-1} \varphi_N\left(\mathbb{P}(R \le i)\right)\right]^{-1}$$

Remark 3. It is possible to have $\mathbb{E}(N) = \infty$, $\mathbb{E}(R) = \infty$ and $\mathbb{P}(V) = 0$. Take $\mathbb{P}(N \ge n) \sim \frac{1}{n}$ when $n \to \infty$ and $\mathbb{P}(R \ge n) = \frac{1}{n \ln n \ln(lnn)}$

Heterogeneous Fireworks

Theorem 24 (Bertacchi and Zucca [2]). *In the heterogeneous case, if*

$$\sum_{n=0}^{\infty} \prod_{i=0}^{n} \varphi_{N_i}\left(\mathbb{P}(R_i < n - i + 1)\right)$$

then $\mathbb{P}(V) > 0$.

Adapting the proof of Theorem 2.3 from Junior et al. [8] we have

Theorem 25. *In the heterogeneous case, if*

(i) $\varphi_{N_i}(\mathbb{P}(R_i < 1)) \in (0, 1)$.
(ii) $\lim_{n\to\infty} \prod_{i=0}^{n-1} \varphi_{N_i}\left(\mathbb{P}(R_i < 2n - 1)\right) = 1$.
(iii) $\lim_{n\to\infty} \prod_{i=n}^{2n-1} \varphi_{N_i}\left(\mathbb{P}(R_i < 2n - 1)\right) > 0$.

then $\mathbb{P}(V) = 0$.

6.2 Reverse Fireworks

Homogeneous Reverse Fireworks. Let us define

$$W = \sum_{n=0}^{\infty} \left[1 - \varphi_N\left(\mathbb{P}(R < n)\right)\right]$$

Theorem 26 (Bertacchi and Zucca [2])

(i) If $W = \infty$ *then* $\mathbb{P}(S) = 1$.
(ii) If $W < \infty$ *then* $\mathbb{P}(S) = 0$.

Theorem 26 can also be obtained as a consequence of Theorem 3.2 from Junior et al. [9] or as a consequence of Theorem 2 from Gallo et al. [4].

Remark 4 (Bertacchi and Zucca [2]). Theorems 23 and 26 admit a similar corollary

(i) For every unbounded random variable R there exists a random variable N such that $\mathbb{P}(V) > 0$ ($\mathbb{P}(S) = 1$). For $\epsilon > 0$ and $\delta \in (0, 1)$ consider N satisfying

$$\mathbb{P}\left(N \ge \frac{\ln(1-\delta)}{\ln(\mathbb{P}(R < n))}\right) \ge \frac{1+\epsilon}{n\delta}.$$

(ii) For every random variable N such that $\mathbb{P}(N = 0) < 1$ there exists a random variable R such that $\mathbb{P}(V) > 0$ ($\mathbb{P}(S) = 1$). Take R satisfying $\mathbb{P}(R \ge n) = p_n$, where $p_n = \inf\{t \ge 0; \varphi_N(1 - t) \le 1 - \frac{2}{n}\}$.

Heterogeneous Reverse Fireworks

Theorem 27 (Bertacchi and Zucca [2]). *In the heterogeneous case,*

$$\sum_{k=0}^{\infty} \left[1 - \varphi_{N_{n+k}}\left(\mathbb{P}(R_{n+k} < k)\right)\right] = \infty, \text{ if and only if } \mathbb{P}(S) = 1.$$

On the other hand, if

$$\sum_{n=0}^{\infty} \prod_{k=1}^{\infty} \varphi_{N_{n+k}}\left(\mathbb{P}(R_{n+k} < k)\right) < \infty, \mathbb{P}(S) > 0.$$

6.3 Galton Watson

Let us define the space of unlabelled GW-trees (the usual GW-trees). Consider a GW-process, with offspring distribution $\mathbb{P}(D = d)$, $0 \leq d < \infty$. We assume that $\mathbb{P}(D = 1) < 1$ (otherwise the resulting random tree is $\mathbb{N}$) and we suppose that $\mu_D := \sum_{d=0}^{\infty} d\mathbb{P}(D = d) > 1$ (the supercritical case). The underlying random graph will be a GW-tree generated by this process. Let $g(s) := \sum_{d=0}^{\infty} s^d \mathbb{P}(D = d)$ be the generating function of D and let $\pi \in [0, 1]$ be the smallest nonnegative fixed point of g. If $\mathbb{P}(D > k) = 0$ for some k we say that the GW-tree has maximum degree k or that it is k-bounded.

Homogeneous Fireworks. In this case, the random number of stations are independent and identically distributed $\mathbb{N}$-valued random variables with common law N. Analogously, The radii of the stations are independent and identically distributed with distribution R (either discrete or continuous random variable).

Definition 8. *We define*

$$\Phi(t) := \varphi_N(\mathbb{P}(R < 1)) + \sum_{n=1}^{\infty} [\varphi_N(\mathbb{P}(R < n + 1)) - \varphi_N(\mathbb{P}(R < n))] t^n.$$

In particular observe that

$$\Phi(0) = \varphi_N(\mathbb{P}(R < 1)) \text{ and the case } N = 1 \text{ a.s.},$$

$$\Phi(t) = \sum_{n=0}^{\infty} [\mathbb{P}(n \leq R < n + 1)] t^n.$$

Theorem 28 (Bertacchi and Zucca [2]). *Consider a Homogeneous Fireworks Process. We have that* *(i)* *If* $\Phi(\mu_D) - 1 > \Phi(0) = \varphi_N(P(R < 1))$ *and* $\mathbb{P}(N = 0) = 0$ *then for the Fireworks process there is survival with positive probability for almost every realization of the environment such that the underlying tree is infinite and there is at least one station at the root.*

(ii) *If* $\Phi(\mu_D) - 1 > \Phi(0) = \varphi_N(P(R < 1))$ *and* $\mathbb{P}(N = 0) > 0$ *then for the Fireworks process* $\mathbb{P}(V|\mathcal{T} = T, N_{\mathcal{O}} = n) > 0$ *for almost every* (T, n) *such that* T *is an infinite (unlabelled) tree and* $n \geq 1$.

(iii) *If the GW-tree is* k*-bounded and* $\Phi(k) \leq 2 - \frac{1}{k}$ *then the Fireworks process becomes extinct a.s. for almost every realization of the environment.*

Homogeneous Reverse Fireworks. In this case, the random number of stations are independent and identically distributed $\mathbb{N}$-valued random variables with common law N, except by numbers of station the root $\mathcal{O}$. For the root, we take $N_{\mathcal{O}} = \min\{n > 0 : \mathbb{P}(N = n) > 0\}$ Besides, the radii of the stations are independent and identically distributed with distribution R (either discrete or continuous random variable).

Definition 9. *We define*

$$\phi_1(t) := \sum_{n=1}^{\infty}[1 - \varphi_N(\mathbb{P}(R < n))]\mu_D{}^n$$

$$\phi_2(t) := \sum_{n=1}^{\infty}[1 - \varphi_N(\mathbb{P}(R < n))]\mu_D{}^n \prod_{j=1}^{n-1} \varphi_N(\mathbb{P}(R < j))$$

Theorem 29 (Bertacchi and Zucca [2]). *Consider a Homogeneous Reverse Fireworks Process. The following hold*

(i) *If* $\phi_1(\mu_D) = \infty$ *then there is survival with probability 1 for the Reverse Fireworks process for almost all realizations of the environment such that the underlying tree is infinite.*

(ii) *If* $\mathbb{P}(N = 0) = 0$, $\phi_1(\mu_D) < \infty$ *and* $\phi_2(\mu_D) > 1$ *then there is survival with positive probability (strictly smaller than 1) for the Reverse Fireworks process for almost all realizations of the environment such that the underlying tree is infinite.*

(iii) *If* $\mathbb{P}(N = 0) > 0$, $\phi_1(\mu_D) < \infty$ *and* $\phi_2(\mu_D) > 1$ *then* $\mathbb{P}(S|\mathcal{T} = T) \in (0, 1)$ *for almost every infinite (unlabelled) tree* T.

(iv) *If* $\phi_1(\mu_D) < \infty$ *and* $\phi_2(\mu_D) \leq 1$ *then there is a.s. extinction for the Reverse Fireworks process for almost all realizations of the environment.*

Definition 10. *We define*

$$\mathcal{M}_c := \left[\limsup_{n\to\infty} \sqrt[n]{1 - \varphi_N(\mathbb{P}(R < n))}\right]^{-1}.$$

Corollary 6 (Bertacchi and Zucca [2]). *There exists a critical value* $\mu_c \in [1, \infty)$, $\mu_c \leq \mathcal{M}_c$ *such that*

(i) $\mu_D < \mu_c$ *implies a.s. extinction for almost all realizations of the environment.*
(ii) $\mu_c < \mu_D < \mathcal{M}_c$ *and* $\mathbb{P}(N = 0) = 0$ *implies survival with positive probability for almost all realizations of the environment such that the underlying tree is infinite.*
(iii) $\mu_c < \mu_D < \mathcal{M}_c$ *and* $\mathbb{P}(N = 0) > 0$ *implies survival with positive probability for almost every infinite (unlabelled) tree.*
(iv) $\mathcal{M}_c < \mu_D$ *implies survival with probability 1 for almost all realizations of the environment such that the underlying tree is infinite.*
(v) If $\mu_D = \mu_c < \mathcal{M}_c$ *then there is a.s. extinction for almost all realizations of the environment.*

7 Open Problems

Some natural extensions for these models are those considering

(i) Fireworks processes (direct and reverse) on $\mathbb{Z}^{d^+}$. An especially interesting case is when $d = 2$ and the *boxes of influence* are distributed as $[0, R_x) \times [0, R_y)$ with R_x independent of R_y and the rumor starting from $(0, 0)$ or from every (x, y) such that $x = 0$ or $y = 0$;
(ii) Fireworks processes on $\mathbb{Z}$. Heterogeneous versions with *radius of influence* non i.i.d. and with individuals being initially placed following a renewal process or a Markovian process.
(iii) Reverse fireworks processes on $\mathbb{Z}$. Individuals throw their *radius of influence* to every direction as in (1) (See Gallo et al. [4]). Gallo et al. [4] believe that conditions for survival will be the same but the final proportion of informed individual will be strictly larger.
(iv) Cone Percolation on Spherically Symmetric and on Galton Watson trees. Obtaining lower and upper bounds for the survival probability and for the extinction time.

Acknowledgments. F.P.M. wishes to thank NYU-Shanghai China and V.V.J. and K.R. wish to thank Instituto de Matemática e Estatística-USP Brazil for kind hospitality.

References

1. Athreya, S., Roy, R., Sarkar, A.: On the coverage of space by random sets. Adv. Appl. Probab. **36**, 1–18 (2004)
2. Bertacchi, D., Zucca, F.: Rumor processes in random environment on N and on Galton-Watson trees. J. Stat. Phys. **153**(3), 486–511 (2013)
3. Bressaud, X., Fernández, R., Galves, A.: Decay of correlations for non-Hölderian dynamics: a coupling approach. Elect. J. Probab. **4**(3), 1–19 (1999)
4. Gallo, S., Garcia, N., Junior, V., Rodríguez, P.: Rumor processes on N and discrete renewal processes. J. Stat. Phys. **155**(3), 591–602 (2014)

5. Gallo, S., Lerasle, M., Takahashi, Y.D.: Markov approximation of chains of infinite order in the d−metric. Markov Process. Relat. Fields **19**(1), 51–82 (2013)
6. Garsia, A., Lamperti, J.: A discrete renewal theorem with infinite mean. Comment. Math. Helv. **37**, 221–234 (1962)
7. Hall, P.: Introduction to the Theory of Coverage Processes. Wiley, New York (1988)
8. Junior, V., Machado, F., Zuluaga, M.: Rumor processes on $\mathbb{N}$. J. Appl. Probab. **48**(3), 624–636 (2011)
9. Junior, V., Machado, F., Zuluaga, M.: The cone percolation on $\mathbb{T}_d$. Braz. J. Probab. Stat. **28**(3), 367–675 (2014)
10. Junior, V., Machado, F., Ravishankar, K.: The cone percolation model on Galton–Watson and on spherically symmetric trees. Preprint arXiv:1510.02821
11. Lebensztayn, E., Rodriguez, P.: The disk-percolation model on graphs. Stat. Probab. Lett. **78**(14), 2130–2136 (2008)
12. Meester, R., Roy, R.: Continuum Percolation. Cambridge University Press, Cambridge (1996)
13. Ravishankar, K., Singh, S.: Broadcasting on $[0, L]$. Discrete Appl. Math. **53**, 299–319 (1994)

Site Percolation on a Disordered Triangulation of the Square Lattice

Leonardo T. Rolla[1,2(✉)]

[1] Argentina National Research Council at the University of Buenos Aires, Buenos Aires, Argentina

[2] NYU-ECNU Institute of Mathematical Sciences at NYU-Shanghai, Shanghai, China

leorolla@nyu.edu

dedicated to Chuck Newman on the occasion of his 70th birthday.

Abstract. In this note we consider independent site percolation in a disordered triangulation of $\mathbb{R}^2$ given by adding one of the two possible diagonals to each face of the usual graph $\mathbb{Z}^2$. The natural conjecture is that $p_c = 1/2$ for every such graph. Bollobás and Riordan proved this for almost every triangulation, in case each diagonal is chosen independently using a given coin. We give an alternative proof to the particular case of a fair coin. The general conjecture remains open.

Keywords: Disordered percolation · Phase transition · Critical parameter

1 Introduction

Let $\widetilde{\mathbb{Z}}^2$ denote the set of 1×1 squares in $\mathbb{R}^2$ having all their corners in $\mathbb{Z}^2$, and define the diagonal configuration space by

$$\Omega = \{ \diagdown, \diagup \}^{\widetilde{\mathbb{Z}}^2}.$$

Let Σ denote the color configuration space

$$\Sigma = \{ \color{blue}\bullet, \color{red}\bullet \}^{\mathbb{Z}^2}.$$

Examples of a diagonal configuration ω and a color configuration σ are

and

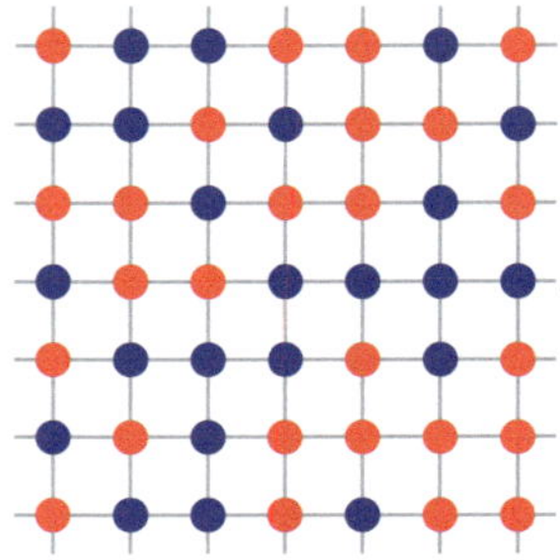

.

V. Sidoravicius (Ed.): *Sojourns in Probability Theory and Statistical Physics - II*, PROMS 299, pp. 228–240, 2019.
https://doi.org/10.1007/978-981-15-0298-9_10

Let P_p denote the probability measure on Σ given by

$$P_p(\sigma_x = {\color{red}\bullet}) = {\color{red}p} \qquad \text{and} \qquad P_p(\sigma_x = {\color{blue}\bullet}) = {\color{blue}\mathbf{1 - p}},$$

independently over $x \in \mathbb{Z}^2$, and let P_p^ω denote the law of the percolation process on the graph obtained by adding the diagonals in ω to the usual graph $\mathbb{Z}^2$. In general, the resulting graph does not have any symmetry, but still there is one critical parameter $p_c(\omega)$ at which the probability of having an infinite red cluster jumps from 0 to 1 (this is a tail event).

Since this graph is a triangulation, site percolation is self-dual, that is, the only way to prevent a given red connection is with a transversal blue connection and vice versa. This is illustrated by

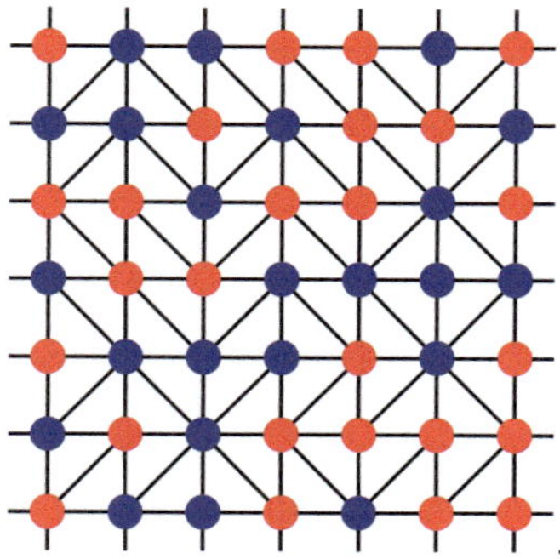
,

where existence of a left-right red crossing prevents a top-bottom blue crossing, and a left-right blue crossing prevents a top-bottom red crossing.

1.1 Statement

Because of self-duality, the obvious conjecture[1] is that $p_c(\omega) = 1/2$. In this note we show that this is true if the diagonal configuration ω is obtained by tossing a fair coin for each square $z \in \widetilde{\mathbb{Z}}^2$.

Let Q denote the probability on Ω given by

$$Q(\omega_z = \boxbslash) = Q(\omega_z = \boxslash) = \tfrac{1}{2},$$

independently over different squares $z \in \widetilde{\mathbb{Z}}^2$. Our sample space will be $\Omega \times \Sigma$, so the process described above is governed by the "quenched measure"

$$P_p^\omega = \delta_\omega \times P_p.$$

The "averaged measure" is given by

$$\mathbb{P}_p = Q \times P_p = \int_\Omega P_p^\omega \, \mathrm{d}Q(\omega).$$

[1] If p_c were smaller, there would be both an infinite blue and an infinite red cluster for any $p \in (p_c, 1 - p_c)$, and if p_c were larger, there would be both infinitely many red and blue circuits surrounding the origin for any $p \in (1 - p_c, p_c)$.

Theorem 1. *For the averaged process,*

$$\mathbb{P}_p(\textit{there is an infinite red cluster}) = \begin{cases} 0, & p \leqslant \frac{1}{2}, \\ 1, & p > \frac{1}{2}. \end{cases}$$

In particular, by Fubini–Tonelli, $p_c(\omega) = 1/2$ *for almost every* ω.

1.2 Related Work and Acknowledgements

Bollobás and Riordan [2] proved a more general result, where the measure Q can be taken as i.i.d. with any parameter. The author was unaware of that work until recently.

The alternative argument presented here is elementary and self-contained. It consists in adapting Smirnov's proof [9] of RSW estimates [6,7] and Kesten's proof of logarithmic expected number of pivotal sites [4]. The argument given in [2] is more general but slightly more involved, as it uses uniqueness of the infinite cluster and requires adapting Menshikov's sausage argument [5].

An important step in the proof is to establish the analogous of Harris–FKG inequality for the averaged measure. Notice that not all events which we would normally call "increasing" are positively correlated, because two such events may have conflicting requirements for the diagonals. To overcome this, we define *robust increasing* events, and prove positive correlations for this type of event. The same idea was already present in [2], and we give a precise description here for the reader's convenience.

Recent work on Bernoulli percolation in $\mathbb{Z}^d$ and Voronoi percolation in $\mathbb{R}^2$ provided new insights for the study of sharp phase transitions [1,3,8]. It would be interesting to apply those ideas to disordered triangulations of the square lattice, and obtain alternative proofs or extend existing results to a more general setting. In particular, it seems that the arguments introduced in [1] to prove a quenched version of RSW estimates for Voronoi percolation would also work for disordered triangulations of the square lattice.

I would like to thank Wendelin Werner for suggesting this problem to me back in 2009, and for inspiring discussions. I also thank him for pointing out the possible connections with [3] and [8]. I thank Vincent Tassion for pointing out reference [2] and the RSW estimates from [1].

2 Robust Increasing Events

An *observable* is a measurable function $f : \Omega \times \Sigma \to \mathbb{R}$.

Definition 1. *We say that an observable* f *is* increasing in σ *if, for each pair* (ω, σ), *switching the color of any site* x *from* ● *to* ● *increases (i.e., does not decrease) the value of* f.

In order to discuss monotonicity with respect to the diagonal configuration ω, we take into account the color configuration σ to see whether it is ⧅ or ⧄ that favors the ${\color{red}\bullet}$'s more than the ${\color{blue}\bullet}$'s, or the other way around.

Given a color configuration σ, each square $z \in \widetilde{\mathbb{Z}}^2$ will be classified as having one of three types, depending on the colors of its four corners.

The first type consists of configurations whose symmetries make it impossible to decide whether ${\color{red}\bullet}$'s and ${\color{blue}\bullet}$'s would prefer ⧅ or ⧄.

–type N: 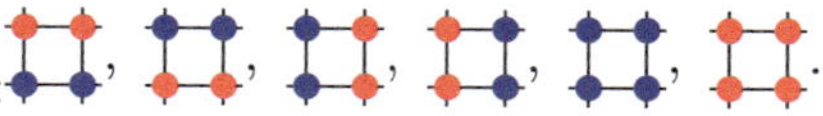

Definition 2. *We say that f is* robust *if flipping the diagonal at any square z of type N does not change the value of f.*

Squares that are not of type N will have type A or B depending on whether it is the ${\color{red}\bullet}$'s or ${\color{blue}\bullet}$'s who prefer ⧄ over ⧅.

– type A: 

– type B: 

Definition 3. *We say that a robust observable f is* increasing in ω *if, for each pair (ω, σ), flipping the diagonal of any square z from ⧅ to ⧄ will increase f for z of type A and decrease f for z of type B. An event $\mathcal{A}$ is called robust, increasing in σ and ω if its indicator function is so.*

We mention that some events that would normally be called "increasing" are not robust. For example, in a 3×3 square (what we call 3×3 contains 16 sites), existence of a red path of length 3 connecting the top-left and bottom-right corners is not robust. Moreover, this event is not positively-correlated with existence of a red path of length 3 connecting the top-right and bottom-left corners: they are in fact mutually exclusive. The above events are not robust because they have requirements for diagonals even when the containing squares are of type N. In the same direction, events requiring existence of disjoint paths are not robust in general.

On the other hand, and that is enough for our needs, for any sets $A, B \subseteq \mathbb{Z}^2$ and any domain $\mathcal{D}$ consisting of a collection of closed squares of $\widetilde{\mathbb{Z}}^2$, the event "A is connected to B by a red path in $\mathcal{D}$" is both robust and increasing.

Lemma 1 (Harris–FKG). *Let f and g be robust non-negative observables, increasing in σ and ω. Then*

$$\mathbb{P}_p(fg) \geqslant \mathbb{P}_p(f)\mathbb{P}_p(g).$$

Proof. Let $\sigma \in \Sigma$ be fixed. For an observable h, consider the projection $h_\sigma : \Omega \to \mathbb{R}$ given by $h_\sigma = h(\cdot, \sigma)$. Observe that, if h is robust and increasing in ω, then h_σ depends only on ω_z for z outside the set $N_\sigma \subseteq \widetilde{\mathbb{Z}}^2$ of squares of type N. Moreover, there is a natural partial order on $\{⧅, ⧄\}^{N_\sigma^c}$ under which h_σ is an increasing function ($\boxbslash \leq \boxslash$ for $z \in N_\sigma^c$ of type A and $\boxbslash \geq \boxslash$ for $z \in N_\sigma^c$ of type B). Since Q induces a product measure on $\{\boxbslash, \boxslash\}^{N_\sigma^c}$, projections of this type satisfy the Harris–FKG inequality with respect to Q. Therefore, if f and g are non-negative, robust and increasing in σ and ω,

$$\mathbb{P}_p(fg) = P_p[Q(f_\sigma g_\sigma)] \geqslant P_p[Q(f_\sigma)Q(g_\sigma)] \geqslant P_p[Q(f_\sigma)]P_p[Q(g_\sigma)] = \mathbb{P}_p(f)\mathbb{P}_p(g).$$

We have used Fubini–Tonelli theorem for the equalities. The first inequality follows from the above observation and the second inequality follows from the standard Harris–FKG inequality, since f and g are increasing in σ. □

3 Proof of Sharp Percolation Threshold

Lemma 2 (Russo–Seymour–Welsh). *In any* $2n \times n$ *rectangle,*

$\mathbb{P}_{\frac{1}{2}}\Big($ 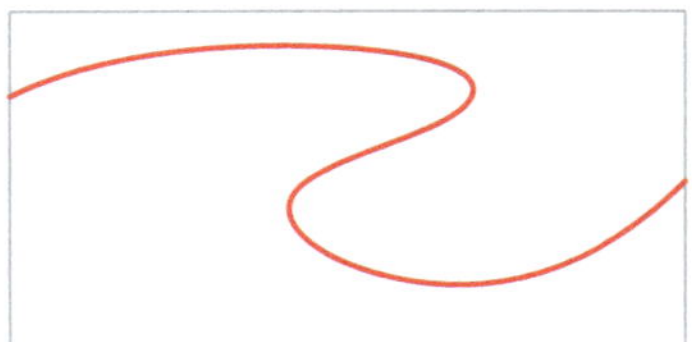 $\Big) \geqslant \frac{1}{16}.$

In the proof we consider an exploration that progressively reveals the color of some sites and the position of some diagonals. Below we show an exploration starts from the top-left corner and targets the bottom-right corner of a rectangle. When the exploration enters a triangle by crossing one of its sides, it looks at the color of the opposite corner in order to decide on where to exit the triangle. When it enters a square, it first reveals the position of the diagonal on that square.

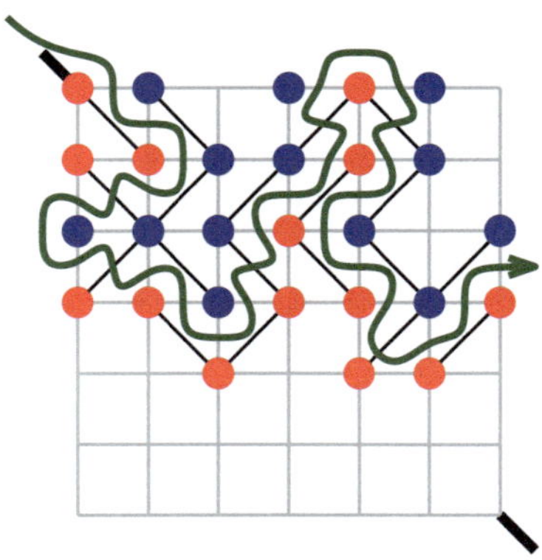

In this procedure, the exploration will leave the rectangle through the right side before the bottom side if there is a left-right red connection, and the bottom side before the right otherwise.

Proof of Lemma 2. We use $\mathbb{P}$ for $\mathbb{P}_{\frac{1}{2}}$. By Harris–FKG we have

$\mathbb{P}\Bigg($  $\Bigg) \geqslant \mathbb{P}\Bigg($ 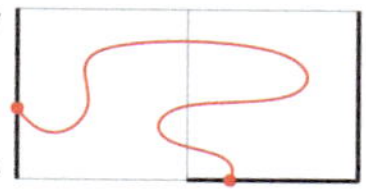 $\Bigg) \times \mathbb{P}\Bigg($ 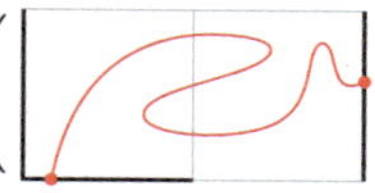 $\Bigg)$,

whence by symmetry it suffices to show that

$\mathbb{P}\Bigg($ 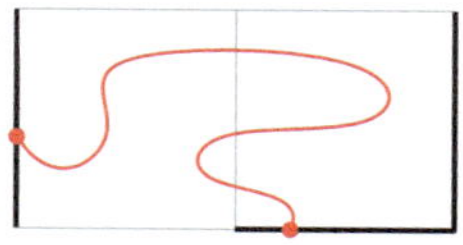 $\Bigg) \geqslant \frac{1}{4}$.

We first try to "bend" the left-side boldface region by starting an exploration path in the left-side square as shown in (1). With probability 1/2 we succeed bending the boldface region until the middle of the rectangle, revealing some diagonals and some blue and red sites like this:

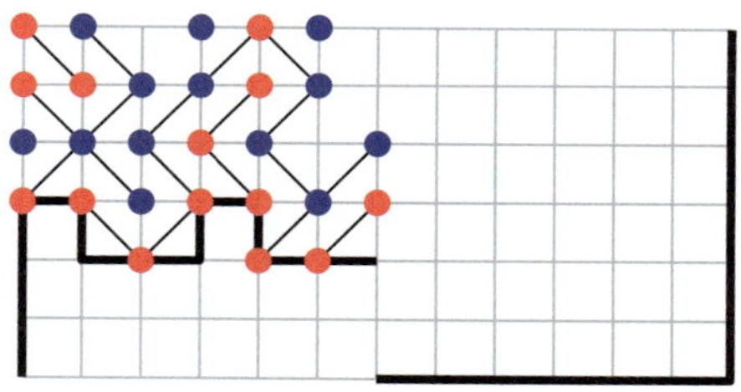

.

Given a partial configuration such as above, the event in (2) is equivalent to a red connection between the two boldface regions. So we need to show that the conditional probability of such red connection is at least 1/2. But such connection is certainly implied by red path connecting two smaller boldface regions contained in a smaller grayed zone given by

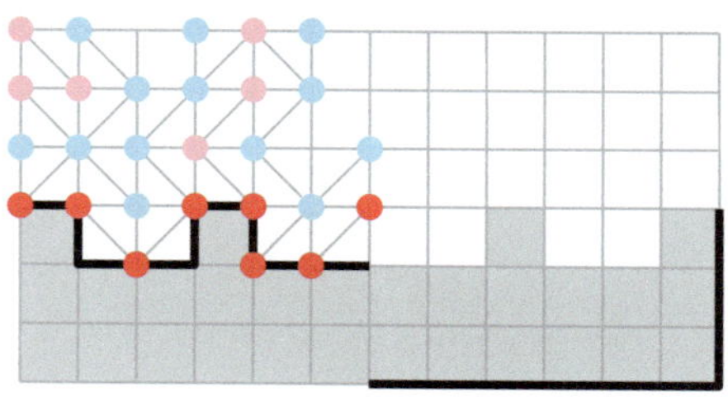

.

Now notice that none of the sites and diagonals revealed so far can interfere with this event, except for some red sites lying on the boldface region. The fact that these sites are red can only help, and the conditional probability of the latter event given that they are red is bounded from below by the probability of the crossing

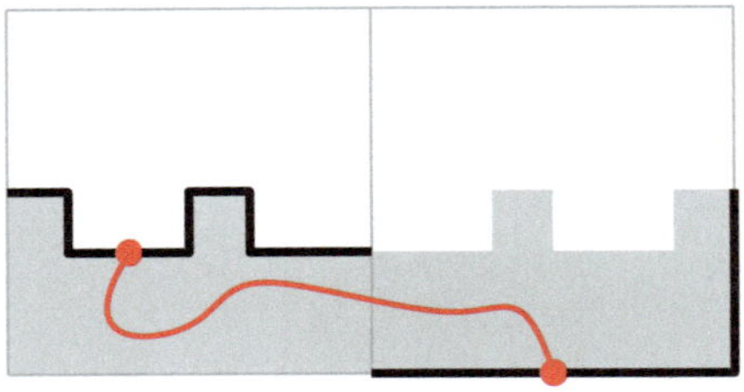

without any conditioning. Finally, the complementary of the latter event is

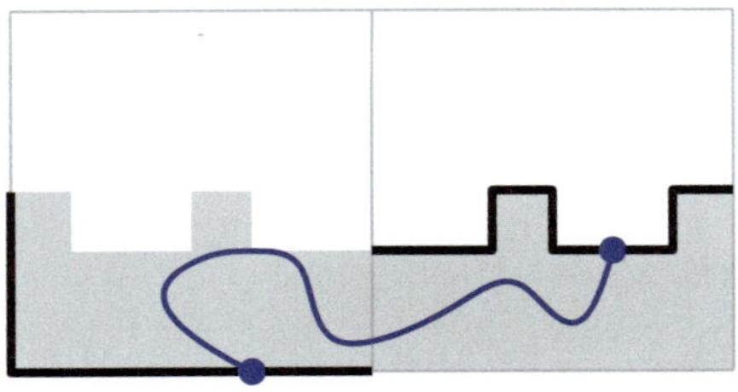

,

which by symmetry has the same probability, concluding the proof. □

As usual, Lemma 2 has the following immediate corollaries.

Corollary 1. *There is $\delta > 0$ such that, in any $8n \times n$ rectangle,*

$\mathbb{P}_{\frac{1}{2}}\Big($  $\Big) \geqslant \delta.$

Corollary 2. *There is $\delta > 0$ such that, in any pair of co-centered squares of size $4n \times 4n$ and $6n \times 6n$,*

$\mathbb{P}_{\frac{1}{2}}\Big($ 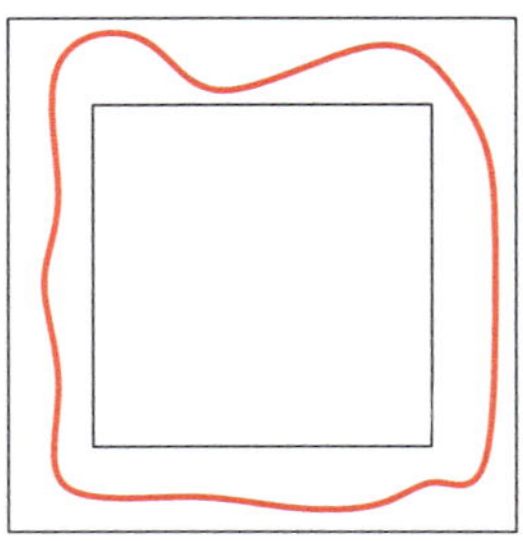 $\Big) \geqslant \delta,$

The last piece in the proof is the following.

Lemma 3 (Kesten). *There is $\beta > 0$ such that, for any $p \geqslant 1/2$, in any $2n \times n$ rectangle, the expected number of pivotal sites satisfies*

$\mathbb{E}_p\Bigg($ 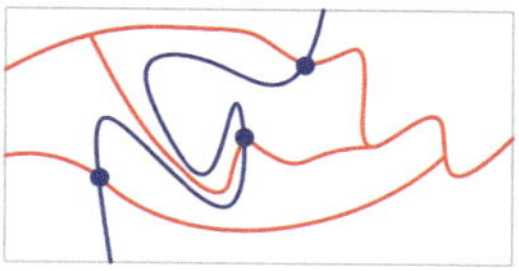 $\Bigg) \geqslant \beta \log_2 n \times \mathbb{P}_p\Bigg($ 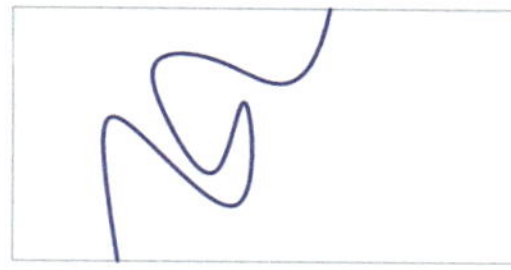 $\Bigg).$

Proof. It suffices to show that

$\mathbb{E}_p\Bigg($ 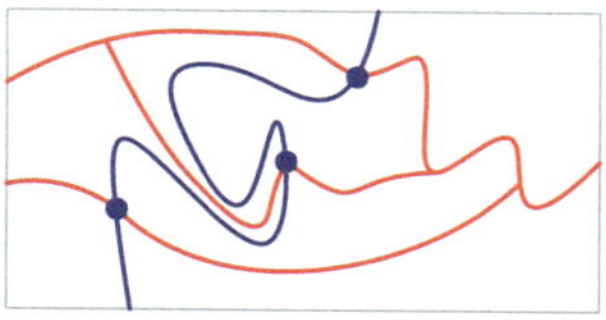 $\Bigg|$ 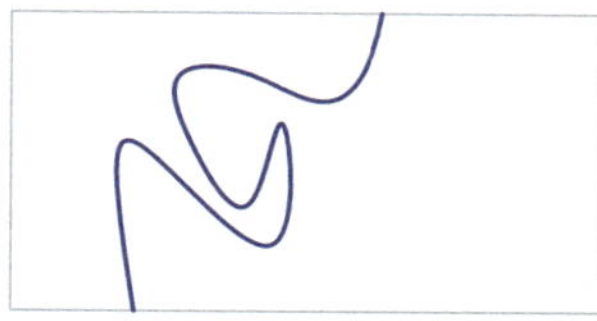 $\Bigg) \geqslant \beta \log_2 n.$

We will determine occurrence of the top-bottom blue crossing using an exploration path that starts at the top-left corner and ends at either the bottom or the right side, as below.

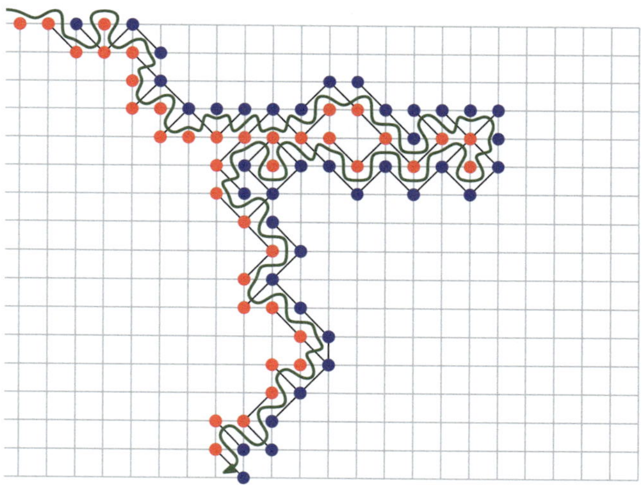

Existence of a top-bottom blue crossing is equivalent to the exploration finding the bottom side before the right side. We want to show that, if it such crossing occurs, then the conditional expectation of the number of pivotal sites given the colors and diagonals revealed in this exploration is greater than $\beta \log_2 n$ for some constant β.

On the above event, there is a self-avoiding blue path that joins the top and bottom sides of the rectangle. What we do now is a little overkilling, but it avoids the hassle of considering all corner cases related to the diagonals. Let us first inflate this self-avoiding blue path to make it squared where it would otherwise use a diagonal, as below.

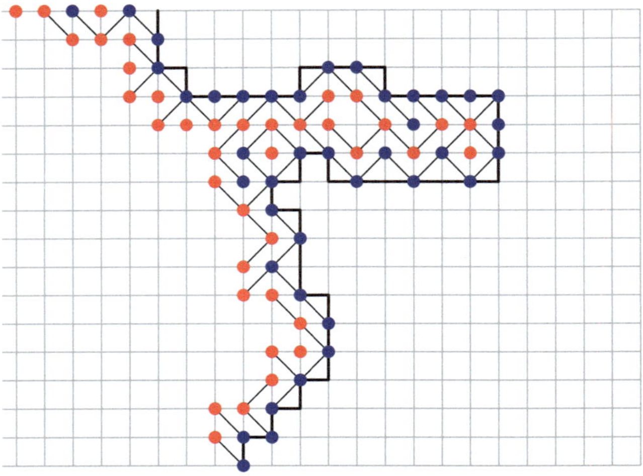

Notice that there are two types of sites in this squared path. The first type consists of blue sites which are adjacent to a red site which is in turn connected to the left side of the rectangle by a red path. The second type consists of sites whose color has not yet been revealed, but which are adjacent to a site of the first type.

Now, in order to find pivotal sites, we consider the domain consisting of the squares that can be reached from the right side of the big rectangle without crossing the squared path, colored in light-gray below.

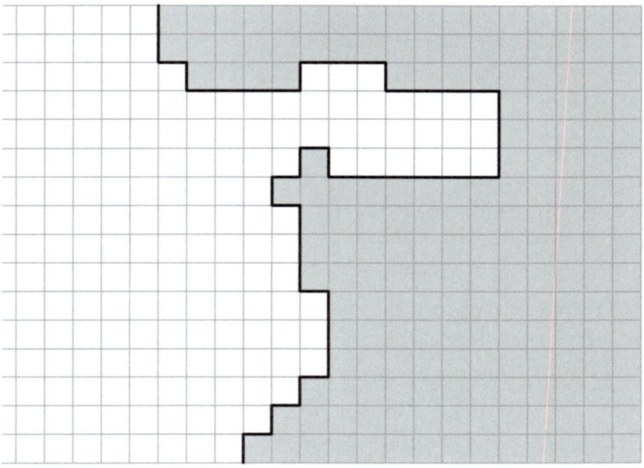

We will "re-sample" the entire process on this domain. Of course we cannot do that, so the red sites that we find on the squared path that turn out to have been previously sampled as blue will remain blue in the end.

We first look for a connection from the right side of the big rectangle to the squared-path that stays in a strip of width $n/4$ using an exploration path as below.

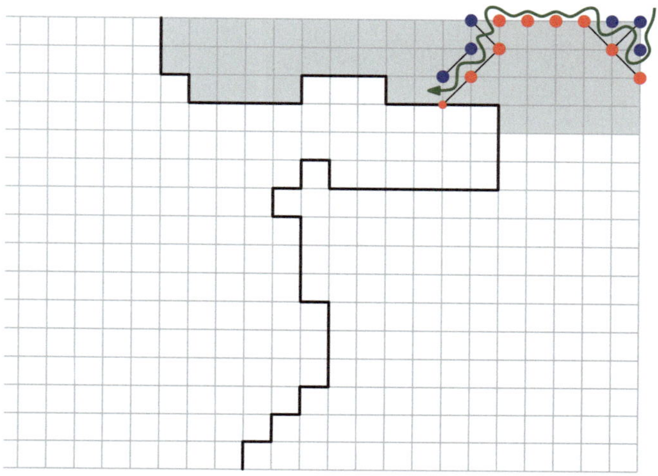

The last red site was drawn smaller to remind us that it may be a site which we already revealed to be blue. By Corollary 1, the probability of finding such a connection is greater than δ.

In case such path is found, we now draw disjoint 4×2^k-sided and 6×2^k-sided squares centered at the point just found, and consider the intersection of the corresponding annuli with the light-gray region in the second previous picture. There are at least $\delta \log_2 n$ such regions that do not go lower than the bottom side of the big rectangle, minus the squares explored in the previous step, as shown below.

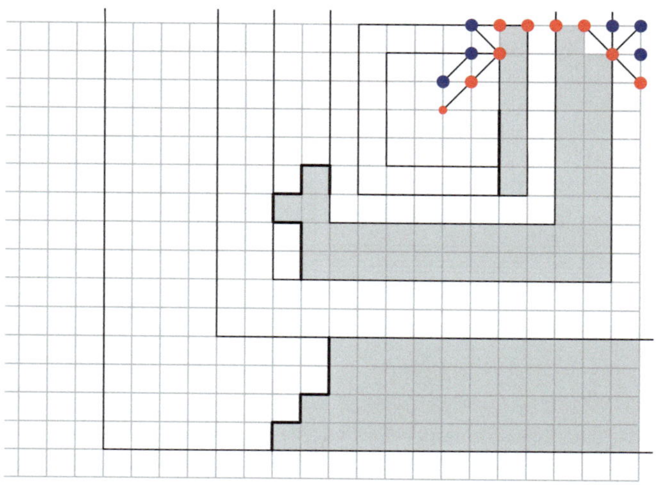

Conditioned on the above picture, by Corollary 2 each of these "tunnels" will contain a red connection with probability at least δ. The conditional expectation for the number of such tunnels that actually contain such connection, given that the previous connection has been found, is thus greater than $\delta^2 \log_2 n$. Therefore, the conditional expectation given existence of a top-bottom blue crossing is greater than $\delta^3 \log_2 n$.

In the example below, two such tunnels ended up providing one red site (drawn smaller), and one of them did not.

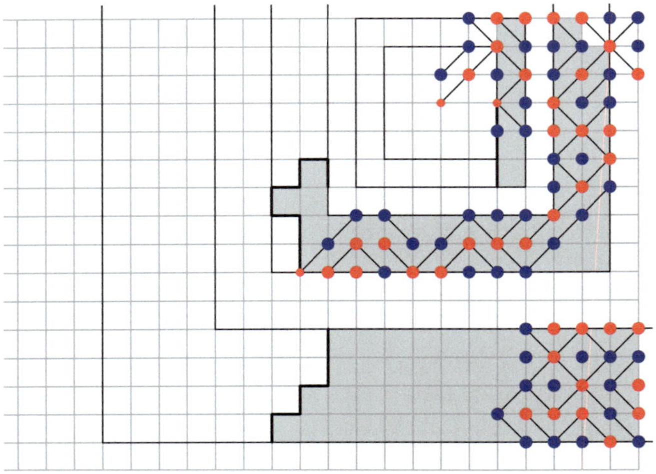

We combine the configuration discovered at this stage with the one previously removed. At this point, a small red site will become a true red site if it had not been revealed in the first exploration, and will be reverted to blue in case it had. The result is highlighted by a light-gray disk below.

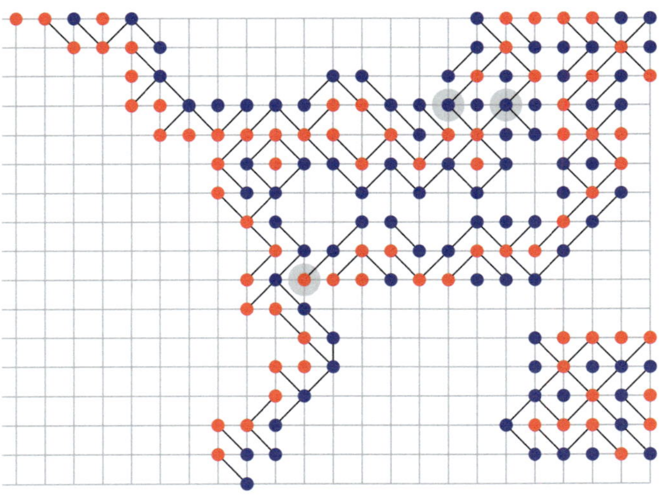

To conclude the proof, notice that each of these highlighted sites is either a pivotal site in case it is blue, or is preceded by a pivotal site in the squared curve in case it is red. □

Proof of Theorem 1. Absence of percolation at $p = 1/2$ follows from Corollary 2 as usual. Define the event

$\mathcal{A}_n =$ 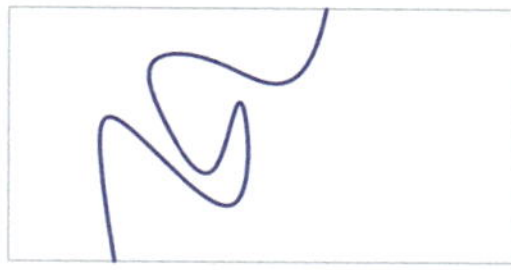

in a $2n \times n$ rectangle. Using Russo's formula and Lemma 3 we get

$$\tfrac{\mathrm{d}}{\mathrm{d}p}\mathbb{P}_p(\mathcal{A}_n) = -\mathbb{E}_p\big[\,\text{pivotal sites}\,\big] \leqslant -\beta \log_2 n\, \mathbb{P}_p(\mathcal{A}_n),$$

which gives

$$\log \mathbb{P}_{\frac{1}{2}+\varepsilon}(\mathcal{A}_n) \leqslant -\varepsilon\beta \log_2 n,$$

and thus, on an $2^k \times 2^{k+1}$ rectangle,

$\mathbb{P}_{\frac{1}{2}+\varepsilon}\Bigg($ 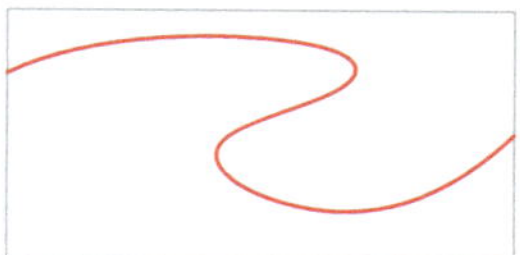 $\Bigg) \geqslant 1 - \mathrm{e}^{-\varepsilon\beta k}.$

To conclude the proof, we arrange $2^k \times 2^{k+1}$ rectangles as

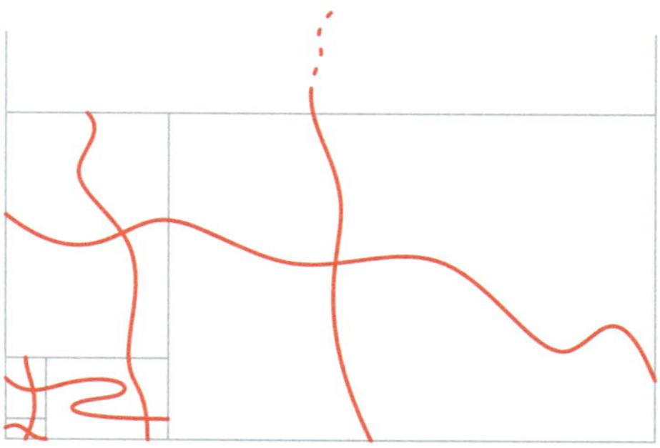

and deduce the second part of the theorem using Harris–FKG inequality. □

References

1. Ahlberg, D., Griffiths, S., Morris, R., Tassion, V.: Quenched Voronoi percolation. Adv. Math. **286**, 889–911 (2016)
2. Bollobás, B., Riordan, O.: Percolation on dual lattices with k-fold symmetry. Random Struct. Algorithms **32**, 463–472 (2008)
3. Duminil-Copin, H., Tassion, V.: A new proof of the sharpness of the phase transition for Bernoulli percolation on $\mathbb{Z}^d$. Enseign. Math. **62**(1–2), 199–206 (2016)
4. Kesten, H.: The critical probability of bond percolation on the square lattice equals 1/2. Comm. Math. Phys. **74**(1), 41–59 (1980)
5. Menshikov, M.V.: Coincidence of critical points in percolation problems. Dokl. Akad. Nauk SSSR **288**, 1308–1311 (1986)
6. Russo, L.: A note on percolation. Z. Wahrscheinlichkeitstheorie und Verw. Gebiete **43**(1), 39–48 (1978)

7. Seymour, P.D., Welsh, D.J.A.: Percolation probabilities on the square lattice. Ann. Discrete Math. **3**, 227–245 (1978). Advances in graph theory (Cambridge Combinatorial Conf., Trinity College, Cambridge, 1977)
8. Tassion, V.: Crossing probabilities for Voronoi percolation. Ann. Probab. **44**(5), 3385–3398 (2016)
9. Werner, W.: Lectures on two-dimensional critical percolation. Preprint arXiv:0710.0856 (2007)

Perturbations of Supercritical Oriented Percolation and Sticky Brownian Webs

Emmanuel Schertzer[1,2] and Rongfeng Sun[3](✉)

[1] Laboratoire de Probabilités, Statistiques et Modélisation (LPSM), Sorbonne Université, CNRS UMR 8001, Paris, France
emmanuel.schertzer@upmc.fr
[2] Center for Interdisciplinary Research in Biology (CIRB), Collège de France, CNRS UMR 7241, PSL Research University, Paris, France
[3] Department of Mathematics, National University of Singapore, 10 Lower Kent Ridge Road, Singapore 119076, Singapore
matsr@nus.edu.sg

To our advisor Chuck Newman, on his 70th birthday.

Abstract. Previously, Sarkar and Sun [13] have shown that for supercritical oriented percolation in dimension 1 + 1, the set of rightmost infinite open paths converges to the Brownian web after proper centering and scaling. In this note, we show that a pair of sticky Brownian webs arise naturally if one considers the set of right-most infinite open paths for two coupled percolation configurations with distinct (but close) percolation parameters. This leads to a natural conjecture on the convergence of the dynamical supercritical oriented percolation model to the so-called dynamical Brownian web.

Keywords: Brownian net · Brownian web · Oriented percolation · Sticky Brownian motion

1 Main Result

Oriented Percolation and the Brownian Web. Following the notation of [13], we consider supercritical oriented percolation on $\mathbb{Z}^2_{\text{even}} := \{(x,t) \in \mathbb{Z}^2 : x+t \text{ is even}\}$, with directed edges leading from (x,t) to $(x \pm 1, t+1)$ for each $(x,t) \in \mathbb{Z}^2_{\text{even}}$. Independently, each directed edge is open with probability p and closed with probability $1-p$ for some $p > p_c$, the critical percolation parameter. A site $z \in \mathbb{Z}^2_{\text{even}}$ is said to be a percolation point if there is an infinite open path of directed edges starting from z, and let $\mathcal{K} \subset \mathbb{Z}^2_{\text{even}}$ denote the set of percolation points. From each $z = (x,t) \in \mathcal{K}$, let $\gamma_z := (\gamma_z(n))_{n \geq t}$ denote the rightmost infinite open path starting from z. When $z = (x,t) \notin \mathcal{K}$, define γ_z to be the rightmost infinite open path starting from $(-\infty, x] \times \{t\}$.

V. Sidoravicius (Ed.): *Sojourns in Probability Theory and Statistical Physics - II*, PROMS 299, pp. 241–261, 2019.
https://doi.org/10.1007/978-981-15-0298-9_11

In [13], Sun and Sarkar showed that after suitable centering and rescaling, the set of paths $\Gamma = \{\gamma_z \ : \ z \in \mathcal{K}\}$ converges to a family of coalescing Brownian motions, known as the Brownian web [3,14,16] and which can also be obtained as the "universal" scaling limit of coalescing random walks [10]. See Fig. 1. More precisely, for $a, b, \epsilon > 0$, let

$$S_{a,b,\epsilon}(x,t) := \Big(\frac{\epsilon}{b}(x - at), \epsilon^2 t\Big), \qquad (x,t) \in \mathbb{R}^2, \tag{1}$$

be a space-time scaling map, and let $S_{a,b,\epsilon}(\gamma_z)$ and $S_{a,b,\epsilon}(\Gamma)$ be defined by identifying each path with its graph in $\mathbb{R}^2$. Then for $p > p_c$, there exist $\alpha(p), \sigma(p) > 0$ such that:

$$S_{\alpha(p),\sigma(p),\epsilon}(\Gamma) \underset{\epsilon \to 0}{\Longrightarrow} \mathcal{W}, \tag{2}$$

where $\Rightarrow$ denotes weak convergence, $\mathcal{W}$ is the *Brownian web*, and both

$$S_{\alpha(p),\sigma(p),\epsilon}(\Gamma) \text{ and } \mathcal{W}$$

are regarded as random variables taking values in $\mathcal{H}$, a suitable space of compact sets of paths. See [13] or the recent survey [14] for further details details on the Brownian web, including the space $\mathcal{H}$ in which the Brownian web takes its value.

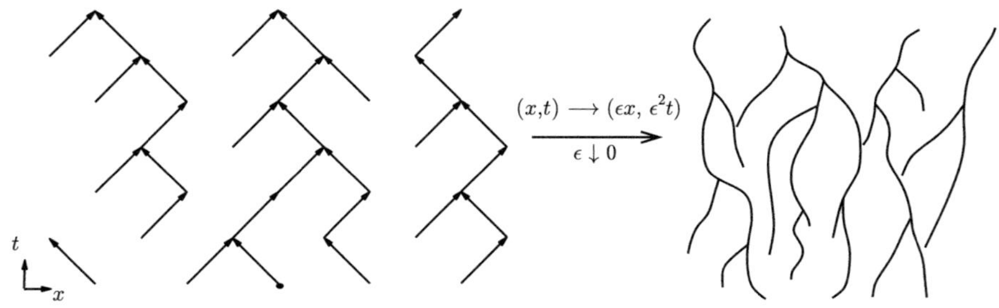

Fig. 1. The Brownian web as scaling limit of coalescing random walks.

Perturbed Percolation Clusters and Sticky Brownian Webs. Let $p > p_c$ and let Ω be the percolation configuration with percolation parameter p. We now describe two natural perturbations of Ω which consists in slightly increasing (resp., decreasing) the fraction of open edges.

In order to do so, we equip each directed edge e of $\mathbb{Z}^2_{\text{even}}$ with an independent uniform random variable ω_e on $[0,1]$. For any $p \in [0,1]$, we declare an edge e to be p-open iff $p < \omega_e$. This construction provides a natural coupling between percolation configurations with different percolation parameters p. For $\epsilon < (p - p_c) \wedge (1-p)$, we consider the percolation configurations $\Omega^{\epsilon,-}$ and $\Omega^{\epsilon,+}$ corresponding to the parameter $p-\epsilon$ and $p+\epsilon$ respectively, and which are coupled in such a way so that $\Omega^{\epsilon,+}$ has a surplus of open edges. Let $\mathcal{K}^{\epsilon,\pm} \subset \mathbb{Z}^2_{\text{even}}$ denote the set of percolation points in the configuration $\Omega^{\epsilon,\pm}$. For every $z \in \mathcal{K}^{\epsilon,\pm}$, let $\gamma_z^{\epsilon,\pm}$ denote the rightmost infinite open path starting from z, and let $\Gamma^{\epsilon,\pm}$ denote the set of all such paths. We note that for ϵ small enough, $\mathcal{K}^{\epsilon,\pm}$ is non-empty since $p > p_c$.

The first aim of this note is to describe the correlation between the sets of right-most paths $\Gamma^{\epsilon,+}$ and $\Gamma^{\epsilon,-}$, thus extending the link between oriented percolation and the Brownian web established in [13]. In order to explain our results, we recall the definition of left-right sticky pair of Brownian webs. This definition is slightly different from the one in Theorem 1.5 [15] (where this object first appeared) and is adapted from the definition of θ-coupled webs in Theorem 4 in [7] or Theorem 76 in [8]. We will comment more on the connection between the definitions in [15] and [7] in the Appendix. We first recall the following from [15, Proposition 2.1].

Definition 1 (Left-right sticky pair of Brownian paths). *Let (l, r) be two paths with respective starting time σ_l, σ_r. Assume first that $\sigma = \sigma_l = \sigma_r$. Then (l, r) is distributed as a left-right sticky pair of Brownian paths with drift b iff the pair (l, r) is distributed as the unique weak solution of the* left-right SDE

$$\begin{aligned} \forall\, t \geq \sigma,\ d\bar{l}_t &= 1_{\bar{l}_t \neq \bar{r}_t} dB_t^l + 1_{\bar{l}_t = \bar{r}_t} dB_t^s - b dt, \\ d\bar{r}_t &= 1_{\bar{l}_t \neq \bar{r}_t} dB_t^r + 1_{\bar{l}_t = \bar{r}_t} dB_t^s + b dt, \end{aligned} \tag{3}$$

(where B^l, B^r, B^s are independent standard Brownian motions) with initial condition $(l(\sigma), r(\sigma))$ at time σ, and subject to the constraint

(C) *$\bar{l}(t) \leq \bar{r}(t)$ for all $t \geq \tau := \inf\{s : \bar{l}(s) \leq \bar{r}(s)\}$.*

If $\sigma_l \neq \sigma_r$, then (l, r) will be called a left-right sticky pair if conditioned on $(l_t, r_t)_{t \leq \sigma_l \vee \sigma_r}$, $(l_t, r_t)_{t \geq \sigma_l \vee \sigma_r}$ solves the left-right SDE (3).

Remark 1. The term *sticky* is motivated by the fact that the process $w = \frac{l-r}{\sqrt{2}}$ is a drifted Brownian motion with sticky reflection at 0 (see e.g. [17]) when l and r have the same starting point.

The following theorem is adapted from [7, Theorem 4] on the characterization of a pair of sticky Brownian webs with zero drift. We will sketch the proof in the Appendix, where we show its equivalence with the characterization given in [15, Theorem 1.5].

Theorem 1 (Left-right sticky Brownian webs). *In distribution, there exists a unique pair $(\mathcal{W}_l, \mathcal{W}_r)$ valued in $\mathcal{H} \otimes \mathcal{H}$ such that*

1. *$\mathcal{W}_l$ (resp., $\mathcal{W}_r$) is a Brownian web with drift $-b$ (resp., b). In particular, for every deterministic z, there is a.s. a unique path l_z (resp., r_z) in $\mathcal{W}_l$ (resp., $\mathcal{W}_r$) starting from z.*
2. *For every deterministic pair $z_1, z_2 \in \mathbb{R}^2$, the pair $(l_{z_1}, r_{z_2}) \in \mathcal{W}_l \otimes \mathcal{W}_r$ is distributed as a left-right sticky pair of Brownian paths with drift b.*

3 (Co-adaptedness) Let $(\mathcal{G}_t; t \in \mathbb{R})$ be the natural filtration induced by the pair of Brownian webs $(\mathcal{W}_l, \mathcal{W}_r)$ (see Definition 2 below for more details). For any deterministic $z_1, z_2 \in \mathbb{R}^2$, the pairs of processes $(r_{z_1}, r_{z_2}), (l_{z_1}, l_{z_2})$ and (r_{z_1}, l_{z_2}) are all Markov with respect to $(\mathcal{G}_t; t \geq 0)$, i.e., for any such pair (l, r), $P((l(s), r(s))_{s \geq t} \in \cdot | \mathcal{G}_t) = P((l(s), r(s))_{s \geq t} \in \cdot | l(t), r(t))$.

We call $(\mathcal{W}_l, \mathcal{W}_r)$ a left-right pair of sticky Brownian webs with drift $b > 0$.

Our main result is the following, which generalizes the main result of [13] from a single Brownian web to a pair of sticky Brownian webs. The constants $\sigma(p)$ and $\alpha(p)$ are as in the convergence statement (2) and as in [13, (2.4)], and were first introduced in [9]. See also (5) below for more details.

Theorem 2. *The function $p \to \alpha(p)$ is differentiable and $\sigma(p) > 0$ on $(p_c, 1]$. Furthermore,*

$$S_{\sigma(p),\alpha(p),\epsilon}(\Gamma^{\epsilon,-}, \Gamma^{\epsilon,+}) \underset{\epsilon \to 0}{\Longrightarrow} (\mathcal{W}^l, \mathcal{W}^r),$$

where $(\mathcal{W}^l, \mathcal{W}^r)$ is a left-right pair of sticky Brownian webs with drift $b := \frac{\alpha'(p)}{\sigma(p)}$, regarded as a random variable taking values in $\mathcal{H} \otimes \mathcal{H}$ (see [14] for further details).

Remark 2. We note that a straightforward extension of [13] will show the convergence of the marginals, namely

$$S_{\sigma(p),\alpha(p),\epsilon}(\Gamma^{\epsilon,-}) \underset{\epsilon \to 0}{\Longrightarrow} \mathcal{W}_l$$

and

$$S_{\sigma(p),\alpha(p),\epsilon}(\Gamma^{\epsilon,+}) \underset{\epsilon \to 0}{\Longrightarrow} \mathcal{W}_r,$$

so most of our work will be dedicated to showing that the two webs have a sticky interaction (conditions 2–3 of Theorem 1).

Discrete Web Perturbations and the Dynamical Brownian Web. We close the introduction with a natural conjecture arising from the previous result. In order to do so, we make a detour, and we consider an alternative (and simpler) model where sticky Brownian webs also emerge naturally. Independently at each edge, declare one of the two randomly chosen out-going edge (or arrow) $(x \pm 1, t+1)$ to be open, whereas the other edge remains closed. At any point $z \in \mathbb{Z}^2_{even}$, there is exactly a single (infinite) path w_z starting from z and we denote by W the infinite collection of paths $\{w_z\}_{z \in \mathbb{Z}^2_{even}}$. This set is often referred to as the *discrete web* that can be described as an infinite set of coalescing random walks. In [3], it was shown that

$$S_{0,1,\epsilon}(W) \underset{\epsilon \to 0}{\Longrightarrow} \mathcal{W},$$

where $\mathcal{W}$ is the Brownian web. Let us now perturb the previous arrow configuration by adding an additional arrow independently at each site $z \in \mathbb{Z}^2_{even}$ with probability $\epsilon \in [0, 1]$. For $\epsilon > 0$, at any point z, there are infinitely many outgoing paths due to the branching of arrows, and we define $\tilde{\Gamma}^{+,\epsilon}$ (resp., $\tilde{\Gamma}^{-,\epsilon}$) to be the the infinite collection of right-most (resp., left-most paths) starting from $\mathbb{Z}^2_{even}$. In [15], it was shown that

$$S_{0,1,\epsilon}(\tilde{\Gamma}^{-,\epsilon}, \tilde{\Gamma}^{+,\epsilon}) \underset{\epsilon\to 0}{\Longrightarrow} (\mathcal{W}_l, \mathcal{W}_r)$$

where $(\mathcal{W}_l, \mathcal{W}_r)$ is a pair left-right sticky Brownian webs with drift 1. In light of Theorem 2, we observe that the left-right sticky pair of webs arises as a perturbation of the underlying path configuration in two distinct models: the discrete web and the supercritical percolation model.

In the discrete web described above, the left-right sticky pair of Brownian webs was obtained by adding some extra branching point in the underlying arrow configuration. We now describe an alternative perturbation of the discrete web W where sticky webs also arise naturally.

Start with the discrete web W and equip each vertex $z \in \mathbb{Z}^2_{\text{even}}$ with an independent Poisson clock with rate ϵ. Every time the clock rings at a given vertex z, switch the direction of the arrow starting from z (i.e., the edge $((x,t),(x\pm 1))$ becomes $((x,t),(x\mp 1))$). This defines a stationary Markov process $(W^\epsilon_s; s\geq 0)$ – called the discrete dynamical web [4] – whose one-dimensional marginal is given by the law of the discrete web. In [11], it was shown that

$$S_{0,1,\epsilon}(W_s; s\geq 0) \underset{\epsilon\to 0}{\Longrightarrow} (\mathcal{W}_s; s\geq 0)$$

where $(\mathcal{W}_s; s\geq 0)$ is a continuum objet called the dynamical Brownian web and the convergence is meant in the sense of finite dimensional distribution. See also [7] for a definition of the continuum object. Not surprisingly, this process is a stationary Markov process with marginal distribution being the law of the Brownian web, and the two-dimensional marginal $(\mathcal{W}_{s_1}, \mathcal{W}_{s_2})$ can be described in terms of a pair of sticky Brownian webs. In this context, the definition of a sticky pair of webs is analogous to Definition 1 with the notable difference that the Brownian motions have no drift and there is no natural ordering of the webs $(\mathcal{W}_{s_1}, \mathcal{W}_{s_2})$ (see condition (C) in Definition 1). See [7,11] for more details.

Dynamical Percolation. Let us now re-consider the supercritical oriented percolation model with parameter $p > p_c$. Analogously to the dynamical web, we can equip every edge with an independent continuous-time Markov chain $(v_z(s); s\geq 0)$ with initial law $\mathbb{P}(v_e(0)=1)=p$ and $\mathbb{P}(v_e(0)=0)=1-p$, and the chain switches from state 0 to 1 at rate ϵp and switches from 1 to 0 at rate $\epsilon(1-p)$. Note that the initial law of the Markov chain is also stationary under the dynamics. For every "dynamical time" $s\geq 0$, we declare an edge to be open if and only if $v_e(s)=1$. This defines a stationary dynamical oriented percolation model, whose marginal distribution at each dynamical time is the (static) oriented percolation model on $\mathbb{Z}^2_{\text{even}}$ with parameter p.

Conjecture 1. Let Γ^ϵ_s be the set of right-most infinite open paths at dynamical time s. There exists a constant $c(p)$ such that

$$S_{\alpha(p),\sigma(p),\epsilon}(\Gamma^\epsilon_s; s\geq 0) \underset{\epsilon\to 0}{\Longrightarrow} (\mathcal{W}_{c(p)s}; s\geq 0)$$

where $(\mathcal{W}_s; s\geq 0)$ is the dynamical Brownian web.

We note that scaling limits of dynamical percolation models have been obtained in the case of standard percolation in $\mathbb{Z}^2$ [1,5,6], but only at criticality (i.e. when $p = p_c$). In the context of oriented percolation, the previous conjecture would show that some interesting large-scale dynamics takes place even in the super-critical regime.

Technical Remarks. As pointed out before Theorem 1, the characterization of a sticky pair of left-right Brownian webs presented in this paper is not new, and is adapted from the one of θ-coupled Brownian webs given in [7, Theorem 4]. However, to the best of our knowledge, this characterization has never been used to prove scaling limit results, and we believe that it could be of interest in other settings. The latter characterization is particularly efficient since convergence boils down to proving that (1) $\mathcal{W}_l$ and $\mathcal{W}_r$ are two Brownian webs, and (2) each pair of paths with deterministic starting points has the required distribution w.r.t. to the *larger filtration* $(\mathcal{G}_t; t \geq 0)$. We will discuss more about this in the appendix.

Finally, we would also like to draw attention to Proposition 2, which provides a minimal characterization of a left-right sticky pair of Brownian paths that could be useful for proving convergence in other contexts.

2 Preliminaries

The main tool used in [13] is the notion of a *percolation exploration cluster*, which we briefly recall here. For every $z = (x, t) \in \mathbb{Z}^2_{\rm even}$, the percolation exploration cluster $C_z := (C_z(n))_{n \geq t}$ consists of a set of sequentially explored edges such that for each time $n > t$, a minimal set of edges $C_z(n)$ before time n are explored in order to find the rightmost open path connecting $(\infty, x] \times \{t\}$ to $\mathbb{Z} \times \{n\}$. See Fig. 2. The percolation exploration cluster C_z provides a good approximation for the rightmost infinite open path γ_z. Let ρ_z be the analog of r_z in [13], i.e., $\rho_z(n)$ is the rightmost position at time n that can be reached by some open path starting from $(-\infty, x] \times \{t\}$. It was shown in [13] that C_z is bounded between the paths γ_z and ρ_z, and furthermore, γ_z and ρ_z converge to the same Brownian motion after proper centering and scaling. The advantage of approximating γ_z by ρ_z is that the latter depends only on the edge configurations explored up to the present and not on the future.

As in [13], we denote by $(\rho(T_i), T_i)_{i \in \mathbb{N}}$ the successive break points along $\rho := \rho_o$. More precisely, the T_i's correspond to the successive times at which $\gamma := \gamma_o$ and ρ coincide. Let

$$\begin{aligned} \tau_1 &:= T_1, & X_1 &:= \rho(T_1), \\ \tau_i &:= T_i - T_{i-1}, & X_i &:= \rho(T_i) - \rho(T_{i-1}) \qquad \text{for } i \geq 2. \end{aligned}$$

It is known from [9] that $((X_i, \tau_i);\ i \geq 2)$ forms a sequence of i.i.d. random variables with τ_2 and X_2 having all finite moments. We will prove the following result.

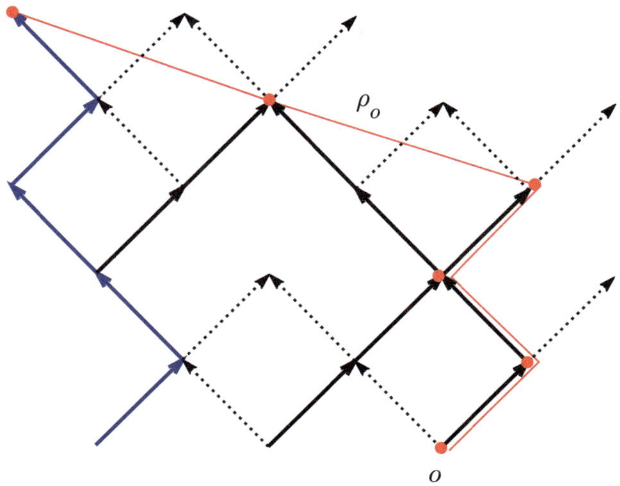

Fig. 2. The path ρ_o and the exploration cluster starting from o. $\rho_o(n)$ is the rightmost position at time n that can be reached by some open path starting from $(-\infty, 0] \times \{0\}$. The exploration cluster associated to o is obtained by exploring the set of vertices connected to $(-\infty, 0] \times \{0\}$ from right to left. See [13] for a precise description of the exploration algorithm to generate the cluster C_o.

Proposition 1. *For every $i, j \in \mathbb{N}$, the function*

$$f_{i,j}(p) := \mathbb{E}_p[X_2^i \tau_2^j] \tag{4}$$

is differentiable on $(p_c, 1)$.

As a corollary, we have

Corollary 1. *$p \to \alpha(p)$ and $p \to \sigma(p)$ are differentiable on $(p_c, 1)$.*

Proof. As shown in [13, (2.4)],

$$\alpha(p) \;=\; \frac{\mathbb{E}_p[X_2]}{\mathbb{E}_p[\tau_2]} \text{ and } \sigma^2(p) \;=\; \frac{\mathbb{E}_p\left[(X_2\mathbb{E}_p[\tau_2] - \tau_2\mathbb{E}_p[X_2])^2\right]}{\mathbb{E}_p[\tau_2]^3}. \tag{5}$$

The result then follows from a direct application of Proposition 1. □

Proof of Proposition 1. Kuczek [9] showed that when $p > p_c$, the distribution of τ_2 has exponential tail. Since $|X_2| \le \tau_2$, it follows that the infinite series

$$f_{i,j}(p) \;:=\; \sum_{(x,n)\in\mathbb{Z}\times\mathbb{N}: x\in[-n,n]} x^i n^j \; \mathbb{P}_p(X_2 = x,\; \tau_2 = n) \tag{6}$$

converges for each $p \in (p_c, 1)$. Let $C_o(n)$ be the set of edges (closed or open) discovered before exploring time horizon n (i.e. before reaching the point $(\rho_o(n), n)$). Again from [9], we know that

$$\mathbb{P}_p(X_2 = x,\; \tau_2 = n) \;=\; \mathbb{P}_p\left(C_o(n) \in A_{x,n}\right)$$

where we say $C_o(n) \in A_{x,n}$ if ρ_o satisfies the following property:

$$o \to (\rho_o(n), n),\ \rho_o(n) = x, \text{ and } (\rho_o(i), i) \not\to (x, n)\ \forall 1 \le i < n. \tag{7}$$

Note that this property is entirely determined by the realization of $C_o(n)$.

To prove the differentiability of

$$f_{i,j}(p) = \sum_{(x,n)\in\mathbb{Z}\times\mathbb{N}:x\in[-n,n]} x^i n^j\ \mathbb{P}_p\left(C_o(n) \in A_{x,n}\right),$$

we will first show that (i) $\mathbb{P}_p\left(C_o(n) \in A_{x,n}\right)$ is differentiable on $(p_c, 1]$ for every pair (x, n); and then show that (ii) the series

$$\sum_{(x,n)\in\mathbb{Z}\times\mathbb{N}:x\in[-n,n]} x^i n^j \frac{\partial}{\partial p}\mathbb{P}_p\left(C_o(n) \in A_{x,n}\right) \tag{8}$$

converges uniformly on every interval $[a, b] \subset (p_c, 1)$.

First note that

$$\mathbb{P}_p\left(C_o(n) \in A_{x,n}\right) = \sum_{C_o(n)\in A_{x,n}} p^{|C_o^{\text{open}}(n)|}(1-p)^{|C_o^{\text{closed}}(n)|}, \tag{9}$$

where $C_o^{\text{open}}(n)$ and $C_o^{\text{closed}}(n)$ denote respectively the subset of open and closed edges in $C_o(n)$. Note that the sum above is finite since under the condition $C_o(n) \in A_{x,n}$, we have

$$C_o(n) \subseteq \{e = ((x,t),(x \pm 1, t+1))\ :\ 0 \le t \le n,\ |x| \le n\}.$$

It follows that $\mathbb{P}_p\left(C_o(n) \in A_{x,n}\right)$ is differentiable with

$$\begin{aligned}
&\frac{\partial}{\partial p}\mathbb{P}_p\left(C_o(n) \in A_{x,n}\right) \\
&\quad = \sum_{C_o(n)\in A_{x,n}} \left(\frac{|C_o^{\text{open}}(n)|}{p} - \frac{|C_o^{\text{closed}}(n)|}{1-p}\right) p^{|C_o^{\text{open}}(n)|}(1-p)^{|C_o^{\text{closed}}(n)|},
\end{aligned}$$

which implies that for all $p \in [a, b]$,

$$\begin{aligned}
&\left|\frac{\partial}{\partial p}\mathbb{P}_p\left(C_o(n) \in A_{x,n}\right)\right| \\
&\le \left(\frac{1}{a} + \frac{1}{1-b}\right) \sum_{C_o(n)\in A_{x,n}} |C_o(n)| p^{|C_o^{\text{open}}(n)|}(1-p)^{|C_o^{\text{closed}}(n)|} \\
&\le cn^2\mathbb{P}_p(C_o(n) \in A_{x,n}) \le cn^2\mathbb{P}_p(\tau_2 = n),
\end{aligned}$$

where we used $|C_o(n)| \le 2n(2n+1)$, and the constant c depends only on a and b. It follows that

$$\sum_{(x,n)\in\mathbb{Z}\times\mathbb{N}:x\in[-n,n]} |x|^i n^j \left|\frac{\partial}{\partial p}\mathbb{P}_p\left(C_o(n) \in A_{x,n}\right)\right| \le c\cdot 3^i \sum_{n\in\mathbb{N}} n^{i+j+3}\mathbb{P}_p(\tau_2 = n).$$

To prove the uniform convergence of the series in (8) for $p \in [a, b]$, we only need to prove the uniform convergence of

$$\sum_{n \in \mathbb{N}} n^{i+j+3} \mathbb{P}_p(\tau_2 = n), \qquad p \in [a, b]. \tag{10}$$

For each $p > p_c$, Kuczek [9] has proved that $\mathbb{P}(\tau_2 \geq n) \leq C_1(p) e^{-C_2(p)n}$, and hence the above series converges pointwise. The uniform convergence follows from the fact that $C_1(p)$ and $C_2(p)$ can be chosen uniformly for $p \in [a, b]$, which also follows from Kuczek's proof. The key ingredient is the following estimate by Durrett [2]:

$$\mathbb{P}_p(o \to \mathbb{Z} \times \{n\}, o \notin \mathcal{K}) \leq c_1 e^{-c_2 n}, \tag{11}$$

where it is easily seen from the proof that c_1, c_2 can be chosen uniformly for $p \in [a, b] \subset (p_c, 1)$. This concludes the proof of Proposition 1. □

3 Invariance Principle for a Pair of Left-Right Paths

Following the notation in [13], let Π denote the space of real-valued continuous functions starting from some time in $\mathbb{R}$, equipped with the topology of local uniform convergence plus convergence of the starting time. Recall from Sect. 1 the definition of the percolation configurations $\Omega^{\epsilon,\pm}$, and for $z \in \mathbb{Z}^2_{\text{even}}$, let $\gamma_z^{\epsilon,\pm}$ and $\rho_z^{\epsilon,\pm}$ be paths defined from $\Omega^{\epsilon,\pm}$ in the same way as γ_z and ρ_z are defined from Ω in Sect. 2. The main result of this section is the following.

Theorem 3 (Convergence of a pair of left-right paths). *Let*

$$z_\pm := (x_\pm, t_\pm) \in \mathbb{R}^2$$

and let $z_\pm^\epsilon := (x_\pm^\epsilon, t_\pm^\epsilon) \in \mathbb{Z}^2_{\text{even}}$ be such that $S_{\alpha(p),\sigma(p),\epsilon}(z_\pm^\epsilon) \to z_\pm$ as $\epsilon \to 0$. Let (l, r) be a left-right pair of sticky Brownian motions with drift $b = \frac{\alpha'(p)}{\sigma(p)}$, starting respectively at the space-time points z_- and z_+. Then

$$\left(S_{\alpha(p),\sigma(p),\epsilon}(\rho^{\epsilon,-}_{z^\epsilon_-}, \gamma^{\epsilon,-}_{z^\epsilon_-}, \rho^{\epsilon,+}_{z^\epsilon_+}, \gamma^{\epsilon,+}_{z^\epsilon_+}) \right) \underset{\epsilon \to 0}{\Longrightarrow} (l, l, r, r), \tag{12}$$

where $\rho^{\epsilon,\pm}$ are defined at non-integer times via linear interpolation, and the scaling map $S_{\alpha(p),\sigma(p),\epsilon}$ is defined in (1).

In the next section, we provide an outline of the proof. We postpone the main technical parts until further sections.

3.1 Outline of the Proof

Until further notice, we will assume that $t_-^\epsilon = t_+^\epsilon$. In the final Subsect. 3.5, we will show how Theorem 3 can easily be deduced from this particular case.

We decompose the proof into several steps. Recall from the introduction that the pair of left-right sticky pair (l, r) with drift b is defined as the unique weak solution of the *left-right SDE*

$$\begin{aligned} dl(t) &= 1_{l_t \neq r_t} dB^l(t) + 1_{l_t = r_t} dB^s(t) - bdt \\ dr(t) &= 1_{l_t \neq r_t} dB^r(t) + 1_{l_t = r_t} dB^s(t) + bdt \end{aligned} \tag{13}$$

subject to the constraint that $l(t) \leq r(t)$ for all $t \geq \tau := \inf\{s : l(s) \leq r(s)\}$, and B^l, B^r, B^s are independent standard Brownian motions. We first need the following characterization of (l, r).

Proposition 2. *Let $b > 0$, and let $z_\pm = (x_\pm, t_\pm)$. The sticky pair (l, r) is the unique process satisfying the following three properties:*

(a) *$l = (l(t))_{t \geq t_-}$ and $r = (r(t))_{t \geq t_+}$ are Brownian motions (defined w.r.t. the same filtration) with unit diffusion constant, respective drift $-b$ and b, and $l(t_-) = x_-$, $r(t_+) = x_+$.*
(b) *$1_{l(t) \neq r(t)}\, d\langle l, r\rangle_t = 0$, where $\langle l, r\rangle$ denotes the cross-variation of l and r.*
(c) *$l(t) \leq r(t)$ for all $t \geq \tau := \inf\{s \geq t_- \vee t_+ : l(s) \leq r(s)\}$.*

Furthermore, $\frac{1}{\sqrt{2}}(r(\tau + t) - l(\tau + t))$ is equal in law to the weak solution of the following SDE:

$$dw(t) \;=\; \sqrt{2}b\, dt + 1_{w \neq 0}\, dB(t), \qquad w(0) = \frac{r(\tau) - l(\tau)}{\sqrt{2}}, \tag{14}$$

where B is a standard Brownian motion.

We postpone the proof till Sect. 3.2.

We first establish the invariance principle for γ and ρ starting from a single point, for which it suffices to consider paths $\rho^{\epsilon,\pm} := \rho_o^{\epsilon,\pm}$ starting at the origin. Denote by $(\rho^{\epsilon,\pm}(T_i^{\epsilon,\pm}), T_i^{\epsilon,\pm})_{i \in \mathbb{N}}$ the successive break points along $\rho^{\epsilon,\pm}$, and let $(X_i^{\epsilon,\pm}, \tau_i^{\epsilon,\pm})_{i \in \mathbb{N}}$ be the successive space-time increments between break points as defined in Sect. 2.

Lemma 1. *For every $n \in \mathbb{N}$, define*

$$W^{\epsilon,\pm}(n) \;:=\; \sum_{i=1}^{n} \left(X_i^{\epsilon,\pm} - \alpha(p)\tau_i^{\epsilon,\pm}\right), \; T^{\epsilon,\pm}(n) \;:=\; T_n^{\epsilon,\pm} = \sum_{i=1}^{n} \tau_i^{\epsilon,\pm}, \tag{15}$$

and for $t \notin \mathbb{N}$, define $W(t), T(t)$ by linear interpolation. Conditional on $o \in \mathcal{K}^{\epsilon,\pm}$,

$$\left(\frac{\epsilon}{\sigma(p)} W^{\epsilon,\pm}(\epsilon^{-2}t)\right)_{t \geq 0} \;\underset{\epsilon \to 0}{\Longrightarrow}\; \left(B(\mathbb{E}[\tau_2]t) \pm \mathbb{E}[\tau_2]\frac{\alpha'(p)}{\sigma(p)}t\right)_{t \geq 0}, \tag{16}$$

$$(\epsilon^2 T^{\epsilon,\pm}(\epsilon^{-2}t))_{t \geq 0} \;\underset{\epsilon \to 0}{\Longrightarrow}\; (t\mathbb{E}[\tau_2])_{t \geq 0}, \tag{17}$$

where B is a standard Brownian motion.

Proof. From [9], conditioned on $o \in \mathcal{K}^{\epsilon,\pm}$, $(X_i^{\epsilon,\pm}, \tau_i^{\epsilon,\pm})_{i\in\mathbb{N}}$ is a sequence of i.i.d random variables with the same distribution as $(X_2^{\epsilon,\pm}, \tau_2^{\epsilon,\pm})$ without conditioning. Furthermore, using Proposition 1 we have

$$\begin{aligned} \mathbb{E}\left[X_2^{\epsilon,\pm} - \alpha(p)\tau_2^{\epsilon,\pm}\right] &= \pm\, \mathbb{E}[\tau_2]\alpha'(p)\epsilon + O(\epsilon^2), \\ \mathbb{E}\left[(X_2^{\epsilon,\pm} - \alpha(p)\tau_2^{\epsilon,\pm})^2\right] &= \mathbb{E}[\tau_2]\sigma^2(p) + O(\epsilon). \end{aligned} \tag{18}$$

The first convergence statement then follows from a standard invariance principle (for triangular arrays). The second convergence statement follows from the law of large numbers (for triangular arrays) and the fact that $T^{\epsilon,\pm}$ is non-decreasing. □

Corollary 2 (Conditional Invariance Principle). *Conditional on $o \in \mathcal{K}^{\epsilon,-}$, resp., $\mathcal{K}^{\epsilon,+}$,*

$$S_{\alpha(p),\sigma(p),\epsilon}(\rho^{\epsilon,-}) \underset{\epsilon\to 0}{\Longrightarrow} l, \quad \textit{resp.}, \quad S_{\alpha(p),\sigma(p),\epsilon}(\rho^{\epsilon,+}) \underset{\epsilon\to 0}{\Longrightarrow} r,$$

where l, resp., r, is distributed as in the left-right pair of sticky Brownian motions with drift $b = \frac{\alpha'(p)}{\sigma(p)}$, defined in Proposition 2.

Proof. We follow closely the proof Proposition 2.2 in [13]. First, we start by replacing the path $\rho^{\epsilon,\pm}$ by $\tilde{\rho}^{\epsilon,\pm}$ where $\tilde{\rho}^{\epsilon,\pm}(T_i^{\epsilon,\pm}) = \rho^{\epsilon,\pm}(T_i^{\epsilon,\pm})$ and for $t \in (T_i^{\epsilon,\pm}, T_{i+1}^{\epsilon,\pm})$, $\tilde{\rho}^{\epsilon,\pm}$ is defined by linear interpolation. More precisely, we write the decomposition:

$$S_{\alpha(p),\sigma(p),\epsilon}(\rho^{\epsilon,\pm}) \;=\; S_{\alpha(p),\sigma(p),\epsilon}(\tilde{\rho}^{\epsilon,\pm}) \;+\; S_{0,\sigma(p),\epsilon}(\tilde{\rho}^{\epsilon,\pm} - \rho^{\epsilon,\pm}).$$

Arguments analogous to the ones in [13] show that the second term vanishes as $\epsilon \to 0$.

From the definition, it is straightforward to see that

$$S_{\alpha(p),0,0}\left(\tilde{\rho}^{\epsilon,\pm}\right)(t) \;=\; W^{\epsilon,\pm}((T^{\epsilon,\pm})^{-1}(t)),$$

and

$$S_{\alpha(p),\sigma(p),\epsilon}(\tilde{\rho}^{\epsilon,\pm}) = S_{0,\sigma(p),\epsilon}S_{\alpha(p),0,0}(\tilde{\rho}^{\epsilon,\pm}) = S_{0,\sigma(p),\epsilon}(W^{\epsilon,\pm}((T^{\epsilon,\pm})^{-1})).$$

Lemma 1 implies that $S_{0,\sigma(p),\epsilon}(W^{\epsilon,\pm})$ converges to $(B(\mathbb{E}[\tau_2]t) \pm \mathbb{E}[\tau_2]\frac{\alpha'(p)}{\sigma(p)}t)_{t\geq 0}$, where B is a standard Brownian motion, and $(\epsilon^2 T^{\epsilon,\pm}(t/\epsilon^2))_{t\geq 0}$ converges to $(\mathbb{E}[\tau_2]t)_{t\geq 0}$. It then follows that $S_{\alpha(p),\sigma(p),\epsilon}(\rho^{\epsilon,\pm})$ converges in distribution to

$$\left(B(t) \;\pm\; \frac{\alpha'(p)}{\sigma(p)}t\right)_{t\geq 0},$$

which is exactly the law of l, resp., r. □

Proposition 3 (Invariance Principle). *Following the same notation as in Corollary 2, we have*

$$S_{\sigma(p),\alpha(p),\epsilon}(\rho^{\epsilon,-}, \gamma^{\epsilon,-}) \underset{\epsilon\to 0}{\Longrightarrow} (l,l), \quad \textit{and} \quad S_{\sigma(p),\alpha(p),\epsilon}(\rho^{\epsilon,+}, \gamma^{\epsilon,+}) \underset{\epsilon\to 0}{\Longrightarrow} (r,r).$$

Proof. The unconditional invariance principle for $\rho^{\epsilon,\pm}$, as well as the fact that $\rho^{\epsilon,\pm}$ and $\gamma^{\epsilon,\pm}$ converge to the same scaling limit, can be established by the same argument as in the proof of [13, Prop. 2.2]. The only modification needed is that we need to use the fact that the tail estimate (11) is uniform in $p \in [a,b] \subset (p_c, 1)$, since the percolation parameters now depend on ϵ. □

To establish Theorem 3, note that Proposition 3 implies that

$$(S_{\sigma(p),\alpha(p),\epsilon}(\rho^{\epsilon,-}_{z^\epsilon_-}, \rho^{\epsilon,+}_{z^\epsilon_+}))_{\epsilon>0}$$

is a tight sequence of $\Pi \times \Pi$-valued random variables. Going to a subsequence if necessary, we can assume the existence of a pair of drifted Brownian motions (l, r) such that

$$S_{\sigma(p),\alpha(p),\epsilon}(\rho^{\epsilon,-}_{z^\epsilon_-}, \rho^{\epsilon,+}_{z^\epsilon_+}) \underset{\epsilon\to 0}{\Longrightarrow} (l, r).$$

In order to prove Theorem 3, it only remains to show that the sub-sequential limit is unique in distribution, i.e., (l, r) satisfies the three conditions of Proposition 2. Condition (a) is satisfied by Proposition 3. Conditions (b) and (c) will follow from the next two results.

Proposition 4. $1_{l(t)\neq r(t)} d\langle l, r\rangle_t = 0.$

Proposition 5. *$l(t) \le r(t)$ for all $t \ge \tau := \inf\{s \ge t_- \vee t_+ : l(s) \le r(s)\}$.*

In order prove Theorem 3, it remains to show Propositions 2, 4, and 5. This is done in the next sections.

3.2 Proof of Proposition 2

It has been shown in [15] that the left-right SDE (13) has a unique weak solution (l, r), and $\frac{1}{\sqrt{2}}(r - l)$ is a weak solution of the Eq. (14). Any such solution (l, r) clearly satisfies conditions (a)–(c) in Proposition 2.

It remains to prove uniqueness of the process satisfying the three properties. Let us assume that there exists a process (l, r) satisfying Proposition 2 (a)–(c). Note that (a) and (b) imply that l and r are independent Brownian motions before they meet at time τ. Therefore we only need to prove uniqueness in distribution from time τ onward, and hence we may assume that l and r both start at 0 at time 0. Define

$$U(t) := \frac{r(t) + l(t)}{2}, \qquad V(t) := \frac{r(t) - l(t)}{2} - bt.$$

By assumption (a), it is straightforward to check that U and V are orthogonal martingales. Furthermore, writing

$$C(t) = \frac{t}{2} - \frac{1}{2}\int_0^t 1_{l(s)=r(s)} d\langle l, r\rangle_s, \tag{19}$$

we have $\langle U\rangle(t) = t - C(t)$ and $\langle V\rangle(t) = C(t)$. By Knight's theorem [12, Theorem V (1.9)], there exist two independent standard Brownian motions B and B' such that

$$U = B'(t - C(t)), \qquad V(t) = B(C(t)).$$

Recovering l and r from U and V, we obtain

$$\begin{aligned} r(t) &= B'(t - C(t)) + B(C(t)) + bt, \\ l(t) &= B'(t - C(t)) - B(C(t)) - bt. \end{aligned} \tag{20}$$

Note that C is increasing. We claim that C is strictly increasing. Assume the contrary that $C(t_1) = C(t_2)$ for some $0 \le t_1 < t_2$. Then by (19), we have

$$\frac{t_2 - t_1}{2} = \frac{1}{2}\int_{t_1}^{t_2} 1_{l(s)=r(s)} d\langle l, r\rangle_s,$$

which is only possible if $l(s) = r(s)$ for all $s \in [t_1, t_2]$, since $d\langle l, r\rangle_s \le ds$. However, by (20), we cannot have both $l = r$ and C being a constant on $[t_1, t_2]$. Therefore $C(t)$ is strictly increasing and admits an inverse C^{-1}. Denote

$$Z(t) := \frac{r(C^{-1}(t)) - l(C^{-1}(t))}{2} = B(t) + bC^{-1}(t).$$

In order to show the uniqueness the law of (l, r), we will show:

(1) Z is $B(t) + 2bt$ Skorohod reflected at 0;
(2) $C^{-1}(t) = 2t - \frac{1}{b}\inf_{s\in[0,t]}(B(s) + 2bs) \wedge 0$.

This would imply that $(C(t))_{t\ge 0}$ is determined by $(B(t))_{t\ge 0}$, and hence the law of (l, r) is unique by (20). We can write

$$Z(t) = B(t) + 2bt + b(C^{-1}(t) - 2t).$$

By (19), it is easily seen that $(C^{-1}(t) - 2t)$ is non-negative, and it only increases when $Z = 0$. Since $Z \ge 0$ by Proposition 2(c), by uniqueness of the solution to the Skorohod equation [12, Lemma VI (2.1)], we deduce that Z is the Skorohod reflection of $B(t) + 2bt$ at 0, and

$$b(C^{-1}(t) - 2t) = -\inf_{s\in[0,t]}(B(s) + 2bs) \wedge 0.$$

This completes the proof of (1) and (2), and thus the proof of uniqueness. □

3.3 Proof of Proposition 4

The following preliminary result will be needed to prove Proposition 4. It shows that two paths $\rho^{\epsilon,\pm}_{z_\pm}$ converge in the scaling limit to two independent Brownian motions before they meet.

Lemma 2. *Let $z_\pm := (x_\pm, t_\pm) \in \mathbb{R}^2$ and let $z^\epsilon_\pm := (x^\epsilon_\pm, t^\epsilon_\pm) \in \mathbb{Z}^2_{\text{even}}$ be such that $S_{\alpha(p),\sigma(p),\epsilon}(z^\epsilon_\pm) \to z_\pm$ as $\epsilon \to 0$. Assume that $z_+ \neq z_-$, and for any $\delta > 0$, define*

$$\tau^\epsilon_\delta = \inf\left\{n \geq t^\epsilon_+ \vee t^\epsilon_- \ : \ \frac{\epsilon}{\sigma(p)}\left|\rho^{\epsilon,+}_{z^\epsilon_+}(n) - \rho^{\epsilon,-}_{z^\epsilon_-}(n)\right| \ \leq \ \delta\right\}.$$

Then

$$\left(S_{\alpha(p),\sigma(p),\epsilon}\left(\rho^{\epsilon,-}_{z^\epsilon_-}(\cdot\wedge\tau^\epsilon_\delta),\, \gamma^{\epsilon,-}_{z^\epsilon_-}(\cdot\wedge\tau^\epsilon_\delta),\, \rho^{\epsilon,+}_{z^\epsilon_+}(\cdot\wedge\tau^\epsilon_\delta),\, \gamma^{\epsilon,+}_{z^\epsilon_+}(\cdot\wedge\tau^\epsilon_\delta)\right),\, \epsilon^2\tau^\epsilon_\delta\right)$$
$$\underset{\epsilon\to 0}{\Longrightarrow} \quad (l(\cdot\wedge\tau_\delta),\, l(\cdot\wedge\tau_\delta),\, r(\cdot\wedge\tau_\delta),\, r(\cdot\wedge\tau_\delta),\, \tau_\delta),$$

where (l, r) is a pair of independent Brownian motions starting respectively at z_- and z_+, with respective drift $-b$ and b and common diffusion constant 1, and

$$\tau_\delta := \inf\{t \geq t_+ \vee t_- : |r(t) - l(t)| < \delta\}.$$

Proof. The proof is the same as in the proof of [13, Prop. 3.3]. The main idea is that the paths $\rho^{\epsilon,\pm}_{z^\epsilon_\pm}$ and $\gamma^{\epsilon,\pm}_{z^\epsilon_\pm}$ can be approximated by their respective percolation exploration clusters, which evolve independently before they intersect. Since the exploration clusters (which are always bounded between ρ_z and γ_z) individually converge to a Brownian motion by Proposition 3, the result then follows. See the proof of [13, Prop. 3.3] for more details, which is more involved than our case since it also considers what happens after the exploration clusters intersect. □

By going to a subsequence if needed, we may assume that

$$S_{\sigma(p),\alpha(p),\epsilon}(\rho^{\epsilon,-}_{z^\epsilon_-}, \rho^{\epsilon,+}_{z^\epsilon_+}) \Rightarrow (l, r).$$

To prove $1_{l(t)\neq r(t)} d\langle l, r\rangle_t = 0$, it suffices to show that for any $\delta > 0$, (l, r) are distributed as independent Brownian motions when $\Delta(t) := r(t) - l(t)$ satisfies $|\Delta(t)| > \delta$.

Define $\kappa_0, \xi_0 = t_- \vee t_+$, and for every $i \geq 0$:

$$\kappa_{i+1} = \inf\{t \geq \xi_i \ : |\Delta(t)| > \delta\}, \qquad \xi_{i+1} = \inf\{t \geq \kappa_{i+1} \ : \ |\Delta(t)| < \delta/2\}, \quad (21)$$

with the convention that $\inf\{\emptyset\} = \infty$. To prove Proposition 4, it then suffices to show that

Lemma 3. *For each $i \in \mathbb{N}$, conditioned on $(l(t), r(t))_{t\leq\kappa_i}$ with $\kappa_i < \infty$, the process*

$$\big(l(s\wedge\xi_i),\, r(s\wedge\xi_i)\big)_{s\geq\kappa_i}$$

is distributed as a pair of independent Brownian motions with unit diffusion constant and respective drift $-b, b$, stopped when their distance reaches $\delta/2$.

Proof. This will be proved by an approximation argument. Recall that we have assumed that

$$S_{\sigma(p),\alpha(p),\epsilon}(\rho^{\epsilon,-}_{z^\epsilon_-}, \rho^{\epsilon,+}_{z^\epsilon_+}) \underset{\epsilon\to 0}{\Longrightarrow} (l, r).$$

Analogous to (κ_i, ξ_i), let us define $\kappa_0^\epsilon = \xi_0^\epsilon = t_-^\epsilon \vee t_+^\epsilon$, and for $i \geq 1$, define

$$\kappa_{i+1}^\epsilon = \inf\Big\{k > \xi_i^\epsilon \ : \ |\Delta\rho^\epsilon(k)| \geq \frac{\sigma(p)}{\epsilon}\delta\Big\},$$
$$\xi_{i+1}^\epsilon = \inf\Big\{k > \kappa_{i+1}^\epsilon \ : \ |\Delta\rho^\epsilon(k)| \leq \frac{\sigma(p)}{\epsilon}\frac{\delta}{2}\Big\},$$

where $\Delta\rho^\epsilon := \rho^{\epsilon,+}_{z_+^\epsilon} - \rho^{\epsilon,-}_{z_-^\epsilon}$. As $[0,\infty]$-valued random variables, $(\epsilon^2\kappa_i^\epsilon)_{i\in\mathbb{N}}$ and $(\epsilon^2\xi_i^\epsilon)_{i\in\mathbb{N}}$ are automatically tight. Therefore by going to a subsequence if necessary, we may assume that

$$\Big(S_{\alpha(p),\sigma(p),\epsilon}\Big(\rho^{\epsilon,-}_{z_-^\epsilon}, \gamma^{\epsilon,-}_{z_-^\epsilon}, \rho^{\epsilon,+}_{z_+^\epsilon}, \gamma^{\epsilon,+}_{z_+^\epsilon}\Big),\ (\epsilon^2\kappa_i^\epsilon, \epsilon^2\xi_i^\epsilon)_{i\in\mathbb{N}}\Big)$$
$$\Longrightarrow \big((l,l,r,r),\ (\bar\kappa_i, \bar\xi_i)_{i\in\mathbb{N}}\big), \quad (22)$$

and by the Skorohod representation theorem, we can assume the above convergence to be almost sure via a suitable coupling. Note that under this coupling, by the definition of $(\kappa_i, \xi_i)_{i\in\mathbb{N}}$ in (21), we must have $\bar\kappa_i \leq \kappa_i$ and $\bar\xi_i \leq \xi_i$ for all $i \in \mathbb{N}$, which will be strengthened to equalities later.

Let us fix $i \in \mathbb{N}$, and to simplify notation, we will omit the subscript i from $\kappa_i^\epsilon, \bar\kappa_i, \xi_i^\epsilon$. Let us consider the composite path $\bar\rho^{\epsilon,\pm}$ defined as follows:

$$\bar\rho^{\epsilon,\pm}_{z_\pm^\epsilon}(t) \ = \ \begin{cases} \rho^{\epsilon,\pm}_{z_\pm^\epsilon}(t) & \text{if } t \leq \kappa^\epsilon, \\ \rho^{\epsilon,\pm}_{w_\pm^\epsilon}(t) & \text{if } t \geq \kappa^\epsilon, \ \text{ where } w_\pm^\epsilon := (\rho^{\epsilon,\pm}_{z_\pm^\epsilon}(\kappa^\epsilon),\ \kappa^\epsilon). \end{cases} \quad (23)$$

We first claim that conditional on $\{\bar\kappa < \infty\}$, these modified paths are good approximations of $\rho^{\epsilon,\pm}_{z_\pm^\epsilon}$, and hence of (l,r). More precisely,

$$S_{\alpha(p),\sigma(p),\epsilon}\big(\rho^{\epsilon,\pm}_{w_\pm^\epsilon}(t) - \rho^{\epsilon,\pm}_{z_\pm^\epsilon}(t)\big)_{t\geq\kappa^\epsilon} \underset{\epsilon\to 0}{\Longrightarrow} 0. \quad (24)$$

Note that conditioned on $\bar\kappa < \infty$, $S_{\alpha(p),\sigma(p),\epsilon}(\rho^{\epsilon,\pm}_{w_\pm^\epsilon} - \gamma^{\epsilon,\pm}_{w_\pm^\epsilon}) \Rightarrow 0$ by Proposition 3. To prove (24), it then suffices to show that

$$S_{\alpha(p),\sigma(p),\epsilon}\big(\gamma^{\epsilon,\pm}_{w_\pm^\epsilon}(t) - \gamma^{\epsilon,\pm}_{z_\pm^\epsilon}(t)\big)_{t\geq\kappa^\epsilon} \underset{\epsilon\to 0}{\Longrightarrow} 0. \quad (25)$$

This is easily seen to hold since the difference between the starting points of these two paths,

$$|\rho^{\epsilon,\pm}_{z_\pm^\epsilon}(\kappa^\epsilon) - \gamma^{\epsilon,\pm}_{z_\pm^\epsilon}(\kappa^\epsilon)| \underset{\epsilon\to 0}{\longrightarrow} 0$$

almost surely on the event $\{\bar\kappa < \infty\}$ by the coupling assumed in (22). In particular, for any $\alpha > 0$, if we denote

$$\widetilde{w}_\pm^\epsilon := (\rho^{\epsilon,\pm}_{z_\pm^\epsilon}(\kappa^\epsilon) - \sigma\alpha/\epsilon,\ \kappa^\epsilon),$$

then except on a set with probability tending to 0, we have

$$\gamma^{\epsilon,\pm}_{\widetilde{w}_\pm^\epsilon}(t) \leq \gamma^{\epsilon,\pm}_{z_\pm^\epsilon}(t) \leq \gamma^{\epsilon,\pm}_{w_\pm^\epsilon}(t) \qquad \text{for all } t \geq \kappa^\epsilon. \quad (26)$$

The advantage of working with $\gamma^{\epsilon,\pm}_{\tilde{w}^\epsilon_\pm}$ and $\gamma^{\epsilon,\pm}_{w^\epsilon_\pm}$ is that except for their starting points, they depend only on the percolation configuration above time κ^ϵ. Therefore by Proposition 3.3 of [13] (modified to take into account the ϵ-dependence of the percolation parameter), $S_{\alpha(p),\sigma(p),\epsilon}(\gamma^{\epsilon,\pm}_{\tilde{w}^\epsilon_\pm}, \gamma^{\epsilon,\pm}_{w^\epsilon_\pm})$ can be approximated by two coalescing Brownian motions started at the same time with distance α apart. By (26) with $\alpha > 0$ chosen arbitrarily small, we then obtain (25), and hence (24).

By going to a further subsequence if necessary, we can now enhance (22) by including the convergence in (24), and by coupling via Skorohod representation, we may assume the convergence to be almost sure, so that

$$S_{\alpha(p),\sigma(p),\epsilon}(\rho^{\epsilon,-}_{z^\epsilon_-}(t), \rho^{\epsilon,+}_{z^\epsilon_+}(t))_{t\leq\kappa^\epsilon}$$

converges almost surely to $(l(t), r(t))_{t\leq\bar{\kappa}}$, and

$$S_{\alpha(p),\sigma(p),\epsilon}(\rho^{\epsilon,-}_{w^\epsilon_-}, \rho^{\epsilon,+}_{w^\epsilon_+})$$

converges almost surely to $(l(t), r(t))_{t\geq\bar{\kappa}}$. On the other hand, we note that conditioned on the percolation configuration up to time κ^ϵ, the law of $(\rho^{\epsilon,-}_{w^\epsilon_-}, \rho^{\epsilon,+}_{w^\epsilon_+})$ depends only on their starting points $w^\epsilon_\pm$. Therefore under the coupling that

$$S_{\alpha(p),\sigma(p),\epsilon}(\rho^{\epsilon,-}_{z^\epsilon_-}(t), \rho^{\epsilon,+}_{z^\epsilon_+}(t))_{t\leq\kappa^\epsilon}$$

converges almost surely to $(l(t), r(t))_{t\leq\bar{\kappa}}$, by Lemma 2, the law of

$$S_{\alpha(p),\sigma(p),\epsilon}(\rho^{\epsilon,-}_{w^\epsilon_-}(t\wedge\xi_\epsilon), \rho^{\epsilon,+}_{w^\epsilon_+}(t\wedge\xi^\epsilon))_{t\geq\kappa^\epsilon}$$

conditioned on $(\rho^{\epsilon,-}_{z^\epsilon_-}(t), \rho^{\epsilon,+}_{z^\epsilon_+}(t))_{t\leq\kappa^\epsilon}$ converges to the law of a pair of independent Brownian motions $(\tilde{l}, \tilde{r})$ with unit diffusion constant and respective drift $-b$ and b, starting respectively at $l(\bar{\kappa})$ and $r(\bar{\kappa})$ at time $\bar{\kappa}$, and stopped when their distance reaches $\delta/2$. Since conditioned on $(l(t), r(t))_{t\leq\bar{\kappa}}$, $(l(t\wedge\bar{\xi}), r(t\wedge\bar{\xi}))_{t\geq\bar{\kappa}}$ has the same distribution as $(\tilde{l}, \tilde{r})$, Lemma 3 would follow once we show that $(\bar{\kappa}_i, \bar{\xi}_i) = (\kappa_i, \xi_i)$.

Indeed, Lemma 2 implies that for each $i \in \mathbb{N}$, $\bar{\xi}_i = \inf\{t \geq \bar{\kappa}_i : |r(t) - l(t)| < \frac{\delta}{2}\}$. Therefore it only remains to show that $\bar{\kappa}_i = \kappa_i$ for every $i \in \mathbb{N}$. We can proceed inductively. We may assume that $|r(\xi_0) - l(\xi_0)| < \delta$. By our definition of κ_1 and the almost sure coupling in (22), we must have $\bar{\kappa}_1 \leq \kappa_1$. On the other hand, we have shown that $(l(t\wedge\bar{\xi}_1), r(t\wedge\bar{\xi}_1))_{t\geq\bar{\kappa}_1}$ is distributed as a pair of independent Brownian motions with initial distance δ at time $\bar{\kappa}_1$. Therefore $\bar{\kappa}_1 = \inf\{t > \bar{\kappa}_1 : |r(t) - l(t)| > \delta\}$, which implies that $\kappa_1 \leq \bar{\kappa}_1$. Therefore $\bar{\kappa}_1 = \kappa_1$. We can now iterate the argument to show that $\bar{\kappa}_i = \kappa_i$ for all $i \in \mathbb{N}$. □

3.4 Proof of Proposition 5

By going to a subsequence if needed, we may assume that

$$S_{\sigma(p),\alpha(p),\epsilon}(\rho^{\epsilon,-}_{z^\epsilon_-}, \rho^{\epsilon,+}_{z^\epsilon_+}) \Rightarrow (l, r),$$

and by the Skorohod representation theorem, we may even assume the convergence to be almost sure.

If $\rho^{\epsilon,-}_{z^\epsilon_-}(t^\epsilon) \le \rho^{\epsilon,+}_{z^\epsilon_+}(t^\epsilon)$, then we have $\rho^{\epsilon,-}_{z^\epsilon_-} \le \rho^{\epsilon,+}_{z^\epsilon_+}$ for all time since $\rho^{\epsilon,\pm}_{z^\epsilon_\pm}$ are defined from the percolation configurations $\Omega^{\epsilon,\pm}$, and almost surely $\Omega^{\epsilon,+}$ contains more open edges than $\Omega^{\epsilon,-}$. It follows that $l \le r$ almost surely.

If $\rho^{\epsilon,+}_{z^\epsilon_+}(t^\epsilon) < \rho^{\epsilon,-}_{z^\epsilon_-}(t^\epsilon)$, then from the definition of $\rho^{\epsilon,\pm}_{z^\epsilon_\pm}$, it is easy to see that $\rho^{\epsilon,+}_{z^\epsilon_+}(n) \ge \rho^{\epsilon,-}_{z^\epsilon_-}(n)$ for all n larger than or equal to

$$\tau^\epsilon_\rho := \inf\{i \ge t^\epsilon : \rho^{\epsilon,+}_{z^\epsilon_+}(i) \ge \rho^{\epsilon,-}_{z^\epsilon_-}(i)\}.$$

Following the same arguments as in the proof of [13, Lemma 3.1], we note that τ^ϵ_ρ is the first time the percolation cluster $C_{z^\epsilon_+}$ in the percolation configuration $\Omega^{\epsilon,+}$ intersects the percolation cluster $C_{z^\epsilon_-}$ in the percolation configuration $\Omega^{\epsilon,-}$. Since the percolation clusters evolve independently before they intersect, and each converges to a Brownian motion with respective drift $\pm b$, it is easily seen (cf. proof of [13, Prop. 3.3]) that $\epsilon^2\tau^\epsilon_\rho$ converges in distribution to $\tau := \inf\{s \ge t_- \vee t_+ : l(s) \le r(s)\}$. Since $\rho^{\epsilon,+}_{z^\epsilon_+}(n) \ge \rho^{\epsilon,-}_{z^\epsilon_-}(n)$ for $n \ge \tau^\epsilon_\rho$, it follows that $r(t) \ge l(t)$ for all $t \ge \tau$, which concludes the proof of Proposition 5. □

3.5 Different Starting Times

Until now, we only considered the case $t^\epsilon_- = t^\epsilon_+$. Let us now show how to extend this result to the case $t^\epsilon_- < t^\epsilon_+$. (The case $t^\epsilon_- > t^\epsilon_+$. can be treated along the same lines.) The proof goes by an approximation argument analogous to the proof Lemma 3. It is sufficient to decompose the path starting from earlier time into two time-windows: before and after reaching the starting time of the later path.

More precisely, we can approximate the path $l(\cdot \wedge t_+)$ by the path

$$S_{\alpha(p),\sigma(p),\epsilon}(\rho^{-,\epsilon}_{z^\epsilon_-})$$

stopped at time t^ϵ_+, and the remaining section of the path (i.e., the path $(l(t); t \ge t^\epsilon_+)))$ by the rescaled exploration cluster starting from the point $\rho^{-,\epsilon}_{z^\epsilon_-}(t^\epsilon_+)$ at time t^ϵ_+, which by definition, only uses the percolation configuration above time t^ϵ_+. Since the percolation configurations before and after time horizon t^ϵ_+ are independent, and since the latter two approximations only depend on the lower and upper part of the percolation cluster at time t^ϵ_+ respectively, it follows that conditioned on $l(t_+)$, the pair of paths $(l(t), r(t); t \ge t^+)$ is distributed as a pair of sticky paths starting from $(l(t^+), r(t^+))$ so that (l, r) is distributed as a left-right pair of sticky Brownian paths.

4 Convergence to the Left-Right Sticky Brownian Webs

In this section, we prove Theorem 2. Before going into the proof, we first need to clarify the definition of the filtration $(\mathcal{G}_t; t \in \mathbb{R})$ alluded to in Theorem 1.

Recall that Π denotes the set of paths with a starting point equipped with a norm d_Π inducing the topology of local uniform convergence plus convergence of the starting time. See e.g. the review article [14] for more details.

For any path with starting point with time coordinate less than t, define the killing operator K_t such that $K_t h(s) = h(s \wedge t)$. Define θ_t the time shift operator such that $\theta_t \circ h(s) = h(s+t)$.

Let us now consider a Brownian web $\mathcal{W}$ and $\mathcal{D}$ a dense subset of $\mathbb{R}^2$. Define the σ-field

$$\mathcal{F}_t = \sigma\big(K_t \circ w_{x_i,t_i} : (x_i,t_i) \in \mathcal{D},\ t_i \le t, w_{x_i,t_i} \in \mathcal{W}\big),$$

where $w_{x,t}$ is the (a.s. unique) path in $\mathcal{W}$ starting from the point (x,t). By a standard path coupling argument, see e.g. [15, Lemma 3.4], it can be proved that $\mathcal{F}_t$ is independent of the choice of the set $\mathcal{D}$. We call $(\mathcal{F}_t; t \ge 0)$ the natural filtration associated to $\mathcal{W}$. The common filtration $\mathcal{G}_t$ referred to in Theorem 1 is defined as follows.

Definition 2. *Let $(\bar{\mathcal{W}}_l, \bar{\mathcal{W}}_r)$ be two (possibly drifted) coupled Brownian webs and let $(\bar{\mathcal{F}}_t^l; t \ge 0)$ and $(\bar{\mathcal{F}}_t^r; t \ge 0)$ be the natural filtration of $\bar{\mathcal{W}}_l$ and $\bar{\mathcal{W}}_r$. We define $\mathcal{G}_t := \bar{\mathcal{F}}_t^l \vee \bar{\mathcal{F}}_t^r$ as the common filtration of the pair $(\bar{\mathcal{W}}_l, \bar{\mathcal{W}}_r)$.*

Proof (of Theorem 2). Let us now consider a supercritical oriented percolation model with $p > p_c$. In [13], it was shown that

$$S_{\alpha(p),\sigma(p),\epsilon}(\Gamma) \underset{\epsilon\to 0}{\Longrightarrow} \mathcal{W},$$

where $\Gamma = \{\gamma_z \ :\ z \in \mathcal{K}\}$ is the set of right-most infinite open paths in the percolation configuration. In previous sections (see Theorem 3), we showed that for every z^ϵ such that $S_{\alpha(p),\sigma(p),\epsilon}(z^\epsilon)$ converges to z, then $S_{\alpha(p),\sigma(p),\epsilon}(\gamma_{z^\epsilon}^{\pm,\epsilon})$ converges to a drifted Brownian motion with drift $\pm b(p)$. Let us denote by $\Gamma^{\pm,\epsilon}$ the set of right-most infinite paths in the percolation configuration $\Omega^{\pm,\epsilon}$. Following the exact same steps as in [13], it can be shown that

$$S_{\alpha(p),\sigma(p),\epsilon}(\Gamma^{+,\epsilon}) \underset{\epsilon\to 0}{\Longrightarrow} \mathcal{W}_r,\ \ S_{\alpha(p),\sigma(p),\epsilon}(\Gamma^{-,\epsilon}) \underset{\epsilon\to 0}{\Longrightarrow} \mathcal{W}_l,$$

where $\mathcal{W}_r$ (resp., $\mathcal{W}_l$) is a drifted Brownian web with drift $b(p)$ (resp., $-b(p)$). (The latter two convergence statements can be seen as a “triangular” extension of the results proved in [13].) In particular, this implies that the sequence of random variables $\{S_{\alpha(p),\sigma(p),\epsilon}(\Gamma^{+,\epsilon}, \Gamma^{-,\epsilon})\}_{\epsilon>0}$ is tight and in order to prove Theorem 2, it remains to prove that any convergent subsequence must be a sticky pair of left-right Brownian webs. According to Theorem 3, it only remains to prove the third condition of Theorem 1, namely that the two Brownian webs are co-adapted. This amounts to proving that for any deterministic $z_1, z_2 \in \mathbb{R}^2$, the processes $(r_{z_1}, r_{z_2}), (l_{z_1}, l_{z_2})$ and (r_{z_1}, l_{z_2}) are Markov processes with respect to the common filtration $(\mathcal{G}_t; t \ge 0)$ induced by $(\mathcal{W}_l, \mathcal{W}_r)$. We only show the Markov property (r_{z_1}, r_{z_2}). The two other processes can be handled by an analogous argument.

Let $m \in \mathbb{N} \cup \{\infty\}$ and f and g be continuous functions on $\Pi^{\otimes 2}$ and $\Pi^{\otimes(2+m)}$ respectively. Let $t \geq 0$ and let $\{z_i\}_{i=1}^2$ and $\{\bar{z}_i\}_{i=1}^m$ be two collections of points in $\mathbb{R}^2$ with time coordinates less than t. According to the definition of $\mathcal{G}_t$, it is enough to prove that conditioned on $(r_{z_1}(t), r_{z_2}(t))$, the law of $(r_{z_1}(s), r_{z_2}(s))_{s\geq t}$ does not depend on $(r_{z_1}(s), r_{z_2}(s), (l_{\bar{z}_j}(s), r_{\bar{z}_j}(s))_{1\leq j\leq m})_{s<t}$ for any finite m This can be proved by an approximation argument analogous to that in Sect. 3.5. Namely, we can approximate each path in $((l_{\bar{z}_j}(s), r_{\bar{z}_j}(s))_{1\leq j\leq m})_{s<t}$ by their rescaled exploration clusters up to time t, and the same for $(r_{z_1}(s), r_{z_2}(s))_{s<t}$, while we approximate $(r_{z_1}(s))_{s\geq t}$ and $(r_{z_2}(s))_{s\geq t}$ by new exploration clusters that start at time t. The desired independence then follows from the independence between percolation configurations above and below time horizon t and passing to the limit as $\epsilon \downarrow 0$. □

Acknowledgement. R. Sun is supported by NUS grant R-146-000-253-114.

A Alternative Constructions of the Left-Right Pair of Sticky Brownian Webs

In this section, we give a proof sketch for Theorem 1. We will show that it is equivalent to the characterization of the left-right Brownian webs given in [15, Theorem 1.5]. We will also discuss the characterization of a pair of θ-coupled sticky webs given in [7, Theorem 4].

In Theorem 1.5 of [15], the pair $(\mathcal{W}_l, \mathcal{W}_r)$ is uniquely characterized by condition 1 in Theorem 1 plus the finite-dimensional distribution, that is, the joint distribution of any finite collection of paths $(l_{z_1}, \ldots, l_{z_m}; r_{\bar{z}_1}, \ldots, r_{\bar{z}_n})$ with deterministic $z_1, \ldots, z_m, \bar{z}_1, \ldots, \bar{z}_n \in \mathbb{R}^2$, where l_{z_i} and $r_{\bar{z}_j}$ are the almost surely unique path in $\mathcal{W}_l$ and $\mathcal{W}_r$, starting at z_i and $\bar{z}_j$, respectively. In [15], the joint law of $(l_{z_1}, \ldots, l_{z_m}; r_{\bar{z}_1}, \ldots, r_{\bar{z}_n})$ is characterized by the properties that: (1) paths evolve independently when they are apart; (2) paths of the same type coalesce when they meet; (3) each pair $(l_{z_i}, r_{\bar{z}_j})$ interact as a left-right pair of sticky Brownian motions as in Definition 1 (see [15] for more details). Therefore to deduce Theorem 1.3, it suffices to show that conditions 1–3 in Theorem 1.3 imply that $(l_{z_1}, \ldots, l_{z_m}; r_{\bar{z}_1}, \ldots, r_{\bar{z}_n})$ satisfy properties (1)–(3) above.

Indeed, the conditions in Theorem 1 imply that, with respect to the common filtration generated by the collection of paths $(l_{z_1}, \ldots, l_{z_m}; r_{\bar{z}_1}, \ldots, r_{\bar{z}_n})$, pairwise, paths of the same type coalesce when they meet, and paths of different types interact as left-right sticky Brownian motions. In particular, for any pair of paths u, v in this collection,

$$1_{u\neq v} d\langle u, v\rangle = 0,$$

where *the cross-variation is taken with respect to the common filtration* generated by the whole collection. This in turn implies that the paths evolve as independent Brownian motions when they are apart. Therefore properties (1)–(3) above hold, and the pair of sticky Brownian webs $(\mathcal{W}_l, \mathcal{W}_r)$ characterized by Theorem 1 is the same as the left-right Brownian web characterized in [15, Theorem 1.5].

The above sketch, in particular, the determination of the joint law of

$$(l_{z_1}, \ldots, l_{z_m}; r_{\bar{z}_1}, \ldots, r_{\bar{z}_n})$$

from conditions 1–3 in Theorem 1, was rigorously implemented by Howitt and Warren in a slightly different context, where they introduced the notion of co-adaptedness. Their argument can be easily adapted to our setting. In [7, Theorem 4], they formulated an analogue of Theorem 1 for a pair of θ-coupled sticky webs $(\mathcal{W}, \mathcal{W}')$, where $\mathcal{W}$ and $\mathcal{W}'$ each is distributed as a standard Brownian web with no drift, and each pair of path $(\pi, \pi') \in (\mathcal{W}, \mathcal{W}')$ with deterministic starting points is distributed as a pair of sticky Brownian motions with stickiness parameter θ. Again, the key is to show that the analogues of conditions 1-3 in Theorem 1 for the θ-coupled sticky webs $(\mathcal{W}, \mathcal{W}')$ uniquely determine the joint distribution of a finite collection of paths $(\pi_1, \ldots, \pi_m; \pi'_1, \ldots, \pi'_n)$ in $(\mathcal{W}, \mathcal{W}')$. More precisely, given $(\pi_1, \ldots, \pi_m; \pi'_1, \ldots, \pi'_n)$, if each pair of paths of the same type are distributed as coalescing Brownian motions with respect to the common filtration, and each pair of paths of different types are distributed as θ-coupled sticky Brownian motions with respect to the common filtration, then the joint distribution of the whole collection is uniquely determined. This is the content of Theorem 76 in Howitt's Ph.D. thesis [8]. The proof is based on martingale characterizations and a decomposition in terms of successive times of coalescence between paths.

References

1. Camia, F., Fontes, L.R.G., Newman, C.M.: Two-dimensional scaling limits via marked nonsimple loops. Bull. Braz. Math. Soc. **37**(4), 537–559 (2006)
2. Durrett, R.: Oriented percolation in two dimensions. Ann. Probab. **12**(4), 999–1040 (1984)
3. Fontes, L., Isopi, M., Newman, C.M., Ravishankar, K.: The Brownian web: characterization and convergence. Ann. Probab. **32**(4), 2857–2883 (2004)
4. Fontes, L., Newman, C.M., Ravishankar, K., Schertzer, E.: Exceptional times for the dynamical discrete web. Stochast. Process. Appl. **119**(9), 2832–2858 (2009)
5. Garban, C., Pete, G., Schramm, O.: Pivotal, cluster, and interface measures for critical planar percolation. J. Am. Math. Soc. **26**(4), 939–1024 (2013)
6. Garban, C., Pete, G., Schramm, O.: The scaling limits of near-critical and dynamical percolation. J. Eur. Math. Soc. **20**(5), 1195–1268 (2018)
7. Howitt, C., Warren, J.: Dynamics for the Brownian web and the erosion flow. Stochast. Process. Appl. **119**(6), 2028–2051 (2009)
8. Howitt, C.J.: Stochastic flows and sticky Brownian motion. Ph.D. thesis, University of Warwick (2007). http://wrap.warwick.ac.uk/56226/
9. Kuczek, T.: The central limit theorem for the right edge of supercritical oriented percolation. Ann. Probab. **17**(4), 1322–1332 (1989)
10. Newman, C., Ravishankar, K., Sun, R.: Convergence of coalescing nonsimple random walks to the Brownian web. Electron. J. Probab. **10**, 21–60 (2005)
11. Newman, C.M., Ravishankar, K., Schertzer, E.: Marking (1, 2) points of the Brownian web and applications. Annales de l'Institut Henri Poincaré, Probabilités et Statistiques **46**(2), 537–574 (2010)

12. Revuz, D., Yor, M.: Continuous Martingales and Brownian Motion, vol. 293. Springer, Heidelberg (1999)
13. Sarkar, A., Sun, R.: Brownian web in the scaling limit of supercritical oriented percolation in dimension 1+ 1. Electron. J. Probab **18**(21), 1–23 (2013)
14. Schertzer, E., Sun, R., Swart, J.M.: The Brownian Web, The Brownian Net, and Their Universality. Advances in Disordered Systems, Random Processes and Some Applications, pp. 270–368 (2017)
15. Sun, R., Swart, J.M.: The Brownian net. Ann. Probab. **36**(3), 1153–1208 (2008)
16. Tóth, B., Werner, W.: The true self-repelling motion. Probab. Theory Relat. Fields **111**(3), 375–452 (1998)
17. Warren, J.: The noise made by a poisson snake. Electron. J. Probab. **7**, 21 (2002)

MIX
Papier aus verantwortungsvollen Quellen
Paper from responsible sources
FSC® C105338

If you have any concerns about our products,
you can contact us on
ProductSafety@springernature.com

In case Publisher is established outside the EU,
the EU authorized representative is:
Springer Nature Customer Service Center GmbH
Europaplatz 3, 69115 Heidelberg, Germany

Printed by Libri Plureos GmbH
in Hamburg, Germany